BEER

TELECOMMUNICATIONS:

CONCEPTS, DEVELOPMENT, AND MANAGEMENT

SECOND EDITION

TELECOMMUNICATIONS:
CONCEPTS, DEVELOPMENT, AND MANAGEMENT

SECOND EDITION

W. John Blyth
THE DMW GROUP, INC.

Mary M. Blyth
DETROIT COLLEGE OF BUSINESS

GLENCOE/McGRAW-HILL
A Macmillan/McGraw-Hill Company
Mission Hills, California

For their commitment to excellence in telecommunications, we are honored to dedicate this book to:

Thomas Muth of Michigan State University
James Koerlin of Golden Gate University
Dan Dennis of MCI Telecommunications

Send all inquiries to:
Glencoe/McGraw-Hill
15319 Chatsworth Street
P.O. Box 9509
Mission Hills, California 91395-9509

ISBN 0-02-680841-2 (Text)
ISBN 0-02-680842-0 (Instructor's Guide)

1 2 3 4 5 6 7 8 9 94 93 92 91 90

CONTENTS

PREFACE

1 INTRODUCTION TO TELECOMMUNICATIONS 1

2 EARLY HISTORY OF TELECOMMUNICATIONS 19

3 STRUCTURE AND REGULATION OF THE TELECOMMUNICATIONS INDUSTRY— AN HISTORICAL PERSPECTIVE 42

5 TELEPHONY 79

6 TELECOMMUNICATION NETWORKS 102

10 PRINCIPLES OF TRAFFIC ENGINEERING 207

11 PRINCIPLES OF TELECOMMUNICATIONS MANAGEMENT 224

14 NEW DIRECTIONS IN TELECOMMUNICATIONS 273

PREFACE

Since telecommunications is a dynamic technology, telecommunications textbooks must change to keep pace with new developments in this exciting industry. Since the first edition was published, the telecommunications industry has undergone an unprecedented period of change. These changes are the result of (1) the introduction of competition into the "natural monopoly" of the telephone industry, (2) the breakup of the Bell System, and (3) advancements in technology. Changes as important as these require us to learn more about them in order to use telecommunications equipment and services efficiently.

Telecommunications: Concepts, Development, and Management provides an overview of the broad field of telecommunications, including voice, data, message, and image communications. It emphasizes breadth rather than depth. Each of the major topics in this book could well be the subject of an entire book — or course — on the subject. In fact, books have been written on many of these topics, but most are written for engineers and specialists rather than for students or business users.

Although it is not possible to adequately cover the rapidly expanding field of telecommunications in one volume, we have included all the standard topics, as well as those of special current interest. Our goal is to provide a balanced overview with representative material selected from each area of telecommunications, as seen by experts in the field.

INTENDED AUDIENCE

This book is designed for an introductory course in telecommunications. Written in nontechnical language, it can be used by students without previous training in telephony, electronics, or computer science.

Students in a variety of college and university programs will be able to learn the basic concepts of telecommunications from this book. Businesspersons will be able to learn about telecommunications products and services and their cost-effective management. The book should also be useful to industry professionals who have detailed telecommunications knowledge of a particular segment of the industry but may lack a broader industry perspective. Furthermore, the reference materials (including the comprehensive glossary; the professional associations; the organizations that sponsor seminars; and the telecommunications periodicals, reports, and newsletters) enhance the book's usefulness to industry professionals.

CHANGES INCORPORATED IN THE SECOND EDITION

Telecommunications: Concepts, Development, and Management, Second Edition, describes the current state of the art in telecommunications. It reflects ideas received from seminar participants, business associates, other instructors, and students. These materials have been field tested by the authors in classroom settings.

This new edition, which builds on the strengths of the previous edition, incorporates the following changes:

- A new chapter has been added to describe the post-divestiture telecommunications industry.
- All appropriate topics have been updated to reflect current status and future trends. There is new or increased coverage on several topics, especially regulation, tariffs, fiber optics, electronic mail, FAX, PBX, teleconferencing, mobile phones, pagers, and videotex.
- Chapter objectives have been added at the beginning of each chapter to help orient readers to the material to be presented.
- A vocabulary section now follows the end of each chapter to call attention to key words and concepts.
- The chapter entitled "Traffic Engineering" has been repositioned so that it precedes the chapters on telecommunications management. Now readers can become familiarized with key traffic engineering terms before they are referred to in operations management.
- A new Appendix C has been added on telecommunications periodicals, reports, and newsletters.
- The bibliography at the end of each chapter has been moved to the back of the book so that all reference materials appear in one place.
- Photographs, bibliography, glossary, and appendices have been updated to reflect the current telecommunications industry.
- A reference list of frequently used acronyms has been added at the end of the book for easy student reference.
- In the Instructor's Guide, new test questions and transparency masters have been added to the existing chapter outlines/lecture notes, answers to review questions, and examination questions.
- A computerized testbank accompanies the Instructor's Guide to provide flexibility in the testing process.

TO THE STUDENT

Welcome to the exciting new world of telecommunications! This book is designed to be your passport into this dynamic field.

ACKNOWLEDGMENTS

We wish to thank the many instructors, students, and professionals who used *Telecommunications: Concepts, Development, and Management,* thereby justifying its revision.

Accurate, up-to-date coverage of a field as broad as telecommunications requires the input of persons with expertise in different segments of the industry. We would like to express our heartfelt appreciation to the following professionals for their contributions and assistance during the preparation of the text: Marvin Pridgeon of Ameritech; William R. Smith of Michigan Bell Telephone Company; Daniel Grove and Scott Davis of Motorola Communications and Electronics; Ralph Carnevale of Compugram Information Network; S. Eric Wachtel of MEDphone Corporation; Timothy Beck of Mitsubishi Electric Sales America; Pamela Debandt of Southwestern Bell Telecom; Drusie Demopoulos of Duffey Communications; Cynthia Milton of Northcom Teleport; Glenn Moore, Byrne Blumenstein, Frank Chuba, and Ben Berman of MCI Telecommunications; David Blyth of Century 21 Programming, Inc.; Dennis Blyth and Shawne Buckley of NCR Corporation; C. Stephen Mayo and Jim Hoogesteger of the University of Michigan; Charles Ploof and Jeff Benson of Northern Telecom; Charles Emery of Telecom Technicians, Inc.; Jeanette Tomaszewski and Gerald Lamphere of Detroit College of Business; and Gini Skwiera of Unisys Corporation.

The following reviewers were very helpful with suggestions and comments on the manuscript, and their contribution is much appreciated: Thomas Dunlop, Ohio University; Ron Kapper, College of DuPage; Stan Kroder, University of Dallas; Robert LaRose, Michigan State University; Ruth Lawrence, Start Technical College; Thomas Muth, Michigan State University; and Donald Reicks, Highline Community College.

To the dedicated publishing professionals, we wish to extend our deepest appreciation for their friendship and constant support. Our original editor, Dennis Gladhill, convinced us that we could make a contribution to telecommunications education by our writing. He applauded our progress and cheered us on as work progressed. Glencoe developmental editors John Gastineau and Patricia Parrott applied their publishing expertise to the manuscript and, with careful attention to detail, guided us along the road of turning the manuscript into a finished product. George Provol, who did such a fine job of marketing the first edition, continued to provide us kindly advice and assistance throughout the project.

W. John Blyth
Mary M. Blyth

TELECOMMUNICATIONS:

CONCEPTS, DEVELOPMENT, AND MANAGEMENT

SECOND EDITION

CHAPTER

1

INTRODUCTION TO TELECOMMUNICATIONS

CHAPTER OBJECTIVES

After completing this chapter, the reader should be able:

- *To define telecommunications.*
- *To identify and describe the four principal types of telecommunication systems available today.*
- *To explain the interrelationship between communications technology and computer technology and describe its effect upon both industries.*
- *To explain why telecommunications is an essential part of today's business environment.*
- *To cite several factors that led to the modification of telecommunications policy and the gradual restructuring of the telecommunications industry in the United States.*
- *To explain why it is important to study telecommunications and have a general understanding of its vocabulary and the role it plays in modern society.*
- *To describe some of the career opportunities available in today's telecommunications industry.*

We are familiar with the telephone. It permits us to communicate with another person nearly anywhere in the world. We know this because we have had firsthand experience with the telephone. When we use the word *telephone*, we can be quite sure that other people know exactly what we mean.

This is not the case, however, with telecommunications. *Telecommunications* is a relatively new word in our vocabularies. In fact, while the word *telephone* has been in existence for over a century, *telecommunications* was not even listed in dictionaries until the mid-1960s. And even today there is misunderstanding regarding its meaning.

THE FIELD OF TELECOMMUNICATIONS

To some people, telecommunications is synonymous with the telephone or voice communications. To others it is synonymous with the computer or data communications. And to still others it is synonymous with broadcasting or any of the other electronic components of the entertainment industry. While each of these areas is an important part of telecommunications, it is only a part. The field of telecommunications is a broad one that includes voice communications, data communications, message (text) communications, and image communications.

THE TERM DEFINED

The prefix *tele* is derived from the Greek root meaning "distant" or "at a distance"; the word *communication* denotes the imparting of information. A literal interpretation suggests that telecommunication is the process of transmitting information over a distance. As the term is currently used, it incorporates the concept of rapid communication. Thus, *telecommunication* is the process of rapidly transmitting information over a distance by electrical or electromagnetic media. Electromagnetic transmission systems include telephone lines, cables, microwaves, satellites, and light beams.

TELECOMMUNICATION SYSTEMS

Two characteristics identify a telecommunication system: (1) communication over a distance, and (2) transmission by electrical or electromagnetic means.

Four principal types of telecommunications systems are available today: voice, data, message (text), and image.

Voice communications is known as *telephony*. Voice communication systems transmit spoken words over telephone networks in the form of electrical energy that varies in amplitude with the sound variations being transmitted. These systems include public and private local and long distance services.

Data communications systems are networks of components and devices organized to transmit data from one location to another—usually from one computer or computer terminal to another. The data is transmitted in coded form over electrical transmission facilities. Data transmission systems try to provide faster information flow by reducing the time spent in collecting and distributing data.

Message (text) systems include computer-based services, such as electronic mail; teletypewriter services, such as telex or TWX; and telegraphic services, such as telegrams. They also include facsimile services, which send exact copies of written text, pictures, sketches, diagrams, and graphs. *Image systems* include video services, such as commercial television, closed circuit-television, and videoconferencing. Data systems, message systems, and image systems all transmit intelligence in coded form by using a stream of pulses that represent digital signals.

COMPUTERS AND TELECOMMUNICATIONS

Communications technology and computer technology developed along parallel lines. Early telephone systems used electromechanical switches to connect calls between telephones. Similarly, early computers used electromechanical equipment controlled by punched cards to perform sorting and computing operations. Both industries contributed to the development of solid-state electronics, and both use it in their present systems. Thus, the technology of these two industries is compatible.

Although computers and telecommunications share the same technical base, the computer industry has been allowed to develop in an unregulated, competitive market. The telecommunications industry, however, has developed as a monopoly under a comprehensive regulatory structure. These two greatly differing philosophies delayed integration of the two industries. They ultimately converged, however, because of the partial deregulation of the telecommunications industry and the blossoming of the microprocessor era.

Today's computer industry relies on telecommunications for moving data from one location to another. In *remote job entry,* raw data is entered into a computer terminal at a site distant from the computer and transmitted over communication lines to the central computer for processing. The processed data is then transmitted back to the entering terminal. This exchange is known as *data communications. Distributed processing* takes the process one step further; some processing of the data is done at the entering site. Data communications is an integral part of distributed processing systems. The computer industry as we know it today could not exist without the telecommunication lines that transmit information from one location to another.

Conversely, computers are penetrating all areas of communications and, as a result, are greatly expanding the use of telephone systems. Computerized components in the telephone system have greatly changed the way of providing telephone service, bringing a higher level of intelligence to network transmission and switching and to network design, management, and control.

Telecommunication switching machines that incorporate computer-controlled hardware are able to provide a wide range of telephone service features, such as abbreviated dialing, call forwarding, call waiting, third-party add-on, and call detail recording.

Computerized telephone exchanges use electronic connections rather than metal contacts, thus enabling them to speed up call completion and improve service quality. Computer technology used in switching devices enables them to process calls more efficiently by automatically selecting the best route for each call.

■ ***TELEMATICS*** The marriage of the computer and telecommunications industries has been one of the more exciting developments of our time. New terms have been coined to describe this merger, such as *information technology* (IT) in England and *Telematique* in France. The latter term was coined by French scientists Simon Nora and Alain Minc; it has been translated into English as *telematics*. Professor Anthony Oettinger of Harvard University coined the word *compunications* to describe the same union; however, this term has not gained widespread acceptance. Regardless of what it is called, the merging of computers and telecommunications into a new and broad discipline has made the information age possible and, with it, exciting opportunities for enhanced productivity and profits.

TELECOMMUNICATIONS IN MODERN SOCIETY

The United States is rapidly becoming an information-based society, and telecommunications is central to this development. The availability of nearly "instant information" made possible by telecommunications technology is changing our jobs, our business organizations, and our personal lives.

TELECOMMUNICATIONS IN BUSINESS ORGANIZATIONS

Communications plays an important role in business organizations, but today the ability to merely communicate is not enough. Communications must be rapid, accurate, and capable of being integrated with other information activities.

Businesses must be productive in order to remain competitive in the world economy. Modern telecommunicatioon systems — with their ability to provide enhanced service features, to switch calls over the most economical route, and to provide call detail recording — save us time and make our work easier. Thus, they make a valuable contribution to the productivity of any organization.

■ ***THE ROLE OF INFORMATION*** Information is the lifeline of modern business; it provides the basis for all business activities. Managers apply judgment to the information available to make their decisions — decisions that have a strong impact upon the success of any enterprise. Recognizing their dependency upon information, businesses began some 25 years ago to focus on the management of this vital resource.

■ ***MANAGEMENT INFORMATION SYSTEMS*** The present trend in information management is to establish formal *management information systems*

(MIS). These systems are generally computer-based and are designed to supply timely information to managers for use in drawing conclusions, making predictions, and recommending courses of action. Some companies have created separate management information systems departments to coordinate the flow of information throughout the organization.

■ ***DATA BASES*** Telecommunications also makes possible the instant availability of information from a centralized *data base* (facts arranged in computer files for access and retrieval). Data bases may be either company-wide or industry-wide. A company-wide data base is a central master file containing information from the major systems of the firm. It provides company-wide access to the master file, thus eliminating the need for duplicate files in each department. With a data base, information can be shared among the various departments of an organization and among remotely located branches. Data bases are protected from unauthorized access by requiring the users to identify themselves with a password or a series of passwords.

In addition, a number of companies have established industry-wide data bases that can be accessed from virtually anywhere via telecommunication lines. These companies provide access to their data bases for a subscription fee. (In addition to the fee, the user pays the telephone company for telephone charges incurred in accessing the data base location.) These information centers are available for many industries and professions, including transportation, banking, investments, publications, law, and medicine.

For example, WESTLAW, one commercial data base system, was created by West Publishing Company, a publisher that has served the legal profession for over a century. The WESTLAW system connects remote terminals located in law offices, courts, and government agencies with a centralized data base by means of telephone lines. Users enter commands or inquiries at a typewriter-like keyboard. Responses are displayed on the cathode ray tube (CRT) or terminal screen. Users can copy information appearing on the screen on paper by means of an associated printer. The system is protected from unauthorized access by a multistep sign-on procedure.

This system has two basic advantages: (1) speed in finding the research material, and (2) search capabilities not possible using books. Lawyers can scan large numbers of cases and identify relevant ones in a few seconds. In addition, they can access case summaries using terms other than those indexed in law books. For example, lawyers can locate a case by entering almost any term associated with that court decision — names of judges, witnesses, companies, and unusual nonlegal terms.

WESTLAW is constantly being updated; as new cases are reported, they are added to the data base. The system has improved legal research by allowing users to accomplish more research in less time and by providing new search capabilities.

INDUSTRIES THAT USE TELECOMMUNICATIONS

Many organizations use telecommunications effectively to enhance their services or to provide new services. There are also industries whose very existence depends upon telecommunications. Among these industries are the airlines, banking, credit card operations, insurance and investments.

■ ***AIRLINES*** All major airlines in the United States use computerized systems for handling reservations. To make a reservation, a person dials a local telephone number that connects to a regional reservation center, probably located in a distant city. The reservation agent at the regional center is linked to a national computer center by telephone lines. This clerk keys in the desired destination on a remote terminal to access the central computer and to obtain information about possible flights. Information displayed on the terminal screen lists the flights on which seats are available and the fares. When the traveler selects a flight, the clerk enters the details of the booking into the system, and the computer's seat inventory is updated. The cancellation of reservations is also handled automatically. The computer cancels the reservation and increases the number of seats available on the flight. In addition, the computer keeps a waiting list of passengers desiring reservations and their telephone numbers. Reservation systems are interlinked with those of other airlines so that connecting flights can be booked.

Telecommunications also permits air traffic controllers to monitor and control air traffic. A computer system identifies each approaching plane and tracks its altitude and speed, enabling the air traffic controllers to give directions for takeoffs, landings, and en route flights.

A number of other industries whose business depends heavily on advance reservations employ telecommunications systems similar to those sed by the airlines. These industries include hotels and motels, car rental agencies, ship lines, and railroads.

■ ***BANKING*** Another major industry dependent upon telecommunications technology is banking. Banks use remote terminals and telecommunication lines to update customers' accounts. Tellers located either at main offices or branch offices insert the customer's passbook into a computer terminal and key in relevant information; the terminal prints the entry in the passbook and transmits the information to update the central computer files.

In some cases, customers can use their pushbutton telephones to perform certain banking transactions, such as paying bills, transferring funds, and determining their bank balances. The telephone functions as an input terminal when it is connected to the bank's central computer by telephone lines. The user communicates with the computer by entering appropriate codes on the telephone keypad.

Banks also use telecommunications to provide automatic teller service. Automatic teller terminals are connected to the bank's central computer via telecommunication lines. To process a transaction, the customer inserts an identifying bank card into the terminal and keys in a personal identification number. After the computer performs appropriate checks on the customer's identify and account status, the customer keys in a transaction code, and the machine completes the transaction. Automatic tellers are located in shopping malls, supermarkets, and places of employment, as well as on bank premises. Many of these terminals accommodate bank cards issued by cooperating banks, credit unions, and savings and loan associations. Automatic tellers offer the convenience of 24-hour operation for the most frequently used services, such as deposits, withdrawals, and payments to utilities and credit card accounts.

■ ***CREDIT CARD SERVICE*** The widespread use of credit cards has resulted in the creation of a new industry to serve as a clearinghouse for credit card transactions. A major function of this service is verifying accounts to reduce fraudulent use of credit cards. When a customer presents a card to be used for payment, the merchant calls the service company to obtain a credit status report. The service company searches its computer files and reports the status of the account. Account verification is a high-speed transaction; the entire process is completed in a few seconds. The service depends upon high-speed telecommunication between a centralized data base and merchants located virtually anywhere in the world.

■ ***INSURANCE*** The use of telecommunications enables insurance companies to operate more efficiently and to offer better service to their customers. Insurance companies with millions of policyholders generally have a home office and a number of branches located throughout the country. Policy records are kept in a central computer located in the home office, and the branch office computers are connected to the home office computer via telecommunication lines. When a policyholder requests information regarding coverage, a clerk in the branch office enters the request on the terminal keyboard. The computer accesses the files and reports the requested information on the terminal screen at the branch office. The entire process takes only a few seconds; thus, the clerk is able to respond to the customer's request without delay.

■ ***INVESTMENTS*** Another industry that depends on the immediate communication of information is the investment business. Both the New York Stock Exchange and the American Stock Exchange — the two major stock exchanges in the country — are located in New York City. All trading of stocks they list is conducted on the floor of the stock exchange. As each transaction is completed, it is recorded on a computer and simultaneously transmitted to brokerage offices all over the country. Some brokerage offices display this information on a continuous tape, thus providing up-to-the-minute details of all trading transactions. Many brokerage offices also have display terminals to obtain current price information on any stock from the computer. The brokerage industry relies heavily on rapid communication since investors' trading decisions are based in part on the latest market prices.

TELECOMMUNICATIONS IN PERSONAL COMMUNICATIONS

Telecommunications has also enhanced personal communications. As our society has become increasingly mobile, telephone services have assumed greater importance in maintaining family and social ties. Long distance telephone calls have virtually replaced letter writing as a means of keeping in touch.

The telephone makes it possible to obtain information, to transact business without leaving home, to summon assistance in times of trouble, and to maintain social contacts.

New service features made possible by sophisticated telecommunications technology have also been favorably received. Many residential customers now enjoy the convenience of such computerized telephone features as call waiting, automatic dialing, add-on, call forwarding, and cordless telephones.

THE RESTRUCTURING OF THE TELECOMMUNICATIONS INDUSTRY

In the United States, the restructuring of the telecommunications industry has taken place gradually over the last two decades; however, the process accelerated rapidly in the early 1980s. It involved technological advancements, the introduction and growth of competition in the industry, and the divestiture of the American Telephone and Telegraph Company (AT&T).

The telecommunications industry in the United States is a private enterprise operating in part as a regulated monopoly. The Federal Communications Commission (FCC), which was created by the Communications Act of 1934, is charged with regulating all interstate and foreign telephone, telegraph, and broadcast communications originating in the United States. The FCC sets public policy and supervises the utilities in the execution of that policy. The Communications Act stipulated that, insofar as possible, the telephone industry be operated to make telephone service available to all the people of the United States at affordable rates.

For half a century, the FCC maintained that the telecommunications industry was a "natural monopoly" and that no geographic location should be served by more than one telephone company. This policy was designed to prevent wasteful duplication of services in high-density areas while fostering the availability of service in low-density areas. The policy resulted in nearly universal, high-quality telephone service at reasonable rates.

Three interrelated factors led to the modification of this policy and the restructuring of the telecommunications industry: (1) technological developments, (2) customer demand for innovative products, and (3) a change in FCC policies and significant court decisions.

TECHNOLOGICAL DEVELOPMENTS

Years of research in electronics — some of which were attributable to the NASA space program — laid the foundation for a variety of practical applications in business and industry. The discovery of the principles of the transistor led to the development of solid-state electronics, resulting in a wide variety of popular consumer products. Today solid-state electronics is used in many industries, such as home appliances, automobiles, and manufacturing, as well as computers and telecommunications.

The many applications of electronics created new products that incorporated both computer and communication capabilities. Telephone switching systems took on many characteristics of computer systems, and telecommunication capabilities became an integral part of computer communications and distributed processing systems. The result was the blurring of boundaries between the telephone and computer industries.

CUSTOMER DEMAND FOR INNOVATIVE PRODUCTS

Until the mid-1950s, telephones and telephone equipment did not reflect any apparent changes even though solid-state technology was available. All telephones were of the same design, and their color was basic black. However, applications of solid-state electronics in the telecommunications industry did result in faster call completion, improved-quality transmission, automatic billing procedures, and similar improvements that were invisible to the user. These technological improvements generated cost savings for the industry that helped keep telephone rates down at a time when other costs were steadily rising. Americans, accustomed to high-quality telephone service, were generally unaware of these improvements; they accepted them as a matter of course.

When "designer" telephones were finally introduced, they met with enthusiastic acceptance in spite of their higher prices. This reaction suggests that the public would probably have been willing to pay more for innovative products much earlier and that the telephone companies should have investigated this possibility before competition forced them to do so.

FCC AND COURT DECISIONS

Through the years the FCC allowed the telephone industry to operate as a friendly monopoly, thus maintaining the comfortable status quo. The first major challenge to this policy was the Carterfone decision of 1968. Since the beginning of the telephone industry, telephone customers had been prohibited from connecting any device not supplied by the telephone company to telephone company lines. The Carterfone decision ended this prohibition, allowing customers to connect telephone equipment manufactured by other companies to telephone company lines.

The next step toward today's competitive market was the Microwave Communications, Inc. (MCI) decision. This 1969 decision allowed MCI, a "nontelephone" company, to build and operate a microwave relay station to provide telephone service between St. Louis and Chicago, in direct competition with the telephone companies. In making this decision, the FCC affirmed that the microwave system would serve the public interest and that competition would not be harmful.

More recent steps toward competition were the Computer Inquiry II decision (1981) and the Modified Final Judgment (1984). Computer Inquiry II investigated whether the computer companies were engaging in communications activities and, therefore, should be regulated. Another consideration was whether AT&T should be permitted to process data during its transmission and thus engage in data processing activities. The Computer Inquiry II decision stipulated that the Bell System would be permitted to engage in data processing activities and that communications services would be divided into two categories: basic and enhanced. Basic services would continue to be regulated; however, enhanced services (services that involve computer processing of the transmitted information) were to be deregulated.

The Modified Final Judgment was undoubtedly the most significant telecommunications regulatory decision in the history of the industry. Its most important stipulation was that the Bell Operating Companies were to be divested (separated) from AT&T. Long distance service was to be provided by AT&T and the other long distance companies. The Bell Operating Companies were allowed to retain monopoly carriage of local telephone service.

A major outcome of the Modified Final Judgment was the reorganization of the world's largest company, AT&T. *Fortune* magazine described this reorganization as "the biggest that American industry has ever witnessed" and as "a coup" that "could fell most governments."

These FCC and court decisions represented a complete reversal of telecommunications policy. After years of protecting the communications industry as a total monopoly, the FCC now embraced competition. As was expected at the time, the new policy encouraged the development of innovative telecommunications products and services and the emergence of new suppliers.

TODAY'S TELECOMMUNICATIONS INDUSTRY

The breakup of the Bell System was a significant event for most Americans. The immediate reaction was confusion and anger. Most residential and many small business customers saw little or no benefit to themselves from the divestiture.

On the other hand, the larger telecommunications users and the equipment and long distance service competitors of AT&T have experienced substantial benefits. Since the divested subsidiaries of AT&T are no longer required to purchase all their equipment from another AT&T organization, previously called Western Electric, the equipment marketplace has become intensely competitive. As a result, more state-of-the-art switching systems and terminal equipment products have become available. Competition in the long distance telephone service market has resulted in improved service quality at lower rates.

USER RESPONSIBILITY

An important outcome of AT&T divestiture has been a change in the patterns of ownership and servicing of telephone systems. Although users had been authorized to buy and install their own telephone equipment for more than a decade, most customers had followed the traditional pattern of obtaining telephones from the telephone company for the payment of a monthly rental fee. In the new era of deregulation, the telephone company no longer has a monopoly on providing telephone systems. Thus, users can now select telephones and telephone systems from any vendor.

The new environment also requires customers to be responsible for the maintenance and repair of their telephones and telephone systems, functions that were formerly handled by the telephone company.

In summary, both individuals and organizations have been required to assume more responsibility for procuring and managing their telephone systems. The new environment presents users with more opportunities and more difficult choices.

GROWING IMPORTANCE OF TELECOMMUNICATIONS

A year-long study of Fortune 1000 companies conducted by the Eastern Management Group in 1988–1989 concluded that not only will telecommunications budgets and personnel grow, but also the charters and mission statements of corporate telecommunication departments will expand.

The study surveyed telecommunications executives, examining communications systems, plans, and expenditures. Ninety percent of the companies interviewed reported that they expected their telecommunications budgets and personnel to grow over the next five years. The largest of the companies expected from 50 to 100 percent growth. Telecommunication's percentage of the overall corporate budget is expected to grow from 2.8 percent to almost 4 percent during this period.

Since divestiture telecommunications is being treated more and more like a mainline activity that contributes significantly to the organization's profitability. Jerry Stern, director of research for the Eastern Management Group, states, "The new age of corporate telecommunications is just arriving. After years of stagnation during the pre–divestiture period, corporate America's commitment to the expanding role of telecommunications is evident."[1]

TELECOMMUNICATIONS AS A DISCIPLINE

As a discipline, or field of study, telecommunications evolved from electronics, specifically, electrical engineering. As with most disciplines, electronics has its own special language. The technical aspects of the language, oriented toward electrical circuitry, often deters laypeople from studying telecommunications. However, the increasing use of electronic communication services, along with the deregulated telecommunications marketplace, has made it essential for all of us to have at least a general knowledge of the field.

WHY STUDY TELECOMMUNICATIONS?

Each of us has some personal contact with telecommunications. Most of us use some form of telecommunications daily, either at home or on the job or both.

■ ***PERSONAL USE OF TELECOMMUNICATIONS*** In today's deregulated marketplace, customers are faced with many choices. The range of equipment, services, and service features seems to be excessive for most requirements. Equipment and service providers vie for our business, each claiming to have the best product, the clearest transmission, or the lowest rates. Choice has become a confusing, time-consuming operation. An understanding of the terminology and a knowledge of the telecommunications services and products available will help us evaluate them in terms of our own needs so that we can make the best decisions.

■ ***BUSINESS USE OF TELECOMMUNICATIONS*** Telecommunications costs represent a significant portion of an organization's operating expenses. Knowledgeable experts agree that with informed management most organizations could reduce their telecommunications expenses by 20–30 percent without decreasing services. To make the best decisions, managers must understand the new telecommunications environment—the new technology, regulatory climate, various supplier offerings, and cost-saving opportunities available in telecommunications—and be able to match them with their organizations' needs. Telecommunications has become so important to all organizations that a knowledge of telecommunications is essential for anyone who wishes to be an effective executive in any industry.

The study of telecommunications will enable all businesspersons to understand the important role that telecommunications plays in the enterprise, to make the most effective use of the organization's telecommunication system, and to make informed decisions regarding its future role in the organization.

■ ***INTEGRATION OF COMPUTERS AND TELECOMMUNICATIONS*** It has often been said that professionals in the various disciplines talk to each other in a language that is only understandable to those in the discipline. Kenneth Sherman, writing in *Data Communications: A User's Guide,* calls attention to the need for computer personnel and telecommunications personnel to be able to communicate effectively. He says, "One of the biggest problems involved with communications is 'communicating.' Different people may use the same term to mean different things."[2]

Today's information industry professionals must be knowledgeable about both the computer and the telephone industry. The study of telecommunications will enable professionals in each field to better understand, and therefore communicate with, the other.

■ ***A POSSIBLE CAREER*** A study of telecommunications will also inform the learner of the many exciting career opportunities available in the telecommunications industry.

CAREERS IN TELECOMMUNICATIONS

Some researchers predict that the telecommunications industry will be the fastest growing, most innovative market in the United States, at least through the end of the century. The phenomenal growth that the industry is experiencing is creating a need for all types of telecommunications personnel. They are in great demand and short supply.[3]

Historically, the principal sources of employment for telecommunications professionals were the telephone and telegraph companies. Within these companies there were jobs as managers, engineers, telephone installers, repairpersons, cable splicers, operators, service representatives, scientists, and research and development personnel.

The introduction of competition in the telecommunications industry has created a wide variety of new jobs in addition to the continuing need for the professionals previously mentioned. Today, career opportunities for both men and women can be found with telephone equipment suppliers, other common carriers, telemarketing organizations, companies that service and repair telecommunication systems, telephone answering services, consulting firms, government agencies, the military, and user organizations.

Within these organizations there are opportunities for employment as managers, analysts, salespersons, technicians, engineers, training specialists, service representatives, cable splicers, telephone installers, repairpersons, and operators.

Because telecommunications plays an important role in the success of any business, all organizations require someone to handle the telecommunications responsibilities. In a small organization, the telecommunications function may be performed by a manager who is also responsible for many other functions or by an administrative secretary. Larger organizations generally have a telecommunications manager and a staff of professionals trained in the various aspects of telecommunications.

There are many ways in which the telecommunications function could be organized. Figures 1.1 and 1.2 illustrate two possible corporation charts. In Figure 1.1 the telecommunications function is in the Management Information Systems group. In Figure 1.2, the telecommunications function is in the Administrative Services group. The responsibilities of the

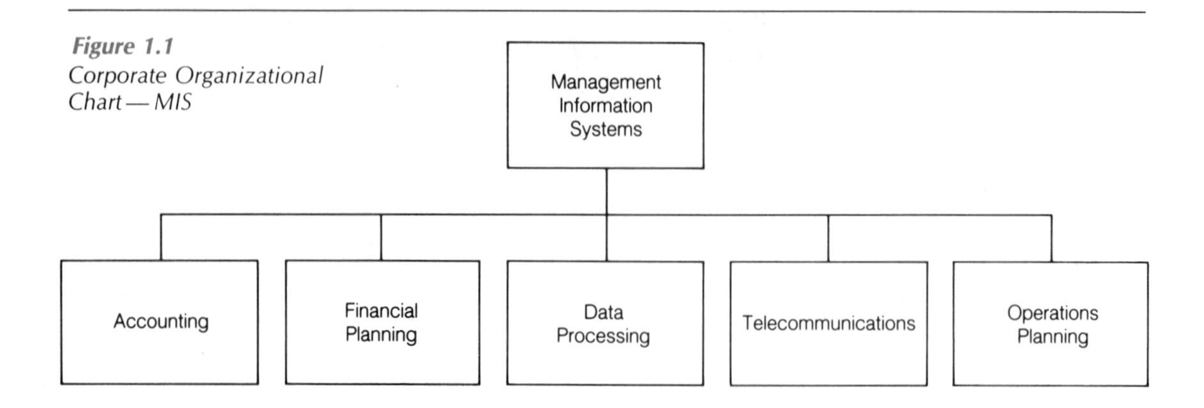

Figure 1.1
Corporate Organizational Chart — MIS

telecommunications manager and the skills required for this position will be examined in Chapter 11.

Figure 1.3 shows a career ladder that depicts the various opportunities available in telecommunications in a large organization. Similar career progressions exist within interconnect companies and operating telephone companies. Figure 1.4 on page 16 illustrates the organization of a typical corporate telecommunications department.

A corporate telecommunications department might include the following telecommunications professionals:

- **Junior Analyst** — Performs a variety of assistance functions (an entry level position).
- **Senior Analyst** — Assists in selecting new telephone systems and upgrading existing systems.
- **Project Manager** — Surveys new technologies and looks for ways to improve service and cut costs; maintains the telephone system.
- **Manager of Voice Communications** — Conducts long-term planning and manages day-to-day operations in voice communications.
- **Manager of Data Communications** — Conducts long-term planning and manages day-to-day operations in data communications.
- **Director of Telecommunications** — Oversees the entire department and has total responsibility for the telecommunications budget.

EDUCATION IN TELECOMMUNICATIONS

For many years, the only way to acquire telecommunications skills, competencies, and understandings was through on-the-job training, either in the telephone industry or in the military. Only recently have colleges and universities recognized the need for telecommunications education and implemented formal courses and programs in the discipline. Even today, however, education lags the field.

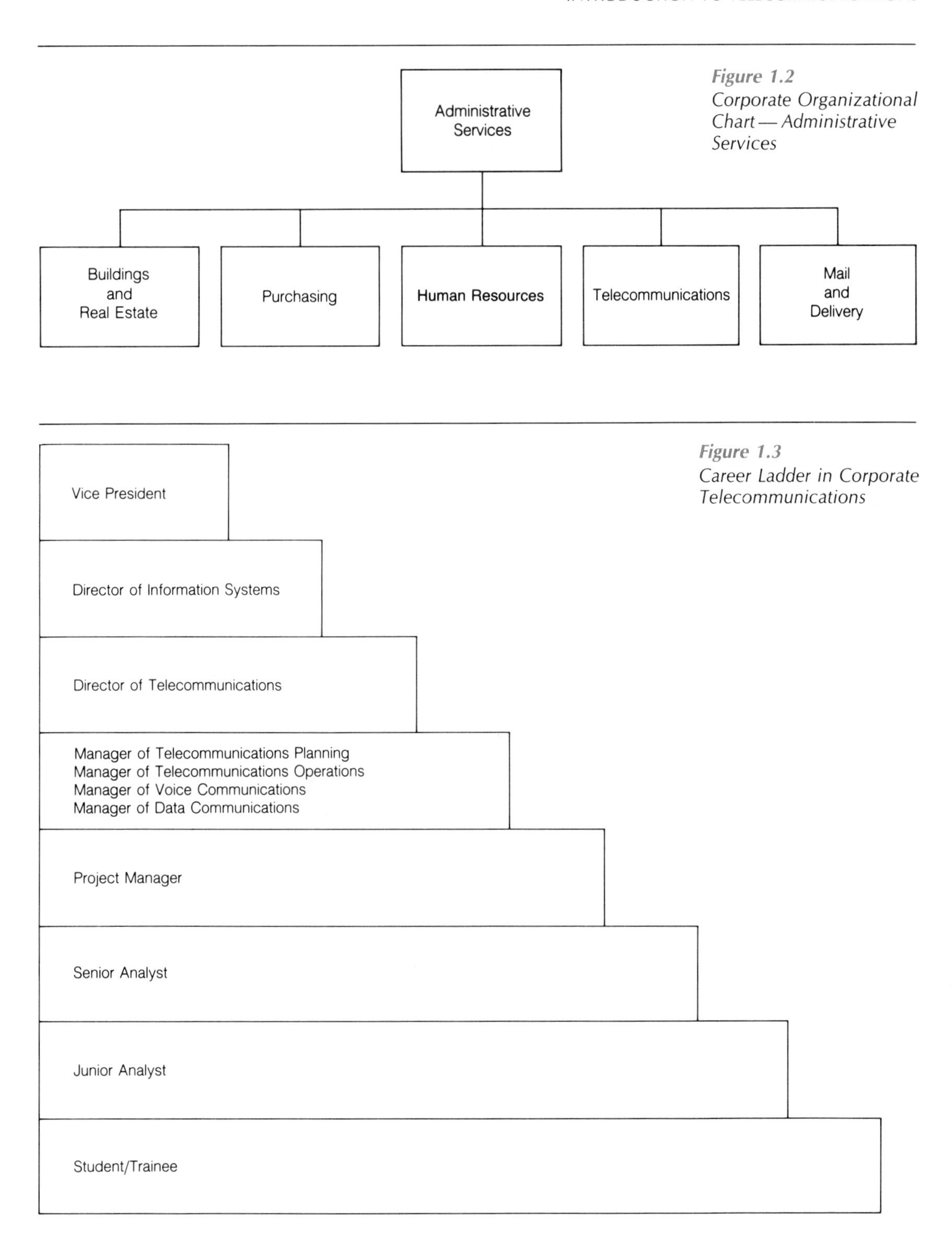

Figure 1.2
Corporate Organizational Chart — Administrative Services

Figure 1.3
Career Ladder in Corporate Telecommunications

Figure 1.4
Organization of a Typical Telecommunications Department

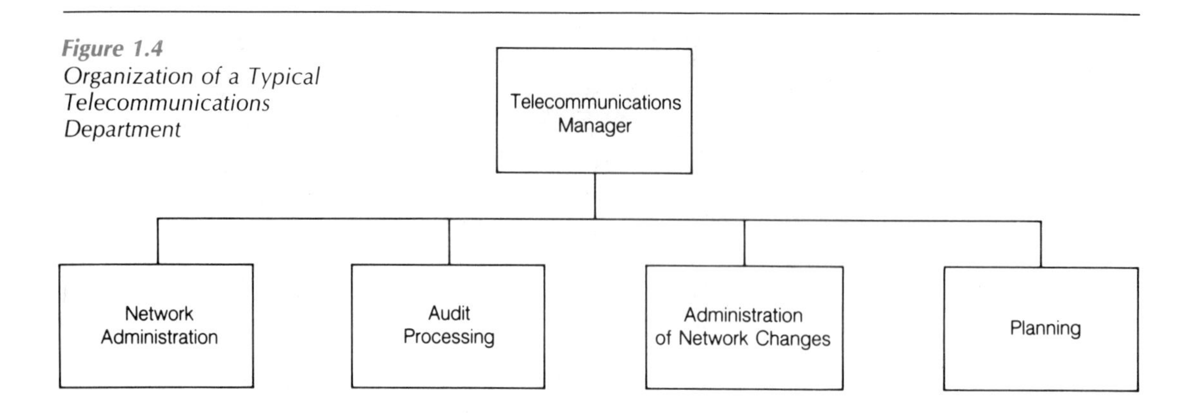

Some of the first telecommunications programs offered were in the engineering curricula, based in electrical engineering, and in radio and television broadcasting. More recently, community colleges have begun to offer two-year Associate Degree programs to train telecommunications technicians. The programs are electronics-based and include the mathematics involved in electronic circuits. These programs are often offered in collaboration with the telecommunications industry and generally include hands-on use of the equipment. Some proprietary vocational schools have also implemented similar programs.

Today, colleges and universities are beginning to address the needs of telecommunications management. Golden Gate University of San Francisco, the first college to offer a telecommunications management program, recently achieved another academic first by establishing a separate school of telecommunications. The aforementioned programs lead to an associate, baccalaureate, or graduate degree.

To meet the needs of professionals who are not interested in degree programs, however, some educational institutions offer certificates for a series of telecommunications courses. Many community colleges, colleges, and universities have also implemented introductory courses in telecommunications in their information management, computer science, instructional technology, office automation, and secretarial curricula.

In addition to course work offered by educational institutions, many organizations provide telecommunications seminars. These short courses cover a wide range of topics specific to the field and are designed for both entry-level and experienced telecommunications personnel. Many organizations regularly send their employees to these seminars. (Appendix C contains a list of organizations that offer such seminars.)

■ ***INDUSTRY RECEPTION OF GRADUATES*** The question asked most frequently by prospective students in telecommunications programs concerns job opportunities for graduates.

Professor Patrick Fitzgibbons of the State University of New York, College of Technology, reported that their 1988 telecommunications graduates have definitely been welcomed by industry. Their employment statistics show 100 percent placement of telecommunications graduates, with 98 percent placement within the first month after graduation. Starting salaries were found to be comparable to those of electrical engineering graduates and slightly higher than either computer science or business graduates.[4]

A study by the International Communications Association (ICA) of telecommunications graduates in industry concluded that, "Telecommunications programs, as represented in the survey, appear to be well-received in the industry, engaging in technology-specific managing and engineering functions and drawing the attendant high salaries."[5]

SUMMARY

This chapter introduced the reader to the field of telecommunications and its role in moving information. The marriage of computers and communications technology has enlarged both fields, making possible data communications, distributed processing, immediate accessing of centralized data bases, improved telephone switching, and a wide array of enhanced telephone service features.

Telecommunications plays a vital role in all business organizations. Additionally, there are industries whose very existence depends upon telecommunications. Among these are airlines, banking, credit-card operations, insurance, and investments.

Through the years the Federal Communications Commission had allowed the telephone industry to operate as a monopoly. Three interrelated factors led to the modification of this policy and the gradual restructuring of the industry: 1) technological developments, 2) customer demand for innovative products, and 3) FCC policy changes and court decisions. Beginning in 1968, a series of FCC and court decisions reversed previous telecommunications policy and gradually introduced competition into the industry, as well as increased user responsibility.

Technological advancements, industry deregulation, and the new competitive environment have promoted industry growth and have increased the importance of telecommunications. We should study this growing field because it has an increasingly vital role in our personal lives and our business world.

There is an urgent need for telecommunications management and telecommunications professionals. Career opportunities for both men and women can be found with interconnect companies, specialized common carriers, consulting firms, government agencies, the military, service companies, telephone answering services, and user organizations, as well as with the operating telephone companies.

In general, telecommunications education lags the field. However, this is beginning to change. Each year finds more telecommunications courses and programs being implemented in colleges and universities.

REVIEW QUESTIONS

1. Name and briefly describe the four principal types of telecommunications systems available today.
2. Briefly describe the relationship between computers and telecommunications.
3. What is a management information system?
4. What role does telecommunications play in data bases?
5. Name several industries whose existence depends upon telecommunications.
6. What three factors led to the gradual restructuring of the telecommunications industry?
7. Cite several reasons for studying telecommunications.
8. What are some of the types of industries in which telecommunications career opportunities can be found?
9. What are some of the jobs available in telecommunications?
10. How are telecommunications graduates faring in industry?

VOCABULARY

telecommunications

voice communications

data communications

message communications

image communications

telephony

telematics

management information systems (MIS)

data base

divestiture

ENDNOTES

1 Don Wiley, "Industry Observations," *Communications News* (February 1989), 2.

2 Kenneth Sherman, *Data Communications: A User's Guide* (Reston, Virginia: Reston Publishing Company, 1981), 11.

3 Mary M. Blyth, "Career Opportunities in Telecommunications," *Business Education Forum* (October 1985), 14–16.

4 Patrick Fitzgibbons, "Telecom Grads Welcomed by Industry," *Network World* (August 1, 1988), 22.

5 International Communications Association, "Major ICA Study Finds Typical Manager," *Communications News* (January 1983), 48–51.

CHAPTER

2

EARLY HISTORY OF TELECOMMUNICATIONS

CHAPTER OBJECTIVES

After completing this chapter, the reader should be able:

- *To discuss the early history of telegraphic communications.*
- *To describe the impact of the telegraph on politics and business.*
- *To trace the early history of the telephone industry.*
- *To describe the formation of the American Telephone and Telegraph Company.*
- *To describe the independent telephone industry.*
- *To discuss the provisions and significance of the Kingsbury Commitment*

The principal form of communication between people throughout the ages has been speech, a process which, until the nineteenth century, was limited by the distance between people. Early civilizations used coded light signals to transmit information over considerable distances, but it was the transmission of speech over distances that stirred people's imaginations. It was not until the nineteenth century, however, that the electromagnetic mode of speech transmission came into use.

Given the historical nature of the telecommunication industry's structure and operating methods, a perspective of its development will be useful in clarifying the dramatic changes taking place in the industry today. This chapter highlights the early history of telegraphic communications, the formation of the Western Union Telegraph Company, the impact of telegraphic communications upon the political and business organizations in the United States, the early history of telephony, the formation of the Bell

System, and the interrelationship between these two industries from the 1880s through the 1920s.

EARLY METHODS OF COMMUNICATION

Humans are social beings; one of our basic drives is to communicate with others. Familiar expressions such as "get it out in the open," "get it off my chest," and "talk it over" reflect our need for communication as a means of making us feel better. Psychiatrists, ministers, and counselors promote mental health as they help us "talk things over."

Businesses use a variety of methods to encourage two-way communication. Periodic consultations between employees and their supervisors; company-sponsored social events such as picnics, bowling teams, and Christmas parties; and business conferences all serve to promote upward and downward communication among employees at various organizational levels.

NONVERBAL COMMUNICATIONS

People have always conveyed information to each other in a variety of ways. In earliest times communication consisted primarily of signs, gestures, and facial expressions. These elementary, nonverbal methods, used originally to beckon, to warn, to approve, or to disapprove, are still effective means of communication. The friend waving to us from across the room, the police officer on the corner signaling traffic, the baseball umpire behind the plate, and the pedestrian hailing a taxicab, all use gestures to relay information to others.

Current interest in body language — a popular term for the large amount of information transmitted through posture, hand and arm movements, facial expressions, and other body-related means — has called attention to nonverbal communications. Researchers in this area claim that as much as 93 percent of a message's effect is derived from nonverbal exchanges. In any case, probably most of us would regard smiling as friendliness, nodding affirmatively as agreement, raising the eyebrows as skepticism, tapping the feet or drumming the fingers as impatience, and frowning as discontent.

Our first use of sound to communicate did not involve the human voice, but a variety of beating or tapping noises to attract attention. These sounds were generally produced by hitting the ground or another surface such as a hollow log. By striking with their fists or with a small mallet, early people could produce sound waves that traveled in all directions from the source. Later, our ancestors progressed to vocal but nonverbal means of communication such as cries, screams, groans, and squeals. These primitive methods, universally recognized as spontaneous expressions of emotion, continue to be used to convey the same meanings today.

The next step — the development of a system of speech — was one of the most important milestones in the history of human communication. Spoken language, with its many words, intonations, and inflections, introduced into the exchange of information a degree of precision previously unknown. Ever since humans first learned to talk, spoken language has been our principal means of communication. Its chief limitation was that the talker and listener had to be "within earshot."

Through the years people developed a number of long-range communication methods, including megaphones, smoke signals, oil lamps, bells, whistles, and semaphores. Many of them are still used today. For example, bells mark the time of day, invite us to worship, and announce football victories. The semaphore, an apparatus for signaling by means of flags, lights, or mechanical arms, is still used to supplement electronic methods in ship, railway, and air communications.

It was the transmission of spoken language over distances, however, that continued to excite people's imagination over the centuries. Finally, with the discovery of electricity in the early nineteenth century, the essential elements for long distance sound transmission were available.

EARLY HISTORY OF TELEGRAPHIC COMMUNICATIONS

The introduction of the electric telegraph of Samuel Findley Breeze Morse revolutionized communications. Morse, who earned his livelihood as a portrait painter and professor of the Literature of Arts and Design at New York University, had always been interested in science and invention. His interest in electricity was heightened when, during a return voyage from Europe, he watched a fellow passenger demonstrate how a piece of iron became magnetized by electric current. This demonstration led to discussions among other passengers of Michael Faraday's recent publication on magnetism.

THE MORSE TELEGRAPH

Upon returning home, Morse constructed telegraph sending and receiving instruments and formulated the principles of his dot-dash-space-code based on the duration or absence of electrical impulses. He developed his code by assigning dots and dashes to the letters of the alphabet, with the most frequently used letter having the simplest code. Thus, Morse code represents the frequently used letter e by a single dot, which requires the least electricity and transmission time; the infrequently used letter x is represented by dot-dash-dot-dot, which requires the most electricity and transmission time. Morse code consists solely of combinations of dots and dashes. A very short burst of electric current represents the dot and a slightly longer burst represents the dash.

Figure 2.1
Morse Sends the First Telegram

(Courtesy Western Union)

Prior to the development of telegraphy, the only long distance communication was by postal service. Early mail service delivered letters by horseback, stagecoach, and steamboat. Delivery schedules were infrequent and often unreliable. The inadequacy of long distance methods of communication in those days was poignantly brought home to Morse in 1825. He had been commissioned by the City of New York to paint a portrait of the Marquis de Lafayette, an appointment that took him to Washington. While there, he received word of the sudden death of his young wife back home in New Haven, Connecticut—word which, because of the slowness of communications, did not reach him until twenty-four hours after her funeral!

On September 2, 1837, Morse demonstrated his telegraph to a group of professors and friends. This exhibition suggested the practicability of the invention and resulted in the enlistment of two partners, chemistry Professor Leonard Gale and a young inventor, Alfred Vail, to help promote the telegraph. Late in September 1837, Morse filed a caveat for his invention in the United States Post Office. (A caveat is a notice of intent to file for a patent.)

In February 1838, Morse demonstrated his telegraph before President Martin Van Buren and his cabinet, hoping to have the telegraph accepted for government use. However, it wasn't until March 3, 1843, that Congress appropriated funds to test its workability. An experimental telegraph line was constructed between Washington and Baltimore. On May 24, 1844 (see Figure 2.1), Morse sent the first public telegram over this 40-mile line, transmitting the message, "WHAT HATH GOD WROUGHT!"

In spite of his successful demonstration, the United States postmaster general decided that the telegraph was a toy that probably would not be successful commercially, and the government withdrew its support.

COMMERCIALIZATION OF THE TELEGRAPH

Morse then enlisted private capital and in May 1845 organized the Magnetic Telegraph Company, the first telegraph company in the United States. By 1851, 50 telegraph companies using Morse telegraph patents were in operation in the United States, each serving a different section of the country. Accordingly, when a user wished to send a telegram over a long distance, it had to be retransmitted from company to company. This procedure not only slowed down delivery but also made it difficult to pinpoint responsibility when poor service resulted.

The various telegraph companies were highly competitive; some prospered while others had difficulty staying in business. In time it became apparent that consolidation of the companies would promote better service. However, petty jealousies blocked attempts at organizing the principal telegraph lines under the Morse patents into an association.

THE TELEGRAPH'S IMPACT ON POLITICS AND BUSINESS

The telegraph soon exerted a strong influence upon the political and economic life of the nation. Its impact upon railway operation and management was substantial. The telegraph provided the railroads with electric train dispatching, informing management of the location of every train on its rail system. The railroads, for their part, provided the telegraph companies with an exclusive right-of-way; telegraph wires were strung on poles beside the railroads. Since each industry had something of value to offer the other, their service contracts with each other became valuable assets.

Another industry closely associated with the development of the telegraph was the newspaper industry. The advent of the telegraph revolutionized the collection and dissemination of news. Since the industry was dependent upon receiving and printing the news as rapidly as possible, it had to employ the only method of rapid communication available at the time—the telegraph. The high cost of individual telegraphic services suggested the need for a cooperative news reporting service. As a result, the New York papers organized the New York Associated Press.

In return for dealing exclusively with certain leading telegraph companies and for promising not to print anything detrimental to these companies, the Associated Press received reduced rates on its business. Telegraph companies, too, benefited from this agreement.

Both state and federal government officials depended upon the telegraph in carrying out their responsibilities. Politicians. and government officials used the telegraph to keep in touch with their constituents, and the

Figure 2.2
Early Telegraph Office

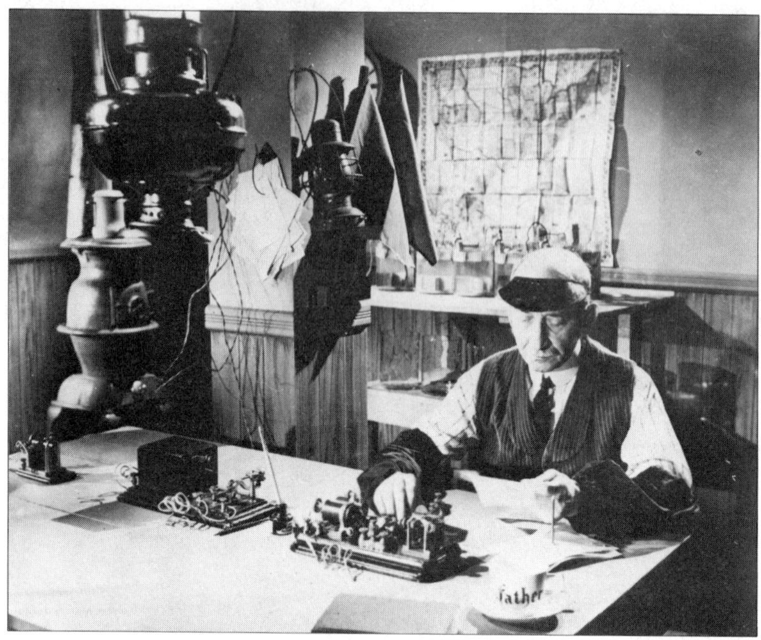

(Courtesy Western Union)

people at home used it to inform their representatives of their wishes with respect to legislative policies.

THE CIVIL WAR YEARS

By the beginning of the Civil War, telegraph lines linked most of the major cities in the nation (see Figure 2.2). The war years saw a dramatic rise in the use of telegraphic communications. The military, faced with assembling troops and supplies and moving them to the battle fronts, found the telegraph invaluable. Businesses, too, deluged the telegraph offices with messages in a desperate effort to get their affairs in order before contacts between the North and the South were broken.

The heavier concentration of railroads and telegraph lines in the northern states gave the Union forces a substantial advantage over the South during the Civil War. Further, the telegraph companies located in the North were relatively secure from the ravages of war, while those whose lines extended from north to south found many of their lines destroyed. Thus, some telegraph companies emerged from the war years in improved economic condition while others were barely able — or unable — to remain in business.

One of the telegraph companies whose business prospered during the war years was Western Union. Originally incorporated under the New

York and Mississippi Valley Printing Telegraph Company, the organization was reincorporated in 1856 as the Western Union Telegraph Company. With its telegraph lines located north of the Mason-Dixon line and the Ohio River, it was situated advantageously to profit from the increase in telegraph business brought about by the war.

When General George B. McClellan in 1861 took over the Department of the Ohio, which included western Virginia, Ohio, Indiana, Illinois, and later, Missouri, he appointed Anson Stager, general superintendent of Western Union, to administer for military purposes all telegraph lines within his department. Stager proceeded to coordinate the operation of the military and commercial telegraph lines within his jurisdiction into a highly effective system.

Western Union took advantage of its good fortune to improve its position within the industry. Gradually, it absorbed all of the competing independent telegraph companies. With the acquisition of the United States and American Telegraph Companies in 1866, Western Union emerged as the nation's largest corporation and its first powerful monopoly.

EARLY HISTORY OF TELEPHONY

On March 7, 1876, U.S. Patent No. 174,465 was issued to Alexander Graham Bell (Figure 2.3) for his invention of the telephone. This patent, often described as "the most valuable single patent ever issued," laid the foundation for an industry that serves nearly every neighborhood and home in the land.

"MR. WATSON, COME HERE. I WANT YOU!"

These historic words, shouted by Alexander Graham Bell to his assistant, culminated years of experimentation in sending electric messages over wires. "I can hear! I can hear the words!" Watson exclaimed excitedly as he rushed into Bell's workshop. The telephone had talked!

Like his father and grandfather, Mr. Bell was a teacher of elocution, or speech transmission. His research in the physiology of speech was no doubt influenced by both his father, the inventor of "Visible Speech," a written code used in training deaf persons to speak vocally, and by his mother, an accomplished musician whose own loss of hearing intensified his interest in the electric transmission of vocal messages. Bell earned his livelihood teaching classes for the deaf; at night he worked on his experiments. Bell postulated that if sound could be converted into electrical signals, it should be possible to transmit speech over a distance electrically. By employing a diaphragm to produce an undulating current, he proved his theory to be right. Thus, he learned the fundamental principle that enabled him to develop the telephone.

Figure 2.3
Alexander Graham Bell in 1876, the Year the Telephone Was Invented

(Reproduced with permission of AT&T)

At the same time that Bell was conducting his experiments, Elisha Gray, an expert electrician and well-known inventor in the field of telegraphy, was independently working on the same problem. Both Bell and Gray attempted to construct a device to transmit a number of telegraph messages over a single wire simultaneously using interrupted tones of different frequencies, a concept know as *harmonic telegraphy*. Bell approached the problem through his knowledge of acoustics; Gray's approach was through electricity. Each concluded that a combination of harmonics could be sent over a wire simultaneously; this discovery led each of them to postulate further that the human voice could generate impulses that could be transmitted over wires.

Bell's application for a U.S. patent was filed in Washington by his attorney on the morning of February 14, 1876. (See Figure 2.4.) For the lack of a better name, he called his invention "an improvement in telegraphy." A few hours later on the same day, Elisha Gray came to the same patent office and filed a caveat — a notice of intent to perfect his ideas to file a patent application within three months — also for an electric telephone. Since Bell's papers were the first to be filed, a patent for the telephone was issued to him on March 7, 1876. However, at this time no one had transmitted a single intelligible sentence by the telephone.

Bell and his assistant, Thomas A. Watson, pressed on with their experiments to test the workability of the theory described in the patent applica-

Figure 2.4
1876 Liquid Telephone

(Reproduced with permission of AT&T)

tion. At Bell's direction, Watson built a variable resistance transmitter that used sulphuric acid as a conductor of electrical current. The transmitter was set up in Bell's workshop (see Figure 2.5) and connected by a wire to a receiver in his bedroom. When the device was ready for testing, Bell adjusted the transmitter while Watson went into the bedroom and put the receiver to his ear. Almost at once he heard Bell's voice saying excitedly, "Mr. Watson, come here. I want you!" Watson rushed down the hall and found that Bell had upset the sulphuric acid, spilling it all over his clothes. Thus, on March 10, 1876, three days after the patent had been issued, Bell had developed a working telephone.

Financial backing for Bell's experiments was provided by the fathers of two of his students, Thomas Sanders, a successful leather merchant, and Gardiner Greene Hubbard, a prominent attorney. Sanders and Hubbard had furnished the money for Bell's experiments in return for an equal share in any patents obtained. However, they were anticipating his developing a harmonic telegraph, not a telephone. They had little interest in the telephone invention and were skeptical about its possibilities for commercial use. In the fall of 1876, they offered to sell all of Bell's patents to Western Union Telegraph Company for $100,000, but the offer was refused.

Figure 2.5
Alexander Graham Bell's Original Laboratory

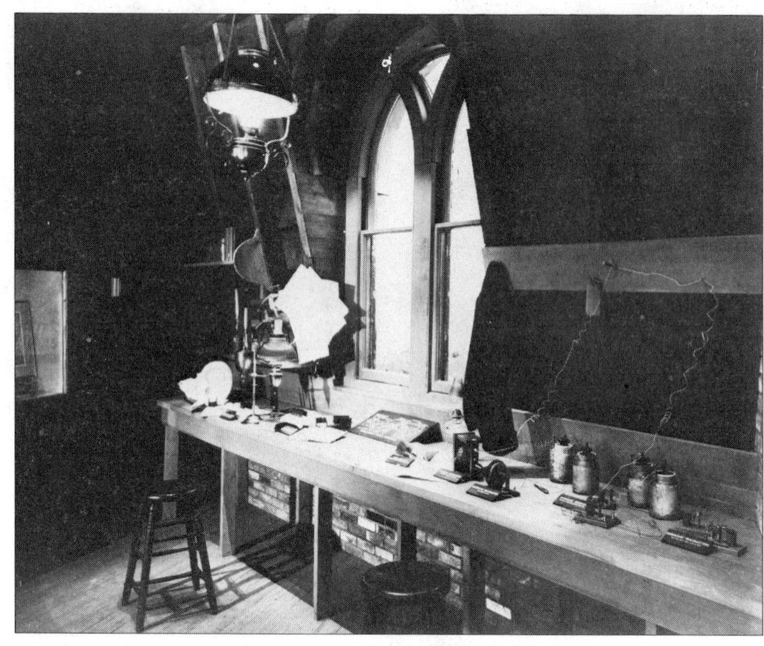

(Reproduced with permission of AT&T)

Meanwhile, Bell continued his experiments to improve telephone performance, hoping to ultimately realize a profit from his invention. On January 15, 1877, he filed an application for a patent for the *box telephone,* an improved version of his original telephone instrument (see Figure 2.6). The patent was issued on January 30, 1877. Bell's assistant, Thomas A. Watson, also continued to experiment and developed further telephone improvements including ringers (bells) and switchboards.

PUBLICIZING THE INVENTION

In the summer of 1876 Bell exhibited his "speaking telephone" at the Centennial Exposition in Philadelphia. It attracted little attention until Bell demonstrated it to Dom Pedro, Emperor of Brazil (Figure 2.7), and Sir William Thompson, British physicist (later Lord Kelvin). "My God, it speaks." said Dom Pedro, and Sir William Thompson, after careful scientific scrutiny, said it was the greatest thing he had seen in America.

On February 12, 1877, Bell gave a lecture before a well-known scientific society, the Essex Institute, at Salem, Massachusetts. The society members there were especially interested in the telephone because Bell had performed his early speech and sound experiments in Salem. This first lecture was free to members of the society, and it created so much interest that Bell was asked to repeat it—this time for an admission fee. Again, the house was filled and the lecture was well received. At the first public

Figure 2.6
The Box Telephone

(Reproduced with permission of AT&T)

Figure 2.7
Bell Demonstrates His Invention to Dom Pedro

(Reproduced with permission of AT&T)

demonstration, a newspaper correspondent from the *Boston Globe* telephoned a report of the lecture to his editor. This was the first use of the telephone in news reporting.

Spurred on by the public interest in his lectures and demonstrations, Bell assembled a presentation resembling a vaudeville show. He rented public halls and, assisted by Watson, entertained audiences by transmitting conversations and songs over the telephone. To add interest, he permitted members of the audience to talk to someone over the telephone. The demand for the lectures was fortuitous for Bell because it solved a temporary money problem. Bell had several urgent reasons for needing money. In addition to establishing a telephone company, he had fallen in love with and wished to marry Mabel Hubbard, Gardiner Hubbard's daughter who had been deaf since early childhood. During the next five months, Bell and Watson delivered lectures in New York and a number of New England cities. On July 11, 1877, Bell and Mabel Hubbard were married. Shortly thereafter, the couple sailed for England, taking with them a complete set of telephones.

THE TRUSTEESHIP

In July 1877 the telephone industry was formally organized by Hubbard with the creation of the Bell Telephone Company. The first organization was a *trusteeship* composed of Alexander Graham Bell, Gardiner Hubbard, Thomas Sanders, and Thomas A. Watson. With Hubbard acting as trustee, the organization began to manufacture and install telephones.

Hubbard, an attorney, had observed that one of his other clients, the Gordon-McKay Shoe Machinery Company, had built a highly successful business organization with its policy of leasing equipment instead of selling it. The company leased the shoe-sewing machines to shoemakers, retaining their title to the machines and receiving a royalty for every pair of shoes sewed with the machines.

By the powers vested in him, Hubbard made a decision that was to have far-reaching effects—the decision to rent telephones instead of selling them. This resulted in the sale of service only, which became the basic principle of the telephone industry. In spite of the critical need for business capital, Hubbard held firmly to the leasing principle as the basis for the development of the telephone business. This policy was a major factor in the financial success of the Bell system; later it became the policy of other telephone companies as well.

The trustees also authorized a system of *franchises,* whereby agents in various parts of the country could provide telephone service by paying a license fee to the trusteeship. In time, the trusteeship was supplanted by a corporate organization.

EARLY TELEPHONE SERVICE

The first telephone instrument bore little resemblance to the telephone that we know today. It was a crude, cumbersome apparatus consisting of a device that served both as a transmitter and receiver, and a connecting length of wire. In order to use the instrument, the user had to shift it back and forth between the mouth and ear. Occasionally, an affluent customer would obtain two instruments in order to use them simultaneously. One of the first telephone's biggest drawbacks was that it had no bell or call signal; thus, there was no way for a person to know that a call was waiting.

By the fall of 1877 there were about six hundred telephone subscribers, all using private lines. Each telephone was connected to another telephone by a direct line consisting of a single iron wire. There were no central exchanges and no switchboards; conversations could take place only between two telephones at each end of the line. On January 28, 1878, the first telephone exchange was opened in New Haven, Connecticut, serving 21 subscribers. The early exchanges served only a few customers, and calls were completed manually by an operator sitting at a switchboard. Calls were "put through" by name rather than by number. The exchange made it possible to connect any telephone with any of the other telephones in the exchange, greatly increasing the usefulness of the telephone. Increased demand for telephone service soon resulted, and within a few months a number of telephone exchanges were opened throughout the country.

On February 12, 1878, the New England Telephone company was formed. This was a licensing, not an operating, company. It held an assignment of rights to the Bell patents for New England and authorized agents to provide telephone service in return for the payment of a license fee.

On March 20, 1879, both the New England Telephone Company and the Bell Telephone Company were consolidated under the name National Bell Telephone Company.

EMERGENCE OF COMPETITION

Meanwhile, Western Union, which had been a successful telegraph company since 1856, reconsidered the prospects of telephony. Already national in scope, Western Union had an extensive network of wires connecting its offices in hotels, railway stations, and other public places that could be used as a nucleus to provide telephone service.

After Bell's patents for his telephone invention had been issued, patent applications for various forms of speaking telephones and transmitters were filed with the United States Patent Office by different individuals. Prominent among these were Elisha Gray of Chicago; Thomas A. Edison of Menlo Park, New Jersey; and Professor Amos E. Dolbear of Somerville, Massachusetts. Western Union purchased the Gray, Edison, and Dolbear patents and organized its own telephone company, the American Speaking Telephone Company. Ignoring the Bell patents, it began offering telephone service to the public. A period of intense competition followed. In many instances, both the Bell Company and the American Speaking Telephone Company established exchanges in the same town, which resulted in two telephone systems that were not interconnected.

Figure 2.8
Theodore N. Vail (about 1885)

(Reproduced with permission of AT&T)

VAIL'S LEADERSHIP

To counter this attack, the Bell Company leaders did two things. First, they hired a professional manager, Theodore N. Vail (Figure 2.8), to manage their organization. Second, they filed a lawsuit against Western Union for infringement of Bell's patents.

Vail had been superintendent of the Post Office's Railway Mail Service, where he was recognized as an outstanding business manager. He left this secure, well-paying job to become the Bell Telephone's first general manager in July 1878 and to face the challenge of directing the struggling young company. When Vail took over, there were only 10,755 telephones in service, competition was intense, and the Bell Company was faced with many technical problems. Vail brought to Bell the management expertise the company so badly needed. His leadership contributed much to the success of the Bell Company and led to the eventual formation of the Bell System.

PATENT LITIGATION

In September 1878 the Bell Company filed a suit in the Circuit Court of the United States, District of Massachusetts, against the giant Western Union Telegraph Company — technically against Peter A. Dowd, agent for the Western Union's telephone subsidiary — for infringement of the Bell patents. Western Union engaged George Gifford, a prominent patent attorney, as its chief counsel in the case.

After investigation, Gifford became convinced that the Bell Company would win and advised settlement of the suit. In November 1879 the two parties reached an out-of-court settlement. The settlement provided that Western Union withdraw from telephone service and sell its network and patents to the Bell Company. In return, Bell agreed to stay out of the telegraph business and to pay Western Union 20 percent of its telephone rental receipts over the 17-year life of the patents. This agreement added 56,000 telephones in 55 cities to the Bell Company. Over the life of the agreement, Bell paid Western Union approximately $7 million.

In March 1880, the American Bell Telephone Company, successor to the National Bell Telephone Company, was formed to carry on the consolidation of Bell and Western Union properties. This company remained parent company of the Bell System until December 30, 1899.

FORMATION OF AMERICAN TELEPHONE AND TELEGRAPH COMPANY

In 1885 a new company, the American Telephone and Telegraph Company (AT&T), was formed to build and operate long lines and render nationwide telephone service. The long distance lines interconnected the regional companies that had developed through the franchise agreements. For the first 15 years of its existence AT&T was a subsidiary of American Bell Telephone Company and was generally called the Long Distance Company. Theodore N. Vail became the first president of the American Telephone and Telegraph Company, a post he held until his resignation in 1887.

In 1900 AT&T absorbed the American Bell Company and became the headquarters company of the Bell System. It continued to provide long distance service through its Long Lines Department.

In 1911 AT&T consolidated the operations of the franchise companies into state or territorial units. These territorial units became the structure known as the Bell Associated Companies. Each of the companies paid a license contract fee to AT&T to cover costs of development of new equipment and improved telephone service. The license fee replaced the royalty fee that franchise companies had paid for the use of the Bell patents.

EARLY TELEPHONE MANUFACTURERS

The first telephones were manufactured in the electrical shop of Charles Williams, Jr., in Boston, where Bell and Watson had conducted many of their early experiments. As the demand for telephones increased, other small shops were licensed to manufacture telephones and related equipment to Watson's specifications. In the next few years telephones and telephone equipment were made by six manufacturers, each producing equipment of differing design and quality. It soon became apparent that a centralized source of high-quality, standardized equipment was needed.

The largest electrical manufacturer in the United States was the Western Electric Manufacturing Company of Chicago, the company that had supplied Western Union's telephone equipment. It had been organized in 1872 as successor to Gray and Barton, manufacturers of electrical equipment, including telegraph apparatus and fire and burglar alarms. From the first, the company gained a reputation for quality workmanship. In 1881 the Western Electric Manufacturing Company was reorganized to enfold some other telephone instrument and switchboard manufacturers.

Since Western Electric had pioneered in electrical equipment and telephone apparatus, the company was well qualified to manufacture Bell telephone equipment. On February 6, 1882, the Bell Company purchased the Western Electric Company and it became Bell's manufacturing unit and the sole supplier of Bell telephone equipment. Ownership of Western Electric gave the Bell System assurance of standardized equipment of high quality, economies of scale, and a dependable supplier.

THE INDEPENDENT TELEPHONE COMPANIES

The early years of the telephone's life saw dramatic improvement in service, rapid growth in the demand for telephones, and the advent of competition. As the value of the telephone became increasingly apparent, many persons appeared who claimed that they had developed instruments that would transmit speech over a distance electrically. Numerous legal battles over patent rights resulted. During the life of the Bell patents, the Bell Company was involved in over 600 lawsuits; it won every one.

The growing interest in telephones caused many people to regard the telephone industry as an attractive business opportunity. As a result, many new telephone companies were formed. Before the expiration of the Bell patents, more than 125 competing companies were in operation. Some of the new companies were organized under a Bell franchise agreement; others began operation "independent" of any Bell affiliation and in direct competition with the franchised companies. The term *independent telephone company* is still used today. Bell's defense against these competing companies took two directions: (1) litigation involving patent infringement and (2) refusal to interconnect independents' subscribers with Bell facilities.

After the expiration of the Bell patents in 1893 and 1894, more independent telephone companies entered the market. By the early 1900s independent telephony became a serious threat to the Bell System. The Bell System generally concentrated on serving the larger cities where potential customers were plentiful. The independents concentrated on serving small communities and rural areas ignored by Bell. However, sometimes two companies established facilities within the same locality. Since the two systems were not interconnected, a customer was limited to calling only those subscribers served by the same telephone company. In order to have contact with a customer of the other telephone company, it was necessary to subscribe to the other service. Businesses, particularly, found this impractical since they were unable to serve an entire community unless they had two telephones and two directories.

■ ***FORMATION OF A NATIONAL TRADE ASSOCIATION*** In June 1897 delegates from independent telephone companies met in Detroit, Michigan, to establish a national trade association for the industry. They adopted the name Independent Telephone Association of America and dedicated their efforts toward addressing mutual problems and strengthening their segment of the industry.

From 1897 to 1915 the association underwent many name changes and mergers, including one with the independent manufacturers' association. Finally, in 1915, the United States Independent Telephone Association (USITA) was chartered, incorporating the previous organizations. During this period of growth and competition, USITA played a vital role in uniting the industry and providing information and advice on many subjects relating to telephony. For many years it has been recognized as the national voice of the independent telephone industry.

In October 1983 USITA dropped the word "independent" from its name and became U.S. Telephone Association (USTA) to attract Bell operating companies (BOCs) to its membership.

■ ***INDEPENDENT MANUFACTURERS*** Prior to the expiration of the Bell patents, many independent experimenters were working on improvements to the telephone. One of the most spectacular developments was the dial system constructed by Almon B. Strowger, a Kansas City undertaker. His original instrument (Figure 2.9) had five pushbuttons lined up in a row. To

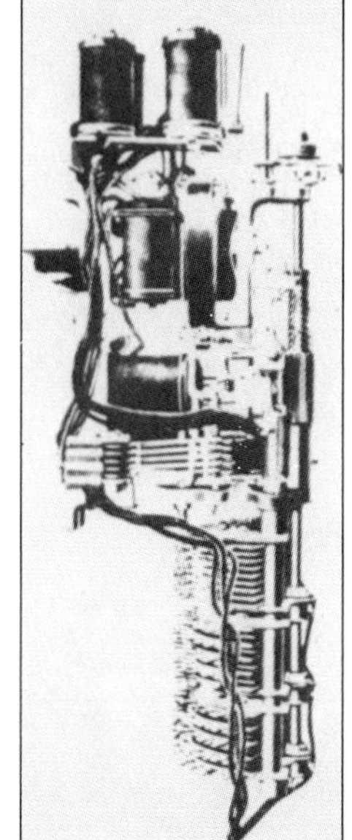

(Courtesy of Bell Laboratories)

Figure 2.9
Strowger Switch

Figure 2.10
Strowger Finger Wheel Dial

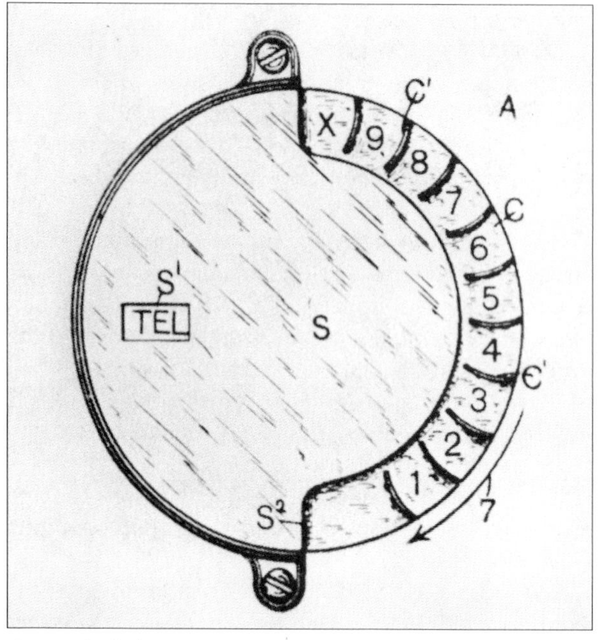

(Courtesy of Bell Laboratories)

call number 21, for example, the user simply pushed the first button twice and the second button once. Strowger later introduced a large dial in place of the buttons (Figures 2.10 and 2.11). Strowger's first unit went into service in La Porte, Indiana, and its use spread rapidly among independent telephone companies. Later, the Bell System also adopted it.

The independent manufacturing and operating companies also pioneered a number of other developments, including a handset telephone, the forerunner of the instrument in use today. It contained a transmitter and receiver in the handle and was connected to the telephone by a cord.

The larger independent telephone companies owned their manufacturing and supply companies. The largest independent telephone company, General Telephone and Electronics (GTE), owned two manufacturing subsidiaries: Automatic Electric and Lenkurt Electric. These subsidiaries manufactured communications equipment for sale not only to GTE companies but also to other firms and governments in the United States and abroad.

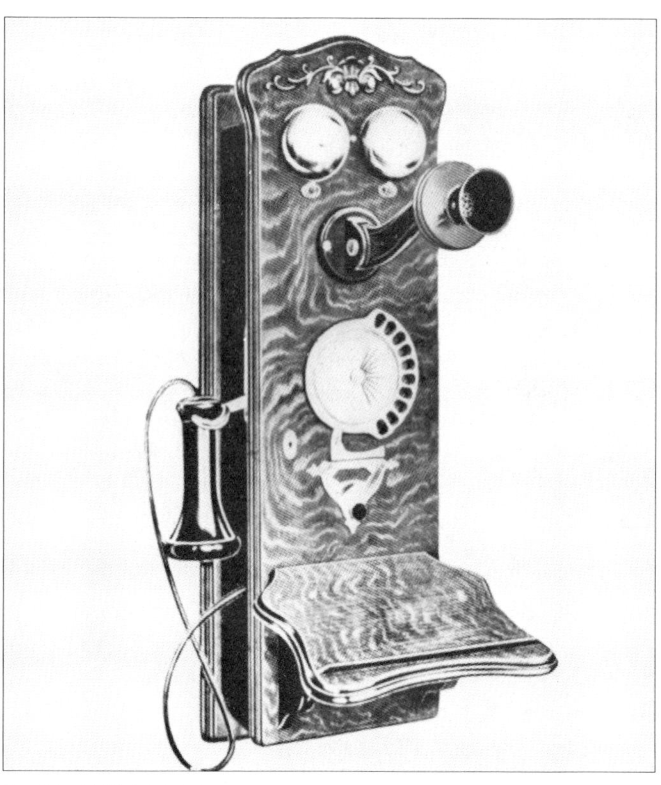

(Courtesy of Bell Laboratories)

Figure 2.11
Strowger Wall Set

VAIL RETURNS

Competition had caused a severe financial drain on AT&T. Between 1902 and 1907 its debt grew from $60 million to over $200 million. In 1907 Theodore N. Vail came out of retirement, at the age of 62, to assume the presidency of AT&T for the second time during a period in which the company was in critical condition. In addition to being on the verge of bankruptcy, its service had deteriorated, and its public image was very poor.

In an effort to make AT&T the sole supplier of telecommunications services in the United States, Vail continued the policy of buying independent telephone companies. He also decided that it would be advantageous for AT&T to get into the telegraph business. Accordingly, in 1909 AT&T bought 300,000 shares of Western Union stock, enough to give it working control. In 1910 Vail became president of Western Union, making him president of both companies.

THE KINGSBURY COMMITMENT

During the first decade of the twentieth century, AT&T had employed these tactics to combat the independents:

1. rate cutting
2. buying out competitors
3. refusing to interconnect

The most powerful weapon was the refusal to interconnect. Since Bell, under patent protection, had developed an extensive long distance network, its service was of more value than that of the independents who had few long distance facilities.

In time, the ruthless competition between telephone companies and the drawbacks of duplication of telephone facilities became apparent to federal and state legislators. In 1910 Congress took the first step toward industry regulation, placing certain aspects of interstate telephone operations under the jurisdiction of the Interstate Commerce Commission. The commission was given jurisdiction over telephone companies for just and reasonable charges, passes and franks, preferences and prejudices, filing contracts, reports to the commission, investigations, furnishing information, joint rates, uniform system of accounts, and preservation of records.

In 1912 the independent companies protested to the Department of Justice that the Bell organization, now controlled by AT&T, was operating in violation of the antitrust laws. In January 1913 the U.S. attorney general responded by advising AT&T that, in his opinion, some planned acquisitions of independent telephone companies were in violation of the Sherman Antitrust Act. That same month the Interstate Commerce Commission began an investigation to determine whether or not AT&T was monopolizing communications in the United States.

Faced with mounting public sentiment against monopolies and a probable government antitrust suit that could break up the company, AT&T reconsidered its position. In December 1913 AT&T vice-president Nathan C. Kingsbury sent a letter to the U.S. attorney general committing AT&T to the following:

1. disposing of its stock in Western Union
2. refraining from acquiring additional independent telephone companies, except as approved by the Interstate Commerce Commission
3. interconnecting its facilities with those of the independents so that the independents could offer nationwide telephone service to their customers

This letter is generally referred to as the *Kingsbury Commitment.* The commitment was a victory for the independents because it prevented their being absorbed by the Bell System and assured them of interconnection with the Bell System. Also, like the Bell operating companies, the independents became monopolies within their respective areas. The commitment established the independents as an integral part of the communications industry and marked the beginning of the two groups' working together toward common goals.

THE GRAHAM-WILLIS ACT

In 1921 Congress passed the Graham-Willis Act, which exempted telephone companies from the provisions of the Sherman Antitrust Act with regard to the acquisition of competing companies. This measure removed the constraints of the Kingsbury Commitment upon AT&T, affording it the opportunity to resume expansion of its territory.

Many persons feared that AT&T would take undue advantage of this opportunity; however, this did not happen. AT&T continued to live up to its commitment, and the 1920s witnessed a unified telephone industry, for the first time.

SUMMARY

The electric telegraph and a code system for its use were developed by Samuel F. B. Morse. After a demonstration of his invention to President Van Buren and his cabinet, Congress appropriated $30,000 to install a telegraph line between Washington, D.C. and Baltimore. On May 24, 1844, Morse sent the first public telegram over this line, transmitting the message, "WHAT HATH GOD WROUGHT!"

Telegraph lines were strung on poles beside the railroad and by 1860 linked most of the major cities in the country. The telegraph was a vital contributor to the development of the railroads, the newspapers, and the New York Associated Press.

The Civil War years saw a dramatic rise in the use of telegraphic communications by businesses and the military. The network of railroads and telegraph lines in the northern states was an important advantage to the Union Army during the Civil War.

The early telegraph system was composed of many small companies. This meant that in order to send a long distance telegram, the message had to be retransmitted from company to company. One of the telegraph companies whose business prospered during the Civil War was Western Union. Gradually, Western Union bought out all of the competing telegraph companies, and by 1866 it was the nation's largest corporation and its first powerful monopoly.

The telephone was invented in 1876 by Alexander Graham Bell, a teacher of speech to the deaf. In 1877 the telephone industry was formally organized with the creation of the Bell Telephone Company. The first organization was a trusteeship. This organization made a decision that was to have far-reaching effects—to rent telephones rather than sell them. Selling only telephone service became a major practice of the telephone industry.

In 1877 Western Union organized the American Speaking Telephone company and began offering telephone service to the public. After a period of intense competition, Bell initiated a lawsuit against Western Union charging patent infringement. Convinced that it would lose if it pursued the matter in court, Western Union reached an out-of-court settlement with Bell providing that Western Union withdraw from the telephone business and sell its telephone network to Bell. Bell, in turn, agreed to stay out of the telegraph business.

In 1885 the American Telephone and Telegraph Company was formed to build and operate long lines for nationwide telephone service. In 1900 AT&T became the headquarters company of the Bell System. Before the expiration of the Bell patents, many new telephone companies were formed. Some of the new companies were organized under a Bell franchise agreement; others began operating "independent" of any Bell affiliation and in direct competition with the franchised companies. After the expiration of the Bell patents, competition between Bell and the independents intensified. Bell employed several tactics to combat the independents, including rate cutting, buying out competitors, and refusing to interconnect with Bell's lines.

In 1909 Bell bought controlling interest in Western Union; and Bell's president, Theodore N. Vail, became president of both companies. In 1911 AT&T consolidated the franchise companies operating under Bell patents into state or territorial units.

In 1912 the independent telephone companies protested to the U.S. Department of Justice that AT&T was operating in violation of the antitrust laws. Faced with a probable lawsuit, AT&T vice-president Nathan Kingsbury sent a letter to the U.S. attorney general committing AT&T to disposing of its stock in Western Union, refraining from acquiring additional independent telephone companies, and interconnecting its facilities with those of the independents. This letter became known as the Kingsbury Commitment.

In 1921 the Graham-Willis Act exempted telephone companies from the Sherman Antitrust Act.

REVIEW QUESTIONS

1. Why was the invention of the telegraph so important?
2. In Morse code, the letter e *is represented by a single dot, and it requires the least electricity to transmit. Why?*
3. Why was it disadvantageous to have many small telegraph companies?
4. How did the telegraph play an important part in the development of the railroads?
5. What was the impact of the Civil War on the telegraph industry? What was Anson Stager's role in uniting the telegraph industry?
6. What was the impact of the Civil War upon Western Union?
7. In the early days of the telephone, how was Western Union able to begin offering telephone service?
8. What did the Bell Company do to counter Western Union's telephone service?
9. What important principle was decided by the out-of-court settlement between Western Union and the Bell Company?
10. Why was it important to the Bell Company to have a single source of supply for telephone equipment?
11. Historically, what has been the role of AT&T in telephony?
12. What is the significance of the word *independent* in the independent telephone companies? What role did these companies play in early telephony?

VOCABULARY

telegraph

harmonic telegraph

trusteeship

franchise

independent telephone company

Kingsbury Commitment

CHAPTER

3

STRUCTURE AND REGULATION OF THE TELECOMMUNICATIONS INDUSTRY—

AN HISTORICAL PERSPECTIVE

CHAPTER OBJECTIVES

After completing this chapter, the reader should be able:

- *To discuss the purpose of regulation and describe its role in the telecommunications industry.*
- *To describe the regulatory process.*
- *To name the two regulatory agencies and define their jurisdiction.*
- *To describe the historical composition of the Bell System.*
- *To describe the organizations that provided common carrier telecommunications services prior to divestiture.*
- *To discuss the major FCC and court decisions leading to the partial deregulation of the telecommunications industry.*

Since 1866 the telecommunications industry in the United States has been subjected to some degree of regulation because the services that it provided were essential to the public welfare. When the industry was developing, it was considered a "natural monopoly," wherein the public would not benefit from competition among telephone companies serving the same area. As the industry matured, this economic concept began to be questioned. Today, local exchange service is allowed to operate as a monopoly, but telecommunications equipment and long distance telephone service markets have been opened to competition.

This chapter examines the regulatory process, common carrier service providers, and the steps taken toward industry deregulation.

GOVERNMENT REGULATION

The Kingsbury Commitment brought an end to competition in the early telecommunications industry, and it was allowed to operate virtually as a monopoly until the interconnect issue arose in the 1950s.

The industry's pre-Kingsbury experience with competition had clearly demonstrated the inefficiencies of running multiple wires along the streets and operating two or more telephone companies that were not interconnected. As a result, popular support for government regulation emerged, forcing the telephone companies to evaluate its options.

The alternatives to competition were government regulation or government ownership. Clearly, the telephone companies did not want to be taken over by the government. Furthermore, they reasoned that if they supported regulation, they could probably influence the form it took. However, if they opposed it, they risked having an unfriendly form of regulation forced on them. Accordingly, both Bell and the independents embraced regulation.

WHY REGULATION?

The rationale for regulation stems from the characteristics of the industry. An industry that supplies essential services to the public welfare is considered a *natural monopoly,* wherein only one organization can operate efficiently in a given market. The government permits such industries to exist as monopolies, but by regulation it ensures that they will not abuse their monopolistic power. This control is accomplished by prescribing strict rules for the conduct of business.

Regulated monopolies enjoy an exclusive franchise to serve a specific geographic area, granted by the state, county, or local governing body. They are sometimes referred to as *franchised monopolies.* The purpose of exclusive franchises is to prevent the duplication of services in high-density areas and the absence of service in low-density areas, thus guaranteeing uniform service to the public over a broad geographic area. Historically, the nature of the telecommunications industry was such that regulation was considered necessary.

THE REGULATORY PROCESS

Regulation is a rule or order that is issued by an executive branch of government and that has the effect of law. Regulation is employed as a substitute for competition where competition either does not exist or where its existence is judged not to be in the public interest.

The purpose of regulation is to assure fair treatment of the rate payers and investors. The process of regulation consists of monitoring and controlling the operation of telephone companies, to ensure their observation of specified policy guidelines. These guidelines, established by the regulatory agency, provide for the following:

1. service on demand
2. uniform policies to all users
3. high-quality service
4. fair rates to customers
5. fair return to stockholders

ROLE OF REGULATION

In all the countries of the world, there is either some type of regulation or the government itself owns and operates the telecommunications system, acting as the sole supplier of telecommunications services.

In many countries, the department responsible for operating the telecommunications system is called the Post, Telephone, and Telegraph (PTT). As the name indicates, the same regulatory agency that is responsible for the postal service is also responsible for the telecommunications service.

In the United States, regulation takes an intermediate position between government services that are administered as public functions and supported through taxation and private enterprises that are subject to no controls and whose prices are determined competitively in the open market. The telephone companies are publicly owned, publicly managed enterprises that sell their services to the general public. However, prices set by the companies are subject to government approval.

The Congress or state legislatures provide only the broad principles of a particular regulatory scheme, leaving their detailed implementation to the agency charged with administering the law (the Federal Communications Commission at the national level and the public utilities commissions at the state level). Modern regulatory agencies are vested with legislative, executive, and judicial powers. They can prescribe regulations having the force of law, police those subject to their authority to ensure that such regulations are not violated, and hear and decide cases involving alleged violations.

STATUS OF REGULATION

Today, the telecommunications industry is undergoing the most challenging period in its history. Advances in technology, coupled with the breakup of the Bell System (described later in this chapter), have changed the telecommunications market considerably. Basic telephone network service is still considered a public utility whose efficient operation is in the public interest. However, the monopolistic aspect of the industry is being debated.

The deregulatory movement in the United States is being watched closely by the rest of the world, and a few other countries are beginning to take steps toward privatization and/or deregulation of their telecommunications industry.

COMMON CARRIERS

In its broadest sense, the term *common carrier* describes any supplier in an industry that undertakes to "carry" goods, services, or people from one point to another for the public. Early common carriers were the pony express and the railroads. In telecommunications, such "carriage" refers to providing transmission capability over the telecommunications network.

A telecommunications carrier that offers communications services to the public is subject to regulation by federal and state regulatory agencies. Before a carrier can construct or operate facilities, it must obtain a license from the regulatory agency. The carrier is accountable to the regulatory agency for providing specified services within its defined geographic area of operation, for applying its policies uniformly to all users, for providing a specified quality of service, and for establishing reasonable rates to users while not exceeding a specified maximum rate of return on investment.

There are two types of telecommunications common carriers — those that are fully regulated and those that are partially regulated. The latter provide long distance service by interconnecting with the facilities of the fully regulated carriers.

TARIFFS

The published rates, regulations, and descriptions governing the provision of communications services are known as *tariffs*. The tariff document does the following:

1. defines the services offered
2. establishes the rate the customer will pay for the service
3. states the obligations of the public utility and the customer in the provision and use of the service

All tariffs must be approved by the appropriate regulatory agency — federal or state — before they can become effective. When approved, they carry the full force of law and serve as a contract between the carrier and the end user. Approved tariffs can also be used by other carriers.

Tariffs must be filed when the following conditions exist:

1. new services are offered
2. rates for existing services are increased or decreased
3. changes are made in existing services
4. services are discontinued

■ ***RATE CASE*** One of the most important aspects of the regulatory process is the approval of rates for the various types of telecommunications services. Rates are established by submitting a tariff document to the appropriate regulatory agency, usually 30 days before the tariff is to take effect. This is called a *rate case*. Written testimony from expert witnesses in both financial and technical areas, justifying and supporting the proposed rate(s), must accompany the tariff document. This information is also distributed to potentially interested parties.

The commission can respond to the tariff request in any one of three ways: (1) it can approve the tariff, (2) it can reject the proposed tariff, or (3) it can delay the request and initiate a public hearing on the request. When a public hearing is held, the public utility must present evidence to show why the petition's request should be granted.

If the rate is approved, a *regulatory order* is issued, thereby making it law. This usually completes the procedure; however, either side has the right to appeal through the appropriate courts.

Tariffs are a function of regulation. To protect the public, all fully regulated common carriers are required to file tariffs. However, as a result of partial deregulation of the telecommunications industry, some telecommunications offerings to the public do no require tariffing. In some cases, certain carriers are required to file tariffs for a service while other carriers are not.

DEVELOPMENT OF REGULATORY AGENCIES

The regulation of telecommunications in the United States began in 1866 with the passage of the Post Roads Act. This act authorized the Postmaster General to fix rates for government telegrams. In 1878 a judicial interpretation of the Commerce Clause of the act included telegraphy as an instrument of commerce subject to federal control. This established a precedent for the subsequent regulation of wire and radio communications.

The Interstate Commerce Act of 1887 established the Interstate Commerce Commission (ICC) to regulate railroads. A provision of the act empowered the ICC to order the interconnection of the lines of telegraph companies in the interest of providing through services.

In 1910 the Mann-Elkins Act extended the authority of the ICC by bringing the regulation of interstate telephone companies under its jurisdiction. The ICC, however, had been created to regulate interstate commerce and focused its attention on regulating transportation rather than communications. It confined its regulation of interstate telephone lines to investigating complaints and prescribing accounting systems rather than investigating the reasonableness of long distance rates and the company's return on investment.

■ ***STATE REGULATORY AGENCIES*** The powers of the individual states to regulate commerce can be found in the Tenth Amendment to the Constitution, which states that "the powers not delegated to the United States by the Constitution . . . are reserved to the states respectively, or to the people." Each state, however, could regulate commerce only within its own borders.

The first steps toward communications regulation were taken by the states. In the competitive era preceding the Kingsbury Commitment, the Bell System had adamantly refused to interconnect its long distance with its competitors' lines. Many states viewed this as an abuse of power that was contrary to the public interest and passed laws requiring the physical interconnection of telephone companies.

The first communications regulatory agencies were established in the states of New York and Wisconsin in 1907. By 1920 more than two-thirds of the states had established regulatory commissions. Today, all 50 states plus the District of Columbia have commissions that, among other activities, regulate intrastate telecommunications.

The state regulatory agencies are called by different names in different states; The two most frequently used names are *public utility commissions (PUCs)* and *public service commissions (PSCs).* In addition to regulating intrastate telecommunications, the state agencies regulate several other industries, including power, gas, and transportation.

■ ***THE FEDERAL COMMUNICATIONS COMMISSION (FCC)*** By the 1930s it was apparent that a revised regulatory structure was needed to centralize authority over telecommunications. Responding to this need, Congress passed the *Communications Act of 1934.* This act initiated the modern regulatory environment for the telecommunications industry by doing the following:

1. creating the Federal Communications Commission and defining its powers
2. establishing as a national goal universal telephone service (a telephone in every household) at affordable rates
3. promoting the construction of the most rapid and efficient telephone system possible

The Federal Communications Commission is an independent government agency, responsible directly to Congress, and charged with regulating interstate and international communications originating in the United States. It is not responsible for regulating intrastate communications — that is the function of the state regulatory agencies.

For many years the FCC consisted of a board of seven commissioners, who were appointed by the President to staggered terms of seven years. In 1983 the membership was reduced to five. The law provides for bipartisanship by requiring that not more than a simple majority — three commissioners — may be from the same political party. One of the commissioners is designated by the President to serve as chairperson.

The FCC regulates interstate and international communications by radio, wire, cable, and television. Its responsibilities include:

1. encouraging the development and operation of broadcast and communications services at reasonable rates
2. licensing and regulating broadcast stations

3. reviewing and evaluating station performance
4. approving changes of ownership and major technical alterations
5. regulating cable television
6. prescribing and reviewing accounting practices
7. regulating and issuing licenses for all forms of two-way radio
8. reviewing applications of telephone and telegraph companies for change in rates and service
9. setting permissible rates of return for communications common carriers
10. reviewing technical specifications of new telecommunications equipment

COMMON CARRIER SERVICE PROVIDERS

AT&T is the largest of the telecommunications common carriers. The following description portrays the Bell System prior to the time the Modified Final Judgment went into effect (1984). The present structure of the telecommunications industry will be discussed in Chapter 4.

THE BELL TELEPHONE SYSTEM

The *Bell System,* often referred to as "Ma Bell," consisted of the companies controlled by AT&T. The units that comprised the Bell System were AT&T; the 22 Bell Associated Companies, later known as the *Bell Operating Companies (BOCs);* Western Electric Company; and Bell Telephone Laboratories. (See Figure 3.1).

AT&T was the headquarters company of the Bell System. Subject to the regulation of the Federal Communications Commission, AT&T interconnected the BOCs by means of its long distance lines; provided a centralized advisory service; controlled Western Electric, the manufacturing and supply unit for the system; and maintained Bell Telephone Laboratories, Inc., an extensive organization devoted to research, development, and design for the communications field.

AT&T and the 22 BOCs comprised the operating units of the Bell System. The BOCs were owned either wholly or in part by AT&T. However, they were operated and managed by local personnel and were responsible for providing telephone service in the areas in which they were established. The Long Lines Department of AT&T linked each regional company with all the others, extending the services given in each locality to every other part of the country. The BOCs operated under state laws and were subject to regulation of the states in which they operated.

To handle the general problems common to all the BOCs and to avoid duplication of expense and effort, the parent company provided the BOCs with services that could best be performed by a centralized organization. These services included manufacturing, engineering, technical research,

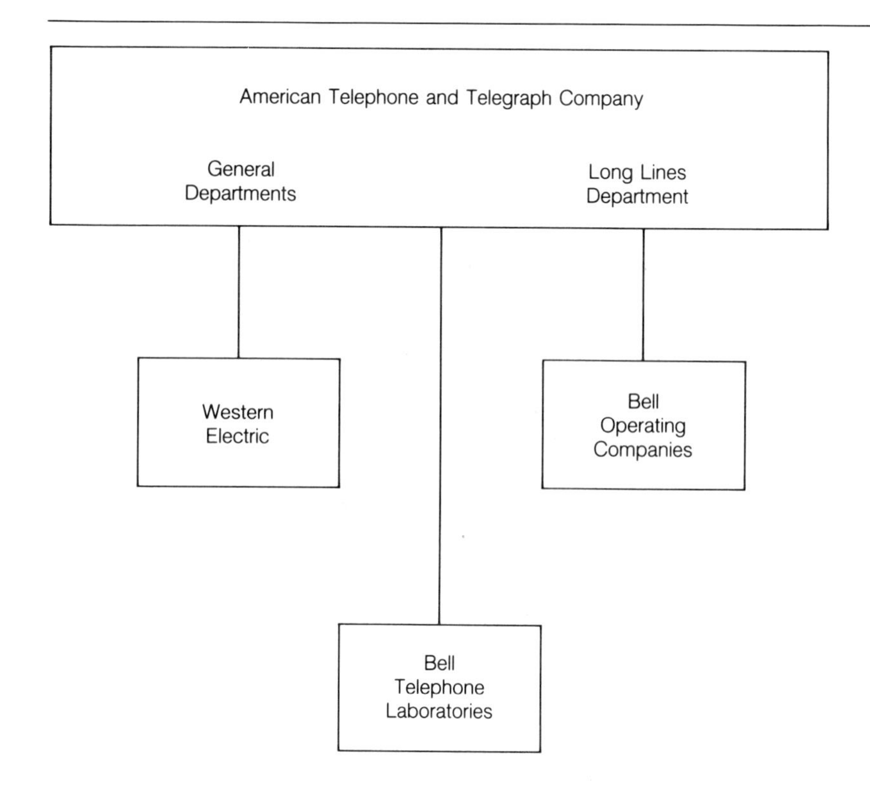

Figure 3.1
Historic Organization of the Bell System

and advice and assistance on operating, legal, accounting, and financial matters. In return for these services, the BOCs paid a license fee to AT&T as part of their operating costs. In addition, they paid all of their net profits to the parent company. These profits were the principal source of AT&T's revenue, enabling it to pay dividends to its stockholders.

WESTERN UNION

Western Union was the first telecommunications company in America. It was founded in 1856 to market the invention of telegraphy. The company grew rapidly and was soon operating a telegraph service to all parts of the United States. In 1943, Western Union acquired Postal Telegraph, Inc., its most serious competitor. Following World War II, the market for telephone services rose sharply, while the market for telegraph services declined steadily. Western Union strengthened its position by offering a teletypewriter service (Telex) and private-line services for voice, data, and facsimile. In 1971, Western Union purchased the Teletypewriter Exchange Service (TWX) from AT&T and began offering this service to its customers.

GENERAL TELEPHONE AND ELECTRONICS CORPORATION

General Telephone and Electronics (GTE) is the second largest domestic communications common carrier and the largest independent telephone company. It serves about 8 percent of American telephones, offering a wide range of voice and data services.

Historically, GTE had two subsidiaries that manufactured telephone equipment: Automatic Electric and GTE Lenkurt, Incorporated. They also owned other subsidiaries, including Sylvania Electric, the British Columbia Telephone Company (Canada), and Telenet, a packet-switching network. In June 1983 they acquired the SPRINT network from Southern Pacific Communications.

INDEPENDENT TELEPHONE COMPANIES

Historically, the non-Bell companies were known as independent telephone companies. They serviced about 20 percent of the telephones in the United States. The independent companies interconnected with long distance carriers to provide service to virtually anywhere. The Bell System was barred from acquiring independent companies by the Kingsbury Commitment.

Most of the independent telephone companies were members of the United States Independent Telephone Association (USITA), which provided an organizational structure to represent them in matters concerning long distance revenues, technical standards, and regulatory matters.

SPECIALIZED COMMON CARRIERS

The emergence and proliferation of computers in the 1960s created a demand for specialized private-line services that were not provided by AT&T. The development of microwave radio transmission technology provided a means of meeting this demand without relying on existing wireline networks. Advancements in technology resulted in the development of a new type of common carrier, the *specialized common carrier (SCC),* which provides long distance service over microwave, satellite, or leased-line facilities. Many of the SCCs' services are designed for customers who transmit large volumes of data. As common carriers, the SCCs are allowed to file tariffs under different regulatory rules, which results in a different pricing structure.

The first SCC licensed by the FCC was Microwave Communications, Inc. (MCI). It began providing long distance telephone service telephone service between St. Louis and Chicago in 1972, relaying telephone calls over microwave stations. Initially, MCI serviced only major cities; however, it now provides service to every city in the United States and to many other countries. Microwave Communications, Inc., the original FCC petitioner, is the parent company of a number of affiliate companies, the largest of which is MCI Telecommunications Corporation, the long distance company. Although considerably smaller than AT&T, MCI is a leading competitor in providing long distance service in the United States.

The first SCC to offer coast-to-coast service was Southern Pacific Communications Company (SPC), which in 1974 began to operate its own private microwave relay system. Originally owned by Southern Pacific Transportation Company, its communications system was designed to support its transportation activities. Services provided to the public resulted in excess capacity in the system. In June 1983 Southern Pacific Communications Company and Southern Pacific Satellite Company were purchased by General Telephone and Electronics. Their long distance service was called Sprint, later to become US Sprint (jointly owned by GTE and United Telecommunications).

VALUE-ADDED CARRIERS

In 1983 another new type of common carrier, known as a *value-added carrier (VAC),* emerged. A VAC leases transmission facilities from existing common carriers and adds computer-controlled services that increase the value of the basic transmission facility. Services provided by the VACs include data transmission, facsimile, electronic mail, and voice mail.

INTERNATIONAL COMMON CARRIERS

International carriers provide a broad spectrum of telecommunications services between nations. These carriers operate only from designated "gateway cities," which include New York, Washington, Miami, New Orleans, Los Angeles, and San Francisco.

STEPS TOWARD INDUSTRY DEREGULATION

By 1934 AT&T was thoroughly established as the controlling firm in the telecommunications industry. It owned approximately 80 percent of the telephones and the principal long distance network in the United States. The remaining 20 percent of the phones were owned by a large number of small independent companies who depended on interconnection with AT&T's network to complete their long distance calls. From 1934 to 1956 the telecommunications industry operated in an environment of regulated monopoly.

CONSENT DECREE OF 1956

In January 1949 the United States Department of Justice brought an antitrust suit against AT&T, alleging violation of the Sherman Antitrust Act. The suit asked that Western Electric be separated from the Bell System and that Western Electric be split into three separate companies, which would compete with other companies for Bell System business.

The suit was settled in 1956 by a consent decree that permitted AT&T to retain its ownership of Western Electric but precluded Western Electric from manufacturing any equipment of a type not sold or leased to Bell System companies. It also limited the Bell System's activities to the telephone business and government projects. This was a most important limitation since it meant that the Bell System would be barred from entering any type of electronic data processing or consumer-related activity that was not limited to common carrier communications.

TERMINAL EQUIPMENT

The equipment at each end of a communications channel that permits the stations to send and/or receive signals is called *terminal equipment*. It is also known as *customer-premises equipment (CPE)* or *interconnect equipment*. The terminal equipment that we are most familiar with is the telephone. Other types of terminal equipment include private branch exchanges (PBXs), modems, teletypewriters, answering machines, and facsimile machines.

From the beginning of the telephone industry, the telephone companies had required their subscribers to use telephones and other terminal equipment provided by them. They accomplished this through tariffs for telephone service, which included the provision of terminal equipment at a monthly rental fee. The tariffs specifically prohibited the use of any equipment that was not provided by the telephone company, which they called *foreign equipment,* and gave the telephone companies the right to remove such equipment and even to discontinue telephone service. This restriction applied not only to telephones but also to such innocuous devices as shoulder rests. The telephone companies claimed that the restriction was necessary in order to protect their network; They argued that the use of any equipment not provided by them could harm the network and/or impair the quality of transmission.

During the late 1950s and early 1960s, many customers bought novelty and color telephones from other suppliers and connected them to the telephone lines without the company's knowledge or approval. If the telephone company were to discover the presence of a foreign attachment, the tariffs permitted them to disconnect the device or terminate service. However, the enforcement of such tariffs proved to be an administrative impossibility.

■ ***THE HUSH-A-PHONE DECISION*** The first successful challenge to the tariff restriction came in the Hush-A-Phone case. The Hush-A-Phone was a cup-like device that was placed on the telephone mouthpiece to direct the speaker's voice into the transmitter. It was a passive, nonelectrical device with the sole purpose of making conversation more private.

After Hush-A-Phone had marketed the device for several years, AT&T warned distributors of the product that the device violated the tariff prohibiting the use of any equipment not furnished by the telephone company. In December 1948 Hush-A-Phone filed a complaint with the FCC, asking that it be allowed to sell the device without the telephone company interference.

In 1956 the matter was finally settled in Hush-A-Phone's favor, and AT&T was ordered to revise its tariffs to permit the use of devices that were "privately beneficial without being publicly detrimental."[1] This was the first decision to chip away at AT&T's absolute control over all telephone equipment.

The case established the very important principle that "some public harm must be shown in order to justify restrictive tariffs, and that it is not the right of the telephone company to control how a subscriber uses his telephone so long as the only impact of the subscriber's use is to affect his own conversation."[2]

■ ***THE CARTERFONE DECISION*** Shortly after the revised tariff went into effect, the Carter Electronics Corporation, a small Dallas-based manufacturing company, again challenged the foreign equipment restriction. The proceeding began in 1961 as a private antitrust suit brought by Carter Electronics against AT&T and GTE in the United States Federal District Court in Texas.

Carter's original objective was quite limited. The company's primary interest was in marketing the Carterfone, a device used to interconnect private two-way radio communication systems with the public telephone network. Carter sought to prove that the tariffs did not apply to the Carterfone since the device did not have an adverse effect on the telephone system. The telephone companies argued that Carter's device did have such an effect.

In June 1968 the FCC decided that the Carterfone could be connected to the public telephone system. The interstate tariffs were amended to allow telephone companies to install a protective device to prevent damage to their equipment.

■ ***THE INTERCONNECT INDUSTRY*** The Carterfone decision changed the structure of the telecommunications industry irrevocably. This decision is considered by many to have been the first step toward divestiture, which opened the way to competition in connection of customer-owned telephone equipment to the Bell and independent telephone company networks. The protective device portion of the tariff was later expanded to require approved FCC registration of any equipment to be connected to telephone company lines. As a result of the Carterfone decision, many companies started to build telecommunications equipment for sale to individuals and businesses. And a new industry, which became known as the *interconnect industry,* was born.

Telecommunications equipment rapidly took on a new look, becoming decorative as well as functional. The variety in telecommunications equipment appealed to subscribers, who enjoyed making their own decisions and expressing their individuality through telephone equipment. Interconnect companies also offered economic advantages. They provided the opportunity for subscribers to reduce their monthly phone bill by owning their own terminal equipment. A number of suppliers soon emerged to fill this need, including ROLM, GTE, General Dynamics, Stromberg Carlson, MITEL, ITT, Northern Telecom, Rockwell International, Nippon Electronics, and OKI Electronics of America.

MCI DECISION

Shortly after the Carterfone decision opened the way for competition in interconnect hardware, the FCC decided that it would be in the public interest to allow "nontelephone" companies to provide common carrier services on a specialized basis in direct competition with existing carriers. The decision was based on a case filed by Microwave Communication, Inc. (MCI), which requested permission to provide private, leased-line telephone service between Chicago and St. Louis via microwave. The ruling, which was known as the MCI Decision, also required existing telephone companies to furnish interconnect service to the new common carriers.

The overall effect of the MCI Decision was to open the way for development of a new segment of the telecommunications industry, the specialized common carriers (SCCs), of which MCI was the first.

SPECIALIZED COMMON CARRIER DECISION

As a result of the MCI decision, a number of other companies filed applications for permission to provide specialized telecommunications service. Rather than deal with each of these applications individually, the FCC issued the Specialized Common Carrier decision in 1971. This action expanded the ruling of the MCI decision to include other specialized carriers. In so doing, the FCC encouraged the offering of new, innovative services not available from the existing common carriers.

■ ***VALUE-ADDED CARRIERS (VACs)*** In 1973 the FCC granted permission to three companies — Packet Communications, Inc. (PCI), Graphnet, and Telenet — to operate value-added networks (VANs) and sell their services, primarily for data transmission. The VAC generally owns no transmission facilities but leases channels from other carriers and enhances the channels or "adds value" by sophisticated computer control to provide new types of service offerings.

The first VAC to receive FCC approval was Packet Communications, Inc., whose plans never materialized. The first operational VAC was Graphnet Systems, Inc. It operated a network to interconnect otherwise incompatible facsimile machines. The third VAC to receive FCC approval was Telenet Communications Corporation, which operated a network to interconnect otherwise incompatible computers.

■ ***RESALE CARRIERS*** In 1981 the FCC approved the resale and shared use of public switched long distance services, and another type of carrier, the *resale carrier,* came into existence. (The resale and shared use of private-line facilities had been allowed by the FCC in 1976).

A resale carrier generally leases transmission facilities from other carriers but owns its own switching equipment. In some cases, however, resale carriers provide services that are completely leased from other carriers, as in the shared use of WATS lines. Resellers are generally found in larger cities, where the market offers more potential customers and therefore greater profits.

■ ***OTHER COMMON CARRIERS (OCCs)*** The SCCs, the VACs, and the resellers, all carriers that provide alternatives to the traditional full-service telephone companies, are collectively referred to as *Other Common Carriers*. The name had its origins in the Bell System practice of referring to the new carriers, somewhat haughtily, as the "other common carriers." The term was first used to describe the SCCs. Later its use was expanded to include other new types of carriers.

THE COMPUTER INQUIRIES

The computer inquiries, launched by the FCC, examined the relationship between the data processing and the telecommunications industries and the possible effects of AT&T's entrance into the data processing industry.

The growing interdependence between the two industries had resulted in the blurring of boundaries between them. Since the communications industry was regulated and the computer industry was not, it became necessary to define the boundaries of each industry in order to determine whether a particular offering was a data processing service or a communications service.

■ ***COMPUTER INQUIRY I DECISION (CI-I)*** In its 1971 ruling, the FCC defined the difference between data processing and data communications. The ruling also stated that the data processing industry was not subject to FCC control. In essence, this meant that the communications carriers could transport data over their networks on a regulated basis, but they could not process it.

■ ***COMPUTER INQUIRY II DECISION (CI-II)*** In 1976 The FCC began a second inquiry into regulatory and policy questions raised by the interdependence of computers and communications. In 1981 this investigation resulted in a decision known as *FCC Computer Inquiry II (CI-II)*, which specified the following:

1. computer companies could transmit data on an unregulated basis
2. the Bell System could engage in data-processing activities
3. enhanced services and customer premises equipment would be deregulated and provided by a fully separate subsidiary of the carriers
4. basic communications services would continue to be regulated

In CI-II the FCC no longer tried to define data communications separately from data processing. Rather, it established two classes of communications services: basic and enhanced.

Basic services were defined as those providing *only* for the transportation of data. This meant that the data traveled unchanged through the network. Basic services would remain regulated.

Enhanced services were those services providing anything beyond basic transportation, such that the information reaching its destination was "different, additional, or restructured" in comparison to the information sent by the source. Enhanced services would not be regulated by the FCC.

The CI-II decision attempted to prevent cross-subsidization of unregulated services by regulated services by requiring the carriers to provide unregulated services through separate companies.

As a result of this decision, AT&T organized American Bell, a fully separated unregulated subsidiary, and renamed the Long Lines Department as AT&T Communications (ATTCOM). Thus, in 1981 the newly aligned AT&T consisted of ATTCOM, American Bell, Western Electric Company, Bell Laboratories, and the BOCs.

The name American Bell, however, proved to be short-lived. As explained in the next section, a stipulation of the Modified Final Judgment (the name given to the settlement in the government antitrust suit of 1974 against AT&T) was that only the Bell operating companies could use the word *Bell* in their company names, with the exception of Bell Laboratories.

■ ***CI-II MODIFICATIONS*** On December 23, 1986, the FCC eliminated the Computer II requirement that AT&T maintain separate subsidiaries for its regulated and unregulated products and services. The adoption of new accounting methods and network disclosure rules eliminated the need for separate companies to prevent cross-subsidization of regulated and unregulated businesses. As a result, AT&T merged AT&T Communications and AT&T Information Systems into a single unit, which it called AT&T Communications and Information Services.

MODIFIED FINAL JUDGMENT

In November 1974 the Justice Department filed an antitrust suit against AT&T. The suit charged monopolization and the conspiracy to monopolize the supply of telecommunications services and equipment and asked that Western Electric be separated from the Bell System. The suit also asked that some or all of the Long Lines Department and perhaps other parts of the Bell System be separated.

The suit dragged on for years; delays were attributable primarily to the hard line taken by AT&T, which spent millions in defense. For years, AT&T argued that divestiture would damage the national defense and ultimately destroy a uniform, quality telephone system thoughout the United States. But the Justice Department remained adamant in its prosecution of the case.[3]

During this time, Congress was studying a variety of legislative proposals that would restructure the telecommunications industry and rewrite the existing statutes. Through all this, certain realities were becoming clear:

- Regardless of what happened in any of these arenas, the marketplace was to be the key instrument for governing the telecommunications industry in the future.
- The Bell System was perceived in some quarters as being too big and too powerful.

Ultimately, however, AT&T reconsidered its options and decided to accept the divestiture solution developed by the Justice Department. Charles L. Brown, Chairman of the Board of AT&T, explained the decision saying, "We were confident that we could ultimately disprove the government's contentions. But we concluded that getting rid of the terrible uncertainty and capitalizing on future market opportunities were more important than vindicating our past behavior in a marketplace that no longer existed, and more important than preserving a corporate structure that would leave our future behavior continually vulnerable to antitrust attacks."[4]

Terms of the original tentative settlement included the following points:

1. AT&T would have to divest itself of all its associated operating companies. It could keep the segments of its business that provided CPE and long distance services.
2. No relationships between the divested companies and AT&T or one another could exist.
3. The divested companies would have to provide equal access to all long distance carriers.
4. The divested companies would be able to provide basic services but would be prohibited from providing long distance services, information services, or any type of nontariffed services. Services would be confined to specified Local Access and Transport Areas (LATAs).

The original decree was modified slightly by Judge Greene and became known as the *Modified Final Judgment (MFJ)*. The modifications included the following:

1. The divested BOCs would be allowed to provide, but not manufacture, CPE.
2. Local operating companies would be allowed to produce and distribute telephone directories.
3. Local exchange companies would be permitted to provide interexchange service and manufacture equipment, provided they could show evidence that their entry into the market would not stifle competition.
4. AT&T would be prohibited from offering electronic publishing services for seven years.
5. AT&T would have to relinquish all rights to the "Bell" name and logo to the operating companies. (This ruling required the renaming of the new American Bell subsidiary; AT&T chose AT&T Information Systems as the new name.)

Because of the magnitude of changes imposed by the judgment, it was agreed that Judge Greene would review the terms of the settlement every three years to ensure that they were still applicable in view of current conditions.

SUMMARY

Competition in the early telecommunications industry in the United States came to an end with the Kingsbury Commitment. Congress established the national policies and regulatory framework, under which the industry was to operate for the next 50 years, by passing the Communications Act of 1934. However, during the last two decades, the telecommunications industry has moved from a completely regulated, monopolistic structure to a largely unregulated, competitive structure.

When the telecommunications industry in the United States was developing, it was considered a "natural monopoly" because the services it provided were essential to the public welfare. The government permits such industries to operate as a monopoly, but by regulation the government ensures that the monopolistic power will not be abused.

The telecommunications industry is regulated by two principal authorities—the Federal Communications Commission and individual state regulatory bodies. The FCC was established by the Communications Act of 1934 to regulate interstate and foreign communications originating in the United States. The state regulatory agencies, collectively referred to as public utility commissions (PUCs) but with different names in different states, regulate the conditions of intrastate communications for the carriers in their jurisdictions.

Tariffs are the published rates, regulations, and descriptions governing the provision of communications services. All tariffs must be approved by the appropriate regulatory agency before they can become effective. Since intrastate tariffs are governed by the individual states, they vary widely from state to state. Interstate tariffs, however, are uniform throughout the United States.

A common carrier is a government-regulated company that provides services to the public. Telecommunications common carrier service providers in the pre-divestiture era included the Bell Telephone System, Western Union, General Telephone and Electronics, specialized common carriers, and international carriers. The largest common carrier was the Bell Telephone System. It consisted of AT&T, the headquarters company; the 22 Bell Operating Companies; Western Electric, the manufacturing unit; and Bell Telephone Laboratories, the research and development organization.

The Consent Decree of 1956 allowed AT&T to own Western Electric, but it prevented the Bell System from entering electronic data processing.

Terminal equipment is the equipment at the end of the communications channel that sends or receives signals. It includes telephones, PBXs, modems, teletypewriters, answering machines, and facsimile machines.

In the 1960s the FCC began making a series of decisions aimed at introducing competition in the telecommunications industry. The first of these was the landmark Carterfone decision, which allowed the interconnection of non-Bell equipment to the telephone network. This resulted in the birth of a new industry, the interconnect industry, which sold customer-premises equipment on a competitive basis.

The next major decision was the MCI decision, which permitted Microwave Communications, Incorporated, a "nontelephone" company to provide private, leased-line telephone service in direct competition with existing carriers. Because this was a specialized type of telephone service, the service provider became known as a specialized common carrier. Soon another new industry, the specialized common carrier industry, came into existence.

In 1981 the Computer Inquiry II decision specified that computer services be divided into two classes: basic and enhanced. Basic services, those providing anything beyond basic transportation, would not be regulated. This decision reversed an earlier decision barring AT&T from engaging in the data processing business. Thus, AT&T was now able to enter the vast field of data processing.

The most momentous decision in recent telecommunications history was AT&T's acceptance of divestiture of the Bell Telephone System. In 1982 the Justice Department and AT&T agreed to settle an antitrust lawsuit brought against AT&T by the Justice Department. The original consent decree was modified slightly by Judge Greene and became known as the Modified Final Judgment. The most important outcome of the MFJ was that on January 1, 1984, the 22 Bell Operating Companies were separated from AT&T. The Bell name was reserved for the use of these companies.

REVIEW QUESTIONS

1. What is the purpose of regulation?
2. What is a telecommunications common carrier?
3. What three functions does the tariff document serve?
4. Name the two regulatory agencies and identify their jurisdiction.
5. What was the significance of the Communications Act of 1934?
6. Who were the principal common carrier service providers prior to divestiture?
7. What FCC decision resulted in the birth of the interconnect industry and what type of products does it provide?
8. Discuss the origin of the specialized common carriers.
9. Historically, what companies comprised the Bell System and what was the function of each company?
10. What was the most important outcome of the Consent Decree of 1956?
11. What was the principal thrust of the Computer Inquiry II decision?
12. What was the principal outcome of the Modified Final Judgment?

VOCABULARY

regulation

natural monopoly

Post, Telephone, and Telepraph (PTT)

common carrier

tariff

rate case

regulatory agency

Public Utility Commission (PUC)

Public Service Commission (PSC)

Federal Communications Commission (FCC)

Communications Act of 1934

Bell System

Bell Associated Companies

Bell Operating Companies (BOCs)

General Telephone and Electronics (GTE)

specialized common carrier (SCC)

value-added carrier (VAC)

Consent Decree of 1956

terminal equipment

customer-premises equipment (CPE)

interconnect equipment

foreign equipment

Hush-A-Phone decision

Carterfone decision

Microwave Communications, Incorporated (MCI)

MCI decision

Specialized Common Carrier decision

resale carriers (resellers)

other common carriers (OCCs)

Computer Inquiry II (CI-II)

basic services

enhanced services

Modified Final Judgment

ENDNOTES

1 Leonard Lewin, ed., *Telecommunications in the United States: Trends and Policies* (Dedham, MA: Artech House, Inc., 1981), 45.

2 Gerald W. Brock, *The Telecommunications Industry: The Dynamics of Market Structure* (Cambridge, MA: Harvard University Press, 1981), 239.

3 Samuel A. Simon, *After Divestiture: What the AT&T Settlement Means for Business and Residential Telephone Service* (White Plains, NY: Knowledge Industry Publications, Inc. 1985), 13–14.

4 Harry M. Shooshan, III, ed., *Disconnecting Bell: The Impact of the AT&T Divestiture* (New York: Pergamon Press, 1984), 4.

CHAPTER

THE POST-DIVESTITURE TELECOMMUNICATIONS INDUSTRY STRUCTURE

CHAPTER OBJECTIVES

After completing this chapter, the reader should be able:

- *To explain the significance of the Modified Final Judgment and its effect on the telecommunications industry.*
- *To describe the reorganized AT&T.*
- *To describe the reorganized BOCs.*
- *To distinguish between the regulated and unregulated portions of the telecommunications industry.*
- *To describe the composition of the post-divestiture telecommunications industry.*
- *To discuss the key issues in the post-divestiture era.*

When the largest corporate divestiture in U.S. history became effective on January 1, 1984, the telecommunications industry, as most Americans knew it, changed dramatically. Prior to this time the average subscriber thought of the industry in terms of the "telephone company," one company that provided terminal equipment, local and long distance service, and installation and repair service. Thus, subscribers had only one telephone company to deal with and received only one monthly telephone bill.

In today's telecommunications environment, subscribers must provide and be responsible for the terminal equipment located on their premises. Thus when a trouble condition develops, the subscriber is responsible for repair, unless the trouble is in telephone company lines. The subscriber must also select a long distance carrier. This results in the subscriber's receiving a monthly telephone bill from each of the companies involved.

INDUSTRY REORGANIZATION

Because AT&T was the dominant organization in the country's telecommunications industry, the reorganization of AT&T was, in effect, the reorganization of the country's telecommunications system.

THE DIVESTITURE

Prior to divestiture, AT&T was the largest company in the world, with assets of $155 billion and one million employees. It was perceived by some as being too big, too powerful, too pervasive. Divestiture split one giant organization into eight smaller, discrete entities.

Subsequent to the divestiture agreement, AT&T and the Justice Department agreed upon a Plan of Reorganization, a blueprint for how the nation's telephone system was to be restructured. The plan addressed a variety of details, the most important of which were the following:

1. The Regional Bell Operating Companies (RBOCs) — the number and composition of the Regional Bell Operating Companies that would be formed by grouping the BOCs.
2. Local Access and Transport Areas (LATAs) — the formation of new geographic areas, LATAs, that would distinguish local telephone service areas from long distance service areas.
3. Equal Access — what steps to take and when to offer the other long distance companies the same type, quality, and price of access to the BOCs' local exchanges as AT&T Long Lines (renamed AT&T Communications after divestiture) enjoyed.

THE REORGANIZED AT&T

Immediately after divestiture, AT&T reorganized as AT&T Communications, the regulated long distance company; and AT&T Technologies, an unregulated corporation combining research, manufacturing, and equipment marketing.

Today's AT&T, however, is composed of AT&T Bell Laboratories, the research and development arm; AT&T Network Systems, the central office and computer equipment marketing arm; AT&T International, the international service and equipment provider; AT&T Technology Systems, the manufacturing arm (formerly known as Western Electric); and AT&T Communications and Information Systems, the provider of long distance service and customer-premises equipment. (See Figure 4.1.)

The existence of the Bell System ended with divestiture. The new AT&T, devoid of its role as "headquarters company," was a much smaller organization. Losing its affiliation with the BOCs meant that AT&T no longer

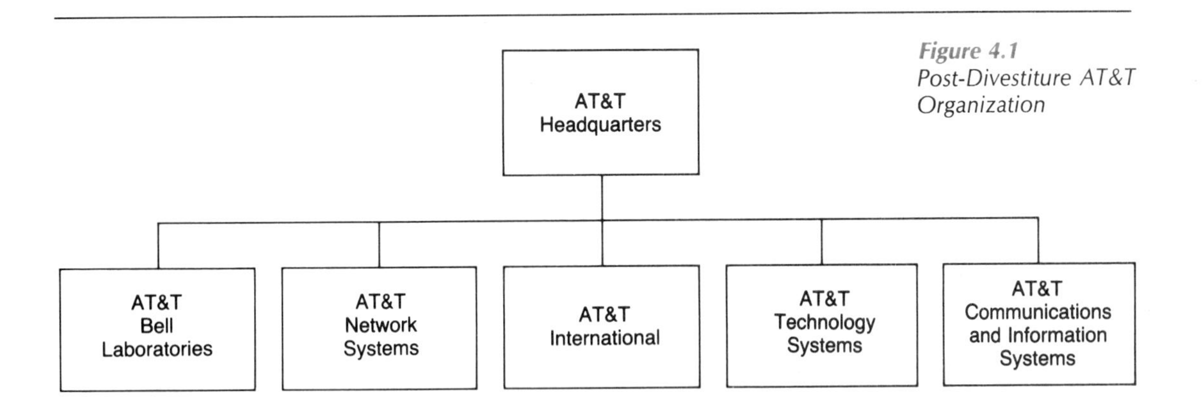

Figure 4.1
Post-Divestiture AT&T Organization

received the BOCs' net profits as its principal source of income. Today, AT&T's income comes from two principal sources: 1) the sale of long distance telephone services and 2) the manufacture and sale of telephone equipment.

THE REORGANIZED BOCs

Following divestiture, the 22 Bell Operating Companies were formed into seven *Regional Bell Operating Companies (RBOCs)* that were entirely independent of AT&T and each other. The seven regions are NYNEX, Bell Atlantic, BellSouth, Ameritech, Southwestern Bell, US WEST, and Pacific Telesis Group. (See Figure 4.2.)

The individual companies that comprise each RBOC are shown in Figure 4.3. It is interesting to note that although the name *Bell* was deemed the sole property of the operating companies, only three regional corporations — Bell Atlantic, BellSouth, and Southwestern Bell — chose to use it as part of their names.

Since the RBOCs no longer had access to the technical support services that had formerly been provided by Bell Labs, Western Electric, and other AT&T groups, they banded together to form a new service organization, which is know as *Bell Communications Research, Inc.* (*BELLCORE*). BELLCORE is funded equally by the seven regional operating companies. In addition to such services as new product evaluation, assignment of new area codes, and special projects requested by the BOCs, BELLCORE serves as a standardization agency for the industry.

■ ***LOCAL ACCESS AND TRANSPORT AREAS (LATAs)*** Local calling areas were mapped into 165 local access and transport areas throughout the United States. The operating companies were empowered to handle calls within their LATAS and to charge all long distance companies, including AT&T Communications, for connecting their customers to the long distance company networks. Only long distance companies were empowered to provide telephone service between LATAs.

Figure 4.2
New Bell Operating Companies

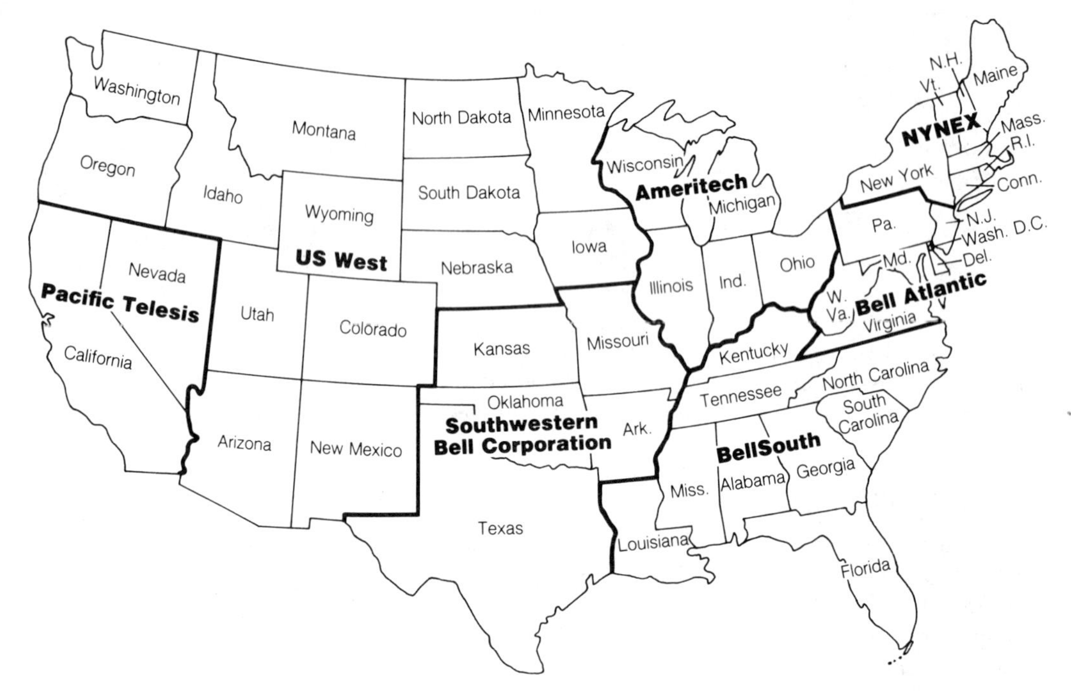

THE REORGANIZED GTE

GTE is the largest non-Bell telecommunications company. Although GTE was not a party to the Modified Final Judgment, it entered into a separate consent decree with the Justice Department, in which it agreed to provide equal access to any long distance carrier desiring it.

In 1987 GTE reorganized its headquarters group in Stamford, Connecticut, into seven operating companies: GTE California, GTE North, GTE Florida, GTE Southwest, GTE South, GTE Northwest, and GTE Hawaii.

Other significant developments occurred in GTE operations, including:

1. the purchase of Airfone Corporation
2. the consolidation of their subsidiaries, manufacturing lighting products, precision materials, and advanced telecommunications equipment, into one major group called *Products and Systems*
3. the sale of US Sprint

Airfone provides air-to-ground telephone service for passengers on commercial airlines. The connection is established from the airplane to one of nearly 70 ground stations located around the country. The call is

Composition of the Regional Bell Operating Companies

NYNEX
New England Telephone Company
New York Telephone Company

Bell Atlantic
New Jersey Bell Telephone Company
Bell Telephone Company of Pennsylvania
Diamond State Telephone Company
Chesapeake and Potomac Telephone Company of Virginia
Chesapeake and Potomac Telephone Company of Maryland
Chesapeake and Potomac Telephone Company of West Virginia
Chesapeake and Potomac Telephone Company of Washington, D.C.

BellSouth
Southern Bell Telephone Company
South Central Bell Telephone Company

Ameritech (American Information Technologies)
Michigan Bell Telephone Company
Ohio Bell Telephone Company
Indiana Bell Telephone Company
Illinois Bell Telephone Company
Wisconsin Bell Telephone Company

Southwestern Bell Corporation
Southwestern Bell Telephone Company

US WEST
Mountain Bell Telephone Company
Northwestern Bell Telephone Company
Pacific Northwest Bell Telephone Company

Pacific Telesis Group
Pacific Bell Telephone Company
Nevada Bell Telephone Company

Figure 4.3
Composition of the Regional Bell Operating Companies

then routed over land lines to the desired telephone. Formerly owned by Western Union and A. F. Holding Company, it was purchased by GTE in early 1987.

In addition to basic telephone operations, GTE produces a variety of consumer and telecommunications products, which are sold to the public and to telephone companies. Consolidation of these activities into one group, the Products and Systems Group, was designed to increase efficiency and therefore profitability.

At one time GTE was the sole owner of GTE Sprint, the third largest long distance carrier in the country. Later, GTE sold half ownership to United Telecommunications. In 1988, GTE sold its remaining interest in Sprint to United Telecommunications, thereby removing itself from the long distance telephone business.

Figure 4.4
Largest Telephone Companies, Including Holding Companies

Largest Telephone Companies, Including Holding Companies

	In Order of Access Lines as of January 1, 1988	Access Lines
1.	Bell Atlantic Corporation	16,056,907
2.	NYNEX Corporation	16,046,014
3.	BellSouth Corporation	15,739,470
4.	Ameritech Corporation	15,094,000
5.	Pacific Telesis Group	12,525,000
6.	US WEST	11,613,000
7.	GTE Corporation	11,559,536
8.	Southwestern Bell Corporation	11,104,974
9.	United Telecommunications, Inc.	3,516,814
10.	Contel Service Corporation	2,377,837
11.	Southern New England Telecommunications Corp.	1,779,204
12.	Centel Corporation	1,422,969
13.	ALLTEL Corporation	982,996
14.	Cincinnati Bell Telephone Co.	747,916
15.	Puerto Rico Telephone Company	696,066

OTHER TELEPHONE COMPANIES

Prior to divestiture, telephone companies that were not part of the Bell System were known as independents. Since divestiture separated the BOCs from AT&T and from each other, they too became independent. Thus, the term *independent* has lost its former significance. As of January 1, 1988, there were 1,384 telephone companies in the United States. Figure 4.4 shows the 15 largest ones.

COMPOSITION OF THE POST-DIVESTITURE TELECOMMUNICATIONS INDUSTRY

The post-divestiture telecommunications industry is composed of interconnect vendors, common carriers, government agencies, associations, and standards organizations.

INTERCONNECT VENDORS

Interconnect vendors provide customer-premises equipment, such as telephone sets, modems, private branch exchanges, speakerphones, answering machines, and related telecommunications equipment. This industry, made possible by the landmark Carterfone decision, was stimulated by the development of the *modular jack,* an interface device which provides easy interconnection of telecommunications equipment to telephone company lines. After initially witnessing an explosion of products and services in the marketplace, the industry has recently experienced a tremendous shake-out of small companies.

Since the interconnect industry is deregulated, there are many vendors in the marketplace in addition to AT&T and the BOCs. A partial list of these companies includes Allied Communications, Contel, Executone, Isotec, ITT, Mitel, NEC, Northern Telecom, ROLM, Siemens, Solid State Systems, Teleco USA, Telenova, TIE/Communications, and Walker Telecommunications.

■ ***AT&T*** AT&T markets customer premises equipment through its Phone Centers. AT&T Phone Centers are located in shopping malls and other high-traffic locations. They carry what is probably the largest selection of telephone sets and telephone accessories available. Their telephone sets may be either purchased or leased. Leasing is available for a monthly charge which includes the following: 1) use of the telephone instrument, 2) maintenance and repair of the telephone instrument, and 3) the option of exchanging the instrument for another style or color.

Telephone sets for purchase come with a manufacturer's warranty against defective parts for a specified length of time. After this time, there is a charge if the telephone instrument develops a problem.

■ ***THE BELL OPERATING COMPANIES*** Prior to divestiture, the BOCs provided telephone sets as part of telephone service. At the time of divestiture, they gave subscribers the option of purchasing the telephone instruments located on their premises. All other telephone instruments were transferred to AT&T.

Since divestiture, the BOCs are no longer responsible for providing maintenance and repair on CPE. They are responsible for the telephone lines connecting the customer to the central office. Some BOCs market CPE through a separate subsidiary. The products are offered for sale only, not for lease.

■ ***MARKET DIFFERENTIATION*** Interconnect vendors may be classified by the market they serve—the residential market or the business market. AT&T serves both residential and business markets; however, their Phone Centers cater primarily to the residential market. Other vendors serving the residential market include department stores, electronic equipment stores, and discount houses.

The BOCs markets CPE through separate subsidiaries. Since each BOC is a separate, autonomous entity, their marketing policies and strategies vary. However, in general, they cater to the business market.

■ ***TELEPHONE REPAIR*** The local telephone companies are responsible for maintaining the telephone lines connecting the subscribers to the central office. They are not responsible for the wiring on the customer's premises or for the terminal equipment.

Thus, when a subscriber has a problem, it is the customer's responsibility to determine whether the trouble is in the telephone instrument, the wiring on the premises, or in the telephone company's facility. Telephone directories provide information on how to determine whether the trouble is in the telephone instrument. If the telephone company sends a repairperson to the customer's premises, there will be a service charge assessed unless the trouble is found to be in the telephone company's facility.

COMMON CARRIERS

As a result of the Modified Final Judgment, the terms *Bell System* and *Independents* disappeared from the telecommunications vocabulary, and new names evolved to describe various segments of the industry. The post-divestiture common carriers are generally segmented into two principal categories: those that furnish local services and those that furnish long distance services.

■ ***LOCAL EXCHANGE CARRIERS (LECs)*** The BOCs and other telephone companies that furnish the local telecommunications service within their franchised areas are known as *Local Exchange Carriers* or LECs. Local telecommunications service includes the transport of telecommunications between customers within the telephone exchange and within the prescribed LATA.

The LEC also provides access to and connection with a long distance carrier. LECs are fully regulated by the state regulatory agency for the intrastate service they provide and by the FCC for the facilites sold to long distance carriers for use in interstate service.

■ ***INTEREXCHANGE CARRIERS (IXCs OR ICs)*** *Interexchange carriers* provide long distance telecommunications services in a competitive environment. They include specialized carriers, value-added carriers, satellite carriers, and resale carriers. IXCs purchase access facilities from the LECS to obtain connections from the customer's premises to the IXC locations known as the *Point of Presence (POP)*. The IXC transmits the information from its POP to a distant POP and then interconnects to the distant customer through LEC access. Figure 4.5 shows the path of a long distance telephone call.

The IXC is regulated by the FCC for interstate business and by the state regulatory agency for intrastate business. There are a few exceptions to this general rule. In some states the IXC is authorized to handle intra-LATA service as well as inter-LATA service, and in other states the IXCs are restricted from providing intra-LATA service.

The major IXCs are AT&T, MCI, and US Sprint, all of which provide nationwide and international service. There are also a number of smaller IXCs. Some have regional networks of their own for all or part of their transmission. They supplement their services by buying or leasing circuits or WATS lines from AT&T or other carriers. Still other IXCs have no

Figure 4.5
Path of a Long Distance Call

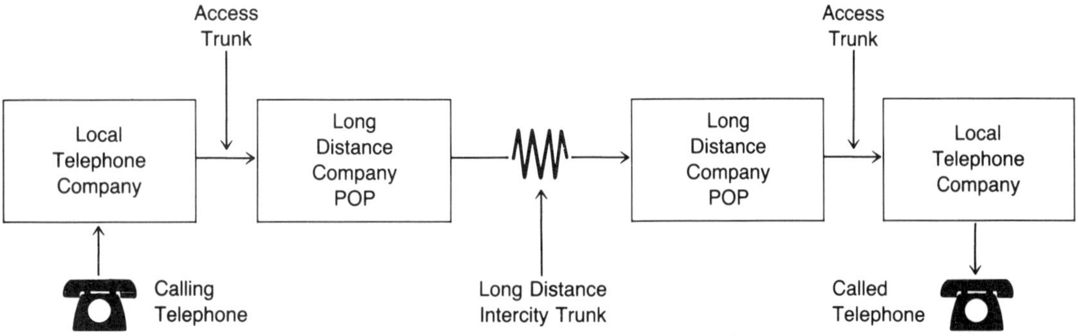

Access Trunk
Local Telephone Company
Long Distance Company POP
Long Distance Company POP
Access Trunk
Local Telephone Company
Calling Telephone
Long Distance Intercity Trunk
Called Telephone

networks of their own; they rely entirely on facilities obtained from other carriers. These IXCs are known as *resellers* or *resale carriers.*

Regulation of the IXCs varies from completely regulated to unregulated, depending on the carrier's degree of industry dominance. Because AT&T is the "dominant carrier" in the industry (has such a large market share that it virtually controls the market) it is fully regulated and will remain so at least until equal access becomes available nationwide. MCI, US Sprint, and some others are partially regulated. The resellers are not regulated.

■ ***INTERNATIONAL CARRIERS*** International carriers provide communications services between countries. The major U.S. carriers providing international service are AT&T Technologies International, Communications Satellite Corporation (COMSAT), International Telecommunications Satellite Organization (Intelsat), ITT World Communications, MCI International (now includes the former RCA Globcom), TRT Telecommunications Corporation, Western Union Telegraph Company, and US Sprint.

Formerly, all international carriers were generally referred to as international record carriers. However, in 1983 the FCC changed its rules for international communications. There now exist two types of international carriers: *International Record Carriers (IRCs)* and *International Carriers.* International carriers are permitted to carry voice and data traffic, while the IRCs can carry voice, data, record (message/ telex), and other traffic, such as facsimile. The FCC no longer restricts the IRCs from domestic markets or domestic carriers from international markets, but International Carriers are still licensed and closely regulated.

GOVERNMENT AGENCIES

Government agencies play an important role in telecommunications policy. These agencies include the Federal Communications Commission, the

National Telecommunications and Information Administration, and the state regulatory agencies.

■ ***THE FCC*** The principal federal regulatory agency continues to be the Federal Communications Commission. Its membership consists of five commissioners appointed by the President to staggered terms of seven years.

The FCC exerts a substantial influence on telephone regulatory matters. There are, however, three other important sources that exert indirect but considerable influence on telecommunications regulation: the Congress, the judiciary, and the White House. Since the FCC was created by Congress and is subject to its authority, Congress wields substantial influence over it. The judiciary's power to review, revise, and possibly reverse FCC rulings makes them a regulatory force. Although the FCC does not deal directly with the White House, it usually reflects its regulatory philosophies, because commissioners appointed by an administration usually reflect that administration's policies in their decisions.

■ ***NATIONAL TELECOMMUNICATIONS AND INFORMATION ADMINISTRATION (NTIA)*** In 1978 the Office of Telecommunications Policy was combined with the Office of Telecommunications of the Commerce Department to form the National Telecommunications and Information Administration.

NTIA advises the President on communications and information policy issues. It manages the radio spectrum allocated to the Federal Government and aids the government in telecommunications planning and research. It has no executive authority; it can issue only recommendations.

■ ***STATE REGULATORY AGENCIES*** Each state has its own regulatory body charged with protecting the public interest in intrastate telecommunications. The National Association of Regulatory Commissioners (NARUC) represents the viewpoints of the state commissions in Washington by bringing matters of common concern to the attention of the FCC and the Congress. All of the states' regulatory commissions are members of NARUC.

Issues that are of common concern in the post-divestiture regulatory environment include depreciation rates, rate-cap (price-cap) regulation, measured service, extended-area service, access charges, and lifeline service.

ASSOCIATIONS

Associations is a general term that includes telecommunications users' groups, vendor groups, trade associations, and special interest groups that have an influence on telecommunications. Although there are many such organizations, some of the major associations and their purposes include the following:

- □ **Association of Data Communications Users, Inc.** Develops procedures for acquiring, reviewing, and disseminating information; takes posi-

tions on governmental and vendor developments; promotes the effective use of data communications; and encourages participation with other existing industry–related groups.

- ☐ **Communications Workers of America (CWA).** A labor union that represents employees of the BOCs and some of the large telecommunication manufacturers in collective bargaining and grievance procedures, such as job displacement because of automation.
- ☐ **International Communications Association (ICA).** Provides for the exchange of information and experience between major users of telecommunications facilities and services; enhances and develops the telecommunications management profession by the establishment of scholarships and aid to universities; and actively promotes a viable domestic and international telecommunications capability.
- ☐ **National Association of Regulatory Utility Commissioners (NARUC).** Represents the state public utility commissions in Washington, D.C.
- ☐ **National Exchange Carriers Association (NECA).** Mandated by the FCC to file interstate access charges for 1600 telephone companies; collects and distributes interstate revenues to its members.
- ☐ **North American Telecommunications Association (NATA).** Principal nationwide trade association of the competitive telecommunications industry; founded in 1970 by Tom Carter to provide legal and legislative regulatory services for the communications equipment industry; representing about 650 vendors, manufacturers, and suppliers of competitive telecommunications equipment for business and residential use.
- ☐ **Organization for the Protection and Advancement of Small Telephone Companies.** Represesnts small telephone companies; informs and educates its members of technological and regulatory developments.
- ☐ **Society of Telecommunications Consultants.** Organization of telecommunications consultants that establishes and maintains standards for the profession.
- ☐ **State Telephone Associations.** Telephone associations in most states, such as Telephone Association of Michigan (TAM).
- ☐ **Tele-Communications Association (TCA).** Telecommunications equipment users, organized to inform members of the telecommunications industry of developments and to protect their interests.
- ☐ **Teiocator Network of America.** Represents independent radio common carriers.
- ☐ **United States Telephone Association.** Represents the interests of all local telephone companies, including the BOCs; was known as the United States Independent Telephone Association before divestiture.

STANDARDS ORGANIZATIONS

Communications standards are established to ensure compatibility among similar communications services. In many cases, the United States developed standards for use in this country ahead of the rest of the world.

Often when the need arose in other countries, U.S. standards were improved, and the improved version became accepted as the international standard. In other cases, international standards groups developed standards that were later adopted in the United States.

■ ***UNITED STATES STANDARDS ORGANIZATIONS*** Standards organizations in this country include the following: American National Standards Institute (ANSI), Electrical Industries Association (EIA), Institute of Electrical and Electronic Engineers (IEEE), and National Institute of Standards and Technology (NIST), formerly National Bureau of Standards.

■ ***INTERNATIONAL STANDARDS ORGANIZATIONS*** The over 100-year-old International Telecommunication Union (ITU) is the "grande dame" of international telecommunications regulation. An agency of the United Nations, the ITU administers most of the international treaties governing radio, satellite, and telephone communications. It also oversees the Consultative Committee on International Telephony and Telegraphy (CCITT), the world's most authoritative telecommunications standards-making body; the Consultative Committee on Radio, which coordinates radio frequencies; and the International Standards Organization (ISO), which coordinates the activities of the various committees.

KEY ISSUES IN THE POST-DIVESTITURE ERA

The key issues in the post-divestiture era include equal access, access charges, federal customer line charges, bypass, teleports, open network architecture, and reassessment of the MFJ.

EQUAL ACCESS

The MFJ required that all BOCs with electronic switching systems and with a market of at least 10,000 access lines offer equal access to all long distance carriers. It also required the access to be phased-in during the period between September 1, 1984, and September 1, 1986.

Although the MFJ equal access requirement applied only to the BOCs, most of the other local exchange carriers elected to provide equal access to their customers at the time they converted to digital equipment.

Equal access is defined as access that is equal in type, quality, and price to that provided to AT&T. Equal access is also called "dial 1" or "1 plus" service. It allows customers to reach the carrier of their choice simply by dialing 1 + area code + telephone number. Previously, this quality service was available only from AT&T; customers using other interexchange carriers had to dial access codes of 12 to 23 digits.

To implement "1 plus" service, subscribers were asked to choose a *primary carrier,* the long distance carrier they wanted to access by dialing "1", just as they did with AT&T. The long distance carriers that chose to participate were listed on a ballot, which the BOCs mailed to their customers asking them to select a carrier. Customers who did not choose a carrier were allocated on a percentage basis, determined by the percentage of customers already having chosen that carrier.

Conversion of a central office to equal access required that the office be equipped with electronic switching systems capable of routing the calls to the desired carrier. This installation was very costly and took considerable time to complete.

The first equal access conversion took place in Charleston, West Virginia, on July 1, 1985. It is estimated that by 1990 about 90 percent of the conversions will have taken place; however, the process will not be completed before 1992.

■ ***ACCESS CHARGES*** Access charges are made to compensate the local exchange carries for providing access to the long distance carrier's network. Access charges are of two types: (1) those levied on the long distance carriers and (2) those levied on residential and business customers. The latter are known as Federal Customer Line Charges.

The local telephone companies are compensated for connecting their customers to the long distance carrier by charging the long distance carrier for the access facilities that are used. These charges are specified in the FCC tariff for interstate use and in the state tariff for intrastate use.

■ ***FEDERAL CUSTOMER LINE CHARGES*** Prior to divestiture AT&T long distance revenues were shared with the local telephone companies based on the local telephone company investment and expenses in providing access from local customers to the AT&T network. This division of revenues resulted in a subsidy of local telephone service by AT&T long distance telephone service. With divestiture, the division of revenues was eliminated. Thus, AT&T, like other long distance companies, could now base its rates on costs of providing the service. However, this also meant that no funds were available to replace the pre-divestiture subsidy to the local telephone companies. The only interstate toll revenues the LECs received were from the disproportionately high access charges paid by the long distance carriers.

A large part of the costs of a long distance company is the charges it pays to the local telephone company for access to its networks. In an effort to reduce this cost of access, the FCC ordered all local telephone companies to add a Federal Customer Line Charge to every line that customers purchase. This line charge has been increased in steps, with accompanying reductions of access rates paid by the ICXs. As a result, long distance subsidy of local service has gradually been eliminated, and the long distance companies have been able to lower their rates. The local telephone companies meet their revenue requirements, even with lower IXC access revenues, with the monies they receive from the Federal Customer Line Charge.

Federal Customer Line Charges were implemented on June 1, 1985, after considerable controversy. The initial monthly charge was $1.00 per line. On June 1, 1986, it was increased to $2.00 per line. On April 1, 1989, the charge was increased to $3.50 per residence and business line.

Some states have also directed local exchange carriers to collect a state customer line charge in addition to the federal charge so that intrastate long distance subsidies can also be eliminated.

BYPASS

Bypass is the use of private communications facilities or services to go around or avoid the local telephone exchanges of the public switched network. Bypass facilities are used to connect two points in a private network or to connect a user directly to an interexchange carrier. Bypass transmission technologies include microwave, fiber-optic cable, satellite, and cellular radio.

A decision to bypass is usually based upon economic factors that justify the expenditure of large sums of money to avoid using the services of the local telephone company. This is attractive to only very large subscribers with heavy traffic volumes, particularly among their own company locations. Improved service is another factor that sometimes encourages companies to bypass. Some organizations are willing to pay higher charges in order to receive higher-quality transmission than that available from the local telephone company.

■ ***TELEPORTS*** One form of bypass that is available in some large cities is teleports. *Teleports* are extensive earth station and satellite antenna complexes constructed to serve large-volume users in metropolitan areas. Teleports usually build their satellite antennas away from densely populated downtown areas and run high-capacity microwave or fiber optic links from customers in the city to the earth station complex. The teleport concept involves concentrating enough long distance voice, data, and video traffic to justify building a set of big dish antennas on one site, enabling users to get bulk discounts on satellite circuits.

Teleports provide communication gateways to metropolitan areas in much the same way that an airport or railroad station serves transportation. The first teleport system (and still the most ambitious) is the New York/New Jersey Teleport. Located on Staten Island, it is an enhanced office park housing the most sophisticated satellite communications facilities linked to a regional fiber-optic network.

Currently, there are about 40 teleports, operational, planned, or under construction in the United States. Worldwide, an estimated 60 teleports exist. The World Teleport Association (WTA) has 63 members from 15 countries. According to a recent study by Frost & Sullivan, it is estimated that by 1995 there will be 200 teleports in the United States alone.

COMPUTER INQUIRY III (CI-III)

Computer Inquiry III examined the degree to which telecommunications carriers could offer enhanced services. In a 1987 decision, the FCC decreed that AT&T and the BOCs could offer unregulated, enhanced services under a set of complex provisions known as *open network architecture (ONA)*. Before the services could be offered, however, plans for implementation of ONA had to be approved by the FCC. Many details remain to be resolved in this matter.

REASSESSMENT OF THE MFJ

In order to ensure that the conditions of the MFJ were working as planned, AT&T and the Justice Department agreed that progress would be reviewed every three years.

■ ***FIRST TRIENNIAL REVIEW*** On September 8, 1987, Judge Greene announced the result of his first triennial review of the restrictions that had been placed on the BOCs.

The most noteworthy restrictions placed on the BOCs were that they were prohibited from the following:

1. providing inter-LATA service
2. entering the telecommunications equipment manufacturing business
3. providing information services

In his report, Greene continued the first two restrictions in their entirety. However, he did relax the restrictions on the BOCs regarding the provision of information services by permitting the BOCs to provide the *transport* of information. They were still restricted from providing actual information. For example, the BOCs could transport videotext information from the suppliers of the information to the customers, but they could not enter into the business of providing the information itself.

SUMMARY

On January 1, 1984, the largest corporate divestiture in U.S. history became effective, and the telecommunications industry was changed irrevocably. Prior to divestiture, AT&T was the largest company in the world, with assets of $155 billion and one million employees. Divestiture split up the one giant into eight smaller organizations.

The post-divestiture AT&T consists of AT&T Bell Laboratories, the research and development arm; AT&T Network Systems, the equipment marketing branch; AT&T International, the international service provider; AT&T Technology Systems, the manufacturing arm; and AT&T Communications and Information Systems, the provider of long distance service and customer-premises equipment.

Following divestiture, the 22 BOCs were formed into seven regional Bell Operating Companies (RBOCs) that were completely independent of AT&T and of each other. The seven regional companies are NYNEX, Bell Atlantic, BellSouth, Ameritech, Southwestern Bell, US WEST, and Pacific Telesis Group. Since the RBOCs no longer had access to the technical support services that had formerly been provided by Bell Labs and other AT&T groups, they banded together to form Bell Communications Research (BELLCORE).

Local calling areas were mapped into 165 local access and transport areas (LATAs) throughout the country. The local exchange carriers (LECs) were empowered to handle intra-LATA calls, and interexchange carriers (IXCs) were empowered to handle inter-LATA calls.

GTE is the largest non-Bell telephone company. In 1987 it reorganized into seven regional operating companies. GTE purchased Airfone, consolidated their subsidiaries, and sold US Sprint.

With the divestiture of the Bell System, the term *independent* became obsolete.

The post-divestiture telecommunications industry is composed of the following: interconnect vendors, who provide customer-premises equipment; common carriers, who provide local, long distance, and international telecommunications service; government agencies, which oversee and regulate the industry; associations, including user groups, vendor groups, and special interest groups, and standards organizations.

Key issues in the post-divestiture era include equal access, bypass, ONA, and the reassessment of the MFJ.

The MFJ required that the BOCs offer equal access to all long distance carriers. Equal access is defined as access that is equal in type, quality, and price to that provided to AT&T. It allows customers to reach the carrier of their choice simply by dialing 1 + area code + telephone number. Previously this service had only been available from AT&T. Customers using other long distance carriers had to dial access codes of 12 to 23 digits.

Access charges are made to compensate the local exchange carriers for providing customers with access to the long distance carrier's network. They are levied on the long distance carriers and on business and residential customers.

Bypass is the use of private communications facilities or services to go around or avoid the local telephone exchanges of the public telephone network. The objective of bypass is to reduce the costs of local telephone service by providing alternative facilities. This alternative is only attractive to very large subscribers with heavy traffic volumes.

Results of the first triennial review of the MFJ were announced on September 8, 1987. The principal change suggested by the review concerned information services. Terms of the MFJ had prevented the BOCs from providing and transporting information services. Judge Greene relaxed this restriction by allowing the BOCs to *transport* information services. However, the restriction on *providing* information services remained in effect.

REVIEW QUESTIONS

1. What happened on January 1, 1984?
2. How has national telecommunications policy changed during the last two decades?
3. What was the most important outcome of the MFJ?
4. What happened to the Bell System as a result of divestiture?
5. Describe the post-divestiture composition of AT&T.
6. What is the relationship of the Regional Bell Operating Companies to AT&T?
7. What is the relationship of the Regional Bell Operating Companies to each other?
8. What is the function of BELLCORE? How is BELLCORE funded?
9. Why did United States Independent Telephone Association (USITA) change its name to United State Telephone Association (USTA)?
10. What are the four categories of organizations/companies that comprise the post-divestiture industry?
11. What is the function of NTIA?
12. What is meant by the term *equal access?*
13. What is meant by the term *bypass?*
14. What is the purpose of the Triennial Review of the MFJ?

VOCABULARY

inter-LATA service

intra-LATA service

Regional Bell Operating Companies (RBOCs)

equal access

AT&T Communications

AT&T Technologies

AT&T Bell Laboratories

AT&T Network Systems

AT&T International

AT&T Technology Systems

AT&T Communications and Information Systems

NYNEX

Bell Atlantic

BellSouth

Ameritech

Southwestern Bell

US West

Pacific Telesis Group

Bell Communications Research, Inc. (BELLCORE)

Airfone

US Sprint

modular jack

local exchange carrier (LEC)

interexchange carrier (ICX or IC)

point of presence (POP)

dominant carrier

International Record Carrier (IRC)

National Telecommunications and Information Administration (NTIA)

Communications Workers of America (CWA)

International Communications Association (ICA)

National Association of Regulatory Utility Commissioners (NARUC)

National Exchange Carriers Association (NECA)

North American Telecommunications Association (NATA)

American National Standards Institute (ANSI)

National Institute of Standards and Technology (NIST)

International Telecommunication Union (ITU)

Consultative Committee on International Telephony and Telegraphy (CCITT)

International Standards Organization (ISO)

primary carrier

access charges

bypass

teleport

Computer Inquiry III (CI-III)

Open Network Architecture (ONA)

CHAPTER

5

TELEPHONY

CHAPTER OBJECTIVES

After completing this chapter, the reader should be able:

- *To describe the principal parts of the telephone and explain the function of each.*
- *To define central office and explain its purpose.*
- *To describe the evolution of telephone switching equipment.*
- *To describe the characteristics of analog and digital signals.*
- *To describe the nationwide and worldwide numbering plans.*
- *To name and describe the principal types of telephone systems.*
- *To describe the principal types of telephone sets and service features.*

The telephone, born in America over a hundred years ago, has become the magic link by which a person can communicate with people across a street, across a city, or across a continent. The telephone is a wonderful device. It accepts the sounds of a human voice, transforms them into signals we cannot see or hear and speeds them along the wires or through space to another telephone. There the sounds come forth, instantaneously delivering a replica of the original voice directly to the listener's ear. Throughout its history, the telephone has been an important force for human betterment. Today, telephones are so much a part of our lives that we take their presence for granted, using them without thought of the sophisticated technology and networks they use.

The telephone is the basic instrument of all communications technology. It provides the greater part of the world's communications business. The four primary types of telecommunications systems in use today are voice, data, message, and image. Voice communication, or *telephony*, refers to the electrical transmission of speech over a distance.

PRINCIPLES OF TELEPHONY

The telephone works as a result of the application of a fundamental physical phenomenon: words spoken into a telephone mouthpiece are converted into electromagnetic impulses and transmitted over telephone lines. At the receiving instrument, these impulses are reconstructed into speech in such a manner that even the voice characteristics of the speaker are recognizable. This conversion phenomenon is basic not only to telephony but to the entire field of telecommunications as well.

THE TELEPHONE INSTRUMENT

All telephones consist of at least three parts: the transmitter, the receiver, and the ringer unit. The transmitter is similar to a microphone; it converts voice vibrations into electrical impulses, which are then transmitted over telephone wires, radio waves, or satellites. The telephone receiver converts the incoming electrical impulses to sounds, and the ringer unit rings or buzzes when activated by an incoming call.

Preautomatic telephone instruments were made up solely of these three parts and required the assistance of an operator to control the completion of the call. Today's modern instruments contain a *control unit*—either a rotary dial or a set of pushbutton keys—that the caller uses to place calls directly, without the assistance of an operator. An auxiliary part of the modern instrument is the *switchhook,* which signals the telephone company central office and incoming callers that the telephone is either idle or in use. The transmitter and the receiver are generally housed in the telephone's handset. The ringer unit and the switchhook are located in its base. When not in use, the handset rests in such a manner that the switchhook is in the "off" position. When a user picks up the handset, the switchhook is released to the "on" position.

TRANSMISSION OF SOUND

Simply stated, a telephone works as follows: sound waves generated by speech or other source of sound are converted to electrical energy at a transmitting instrument and are sent over wires to a receiving instrument, where they are reconstructed as sound. The voice passes over the wire in the form of an *analog signal,* that is, in a continuous flow of electrical energy that oscillates in proportion to the frequency and intensity of the sound being transmitted. Sound that can be heard by the human ear consists of frequency ranges between about 30 Hz (Hertz) and 15,000 Hz. (*Hertz* is a standard unit of frequency that has replaced, and is identical to, the older unit of measurement called "cycles-per-second.") Telephone transmission equipment is designed to use frequencies varying from 300 Hz to 3,400 Hz—a range that is satisfactory for telephone transmission because it produces voice characteristics that are both intelligible and recognizable.

ANALOG VS. DIGITAL TRANSMISSION

Digital transmission of sound uses a different physical process than analog transmission. In digital systems a stream of discrete "on" and "off" pulses, called *bits,* are sent over the transmission facility. Since analog signals can be converted to digital signals and digital signals converted to analog, any type of information can be transmitted in either analog or digital form. The conversion from analog to digital and/or from digital to analog is performed by a unit of equipment called a *modem,* which performs the functions of a modulator and demodulator. The modulator is used in transmission, and the demodulator in reception. The Bell System term *data set* is synonymous with the word *modem,* which is used industrywide.

The traditional telephone line consists of one pair of wires and can carry one analog signal — a continuous stream of frequencies. A *trunk line,* or talking path between two central offices, may consist of one pair of wires carrying one analog signal at a time. Trunk lines can also be constructed to carry digital signals by providing special equipment that breaks down the analog signal into discrete bits or pulses through a technique known as *pulse code modulation (PCM).* By converting the analog signals to digital signals, the unwanted noise that is characteristic of analog transmission is virtually eliminated. The result is improved sound quality because signal strength can be amplified without amplifying the noise and distortion. Digital transmission is also extremely efficient because the discrete bits can be transmitted at exceptionally high speeds. Although it is not economically feasible to replace transmission systems that are in good working order, when new systems are required, digital systems are clearly the first choice.

TELEPHONE LINES

Each telephone line consists of a pair of wires. Early telephones were equipped with batteries so that they could provide their own electrical current. The telephone contained no flow of electrical current except when they were in use. Each telephone also contained a hand-operated electrical generator called a *magneto.* By cranking the handle on this generator, the user activated a bell or light that signaled an operator. When the operator answered the signal and manually connected the caller's line to the line being called, the electrical circuit was completed, current flowed, and the caller could talk to the party called. Modern telephone lines employ *common-battery operation,* in which a central source of electricity is always available for use.

Telephone lines connect individual instruments to a central point, known as a *central office.* Lines connecting a customer's telephone to a central office operator originally consisted of a single wire, over which current flowed in one direction; the ground served as a means of completing the electrical circuit. Because these single-wire lines provided extremely poor transmission, they were soon replaced with two-wire lines that formed a "loop" between the telephone and the central office. The

Figure 5.1
Broadway and Maiden Lane, New York City

(Reproduced with permission of AT&T)

wires were strung through the air on poles with crossarms to hold them apart. As the number of telephones increased, the poles and wires became so numerous and unsightly that their continued use was impractical. (See Figure 5.1.) Several techniques were developed to compact the lines by placing them in pipes or cables. Today, wires are packed into cables. Each wire is insulated from the other wires and covered with a lead or plastic sheath that is sealed to prevent water from entering the cable. The cables may be placed either overhead or underground.

The wires connecting the customer premise's equipment to the local telephone office is known as a *local loop* or *subscriber loop*. The local loop consists of a *drop wire*, a *distribution cable*, and a *feeder cable*. Each loop consists of a pair of drop wires. The drop wire runs from a residence or building to a telephone pole, where it is connected to a distribution cable. Several distribution cables are connected to a much larger feeder cable, which extends to the central office. Frequently the feeder cable is placed in underground ducts, reducing the need for unsightly poles (see Figure 5.2).

Traditional local loops consisted of wire-pair transmission facilities. Today, some telephone companies are experimenting with fiber optic transmission systems to replace the wire-pair loops.

A second type of telephone line connects central offices to each other. These lines are known as *trunks*. In the past they were usually made up of

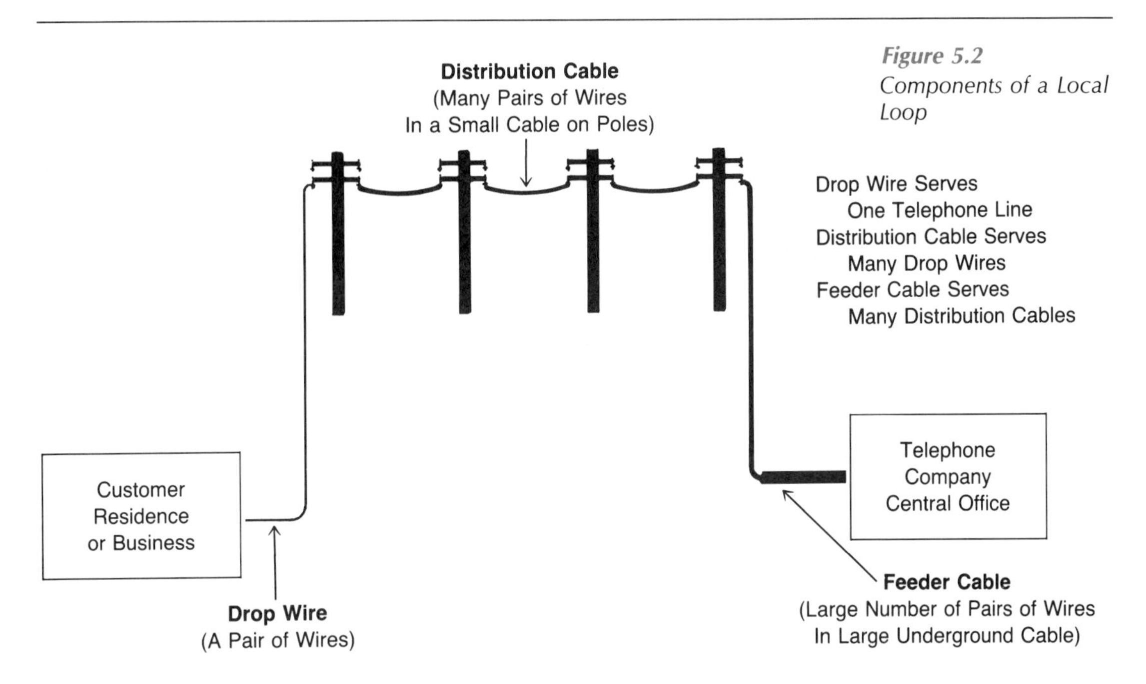

Figure 5.2
Components of a Local Loop

wires in large cables that were almost always carried underground in cable ducts. Today fiber optics is being widely used in trunks.

CENTRAL OFFICES

Bell's experiments were conducted at the electrical shop of Charles Williams, Jr., in Boston. Appropriately, when the first permanent outdoor telephone line was strung on April 4, 1877, it was placed between the electrical shop and Williams's home, a distance of about three miles. Shortly thereafter, the first telephones rented from Bell for business use were connected by a line from the Boston office of a young banker to his home in Somerville. These first telephone lines were strung directly from one individual telephone to another; no "central" point intervened.

As more telephones came into use and more and more wires were strung, it became apparent that stringing direct lines between separate instruments was no longer feasible. Moreover, people realized that the value of the telephone would be substantially increased if it were connected to more than one other telephone. Thus, the process of transferring a connection from one telephone circuit to another by interconnecting the two circuits was developed. This process is known as *switching*. The central office, or *telephone exchange* as it is sometimes called, came into being as a switching center—a point at which two circuits could be interconnected to make a talking path between two telephones. Originally, "central" was used to identify the operator who manually connected telephone lines to each other. In time, automatic switching equipment performed this function, but the term *central office* is still used today to describe the place where call switching is done. Besides acting as a focal point of many telephone lines, the central office provides a common source of electricity to a number of telephones.

Each central office serves a specific geographic area, functioning as a switching center for the individual telephone lines that terminate there. The size of the area served depends on the number of telephone lines; large metropolitan areas require many central offices. Each office contains trunks to other central offices and to long distance switching offices, so that any telephone line can be connected to any other telephone line anywhere. These connections are implemented by the *central office switching equipment,* the general industry name for the mechanical, electromechanical, or electronic equipment that routes a call to its ultimate destination. During the routing process, the call may be switched once or any number of times. The greater the distance a call has to travel, the more likely it is to require multiple switchings.

Each central office has the capacity to handle a certain number of calls at one time. If too many calls come into the central office switching center simultaneously, the system becomes overloaded and the customer either fails to get a dial tone or gets a fast busy signal before the dialing has been completed. This condition is known as *blocking*. It will occur occasionally because telephone companies have found that it is not economically possible to provide each central office with enough equipment to meet every calling load and condition during peak times.

AUTOMATIC SWITCHING

In the early days when all switching was done manually, operators interconnected lines by inserting a cord with jack plugs into each of the two lines to be connected (Figure 5.3). As the use of telephones spread, manual call switching quickly became obsolete, partly because of the ever-increasing numbers of operators that would have been required. Thus, in the 1920s, manual switching began to be replaced by automatic exchanges, in which the caller dials the desired number directly to effect the connection between the two telephones. The equipment that makes the connections consists of banks of relays and switches mounted on rows of frames that extend from floor to ceiling and occupy entire buildings.

The story of the invention of automatic switching by a Kansas City undertaker, Almon B. Strowger, provides an interesting sidelight in the history of the telephone. According to tradition, Strowger suspected the local telephone operator of misdirecting calls intended for him to his competitor who was a relative of the operator. Accordingly, he vowed to find a way to eliminate the need for operators and went to work to develop an automatic switching device that would make his competitor's accomplice obsolete. In 1891 a patent was issued to Strowger for such an invention: a two-motion (vertical and horizontal) switch that was subse-

(Reproduced with permission of AT&T)

Figure 5.3
Early Telephone Switchboard

quently named the *Strowger switch*. The principle on which Strowger's invention was based was so fundamental to automatic telephone systems that for many years telephone companies had to obtain a license from him before updating their equipment.

A central office switching system receives signals generated by a calling telephone, registers the signals in an electrical buffer, and translates them into a series of equipement operations that completes the call. (The signals generated by rotary-dial telephones are dial pulses. Signals generated by pushbutton telephones are musical tones.)

The earliest automatic central offices were equipped with step-by-step systems (Figure 5.4), which offered a means of selecting a series of paths through which the circuit from the calling party to the called party could be progressively established. Each digit dialed activated one switch in the central office, and the sequence of digits selected the called telephone number for each connection one digit at a time.

Another type of electromagnetic switching equipment, the *panel system,* derived its name from the way groups of numbers were arranged on frames resembling panels. In this system all the digits dialed were stored in

Figure 5.4
Step-by-Step Switching Equipment

(Reproduced with permission of AT&T)

a control device, and the call advanced at the completion of dialing, based upon the predetermined logic wired into the equipment.

Both step-by-step and panel systems were limited to 10,000 telephone lines. Although extremely reliable, these systems were slow by today's standards because of their mechanical nature. In addition, maintenance proved costly because the mechanical equipment required frequent cleaning, lubrication, and adjustment. A subsequent development, the *crossbar system,* which derived its name from the way connections were established, consists of series of switches with vertical and horizontal *talking paths.* (See Figure 5.5.) At the intersecting points between these paths, or crosspoints, contacts are made on platinum or another precious metal that produces virtually noise-free connections. An important feature of crossbar equipment is its pretesting of trunk and line availability before it establishes connections. Moreover, the circuit connections in this type of equipment are established with fewer mechanical movements than a step-by-step switch, resulting in less wear and easier maintenance. Crossbar systems have a capacity of up to 30,000 telephone lines or three number groups of 10,000 lines each. (See Figure 5.6.)

Although both step-by-step and crossbar equipment can still be found in local exchanges and private exchanges throughout the United States, these traditional systems are increasingly being replaced by *electronic switching systems (ESSs),* which establish connections at phenomenal speeds and have a capacity of up to 100,000 lines (Figures 5.7 and 5.8). Electronic switching equipment is called *stored program control* switching because it can be programmed to perform a variety of functions in addition to

Figure 5.5
Crossbar Switching Equipment

(Reproduced with permission of AT&T)

Figure 5.6
Crossbar Switching Equipment

(Reproduced with permission of AT&T)

Figure 5.7
Electronic Switching System Equipment

(Reproduced with permission of AT&T)

Figure 5.8
Digital Central Office Switching System, DMS-10

(Courtesy of Northern Telecom)

conventional call completion. Although relatively expensive to purchase, it is economically attractive because of its low maintenance costs and revenue-producing service features.

TELEPHONE NUMBERS

Telephone numbering system have been established for not only the United States but also the rest of the world.

NATIONWIDE NUMBERING PLAN

The telephone industry has divided the United States into geographic areas called *numbering plan areas (NPAs)*. Each NPA is identified by a unique three-digit code, called the *area code,* which precedes the seven-digit telephone number. (See Figure 5.9.) Thus, seven digits are required to place a call to any telephone within the same NPA, and ten digits are required if the telephone call is to terminate at a telephone in a different NPA. Because the digit "0" is reserved for operator calls and the digit "1" is reserved for identifying long distance calls, only the digits "2" through "9" are available for use as the first digit of a seven-digit telephone number. Mathematically, then, there can be no more than 8 million telephone numbers in a given geographical area.

Most telephone companies require that customers prefix the digit "0" on calls that require the assistance of an operator. The digit "1" is frequently called a *long distance access code*. Dialing the extra "1" helps remind the user that there will be a charge for the call and improves the credibility of the telephone bill. Access codes other than the "1" are generally not assigned in public telephone systems because even the use of a single digit as an access code would invalidate an entire block of one million digits for general telephone number use. Private telephone systems employ access codes such as the digit "9" to reach the public network, and other codes to meet the specific design of the private system. The use of these access codes does not interfere with the public telephone system and is compatible with the overall numbering plan.

WORLD NUMBERING PLAN

The International Telecommunication Union (ITU), an agency of the United Nations, was established to provide worldwide standards for communications. The union's Consultative Committee on International Telegraphy and Telephony (CCITT) is the medium for recommendations for international communication systems. By following CCITT recommendations, the United States and other United Nations countries achieve compatability between their telecommunication systems and make international communications possible. One of the CCITT's most important recommendations concerned a world-wide numbering plan that assigns each customer a unique international telephone number.

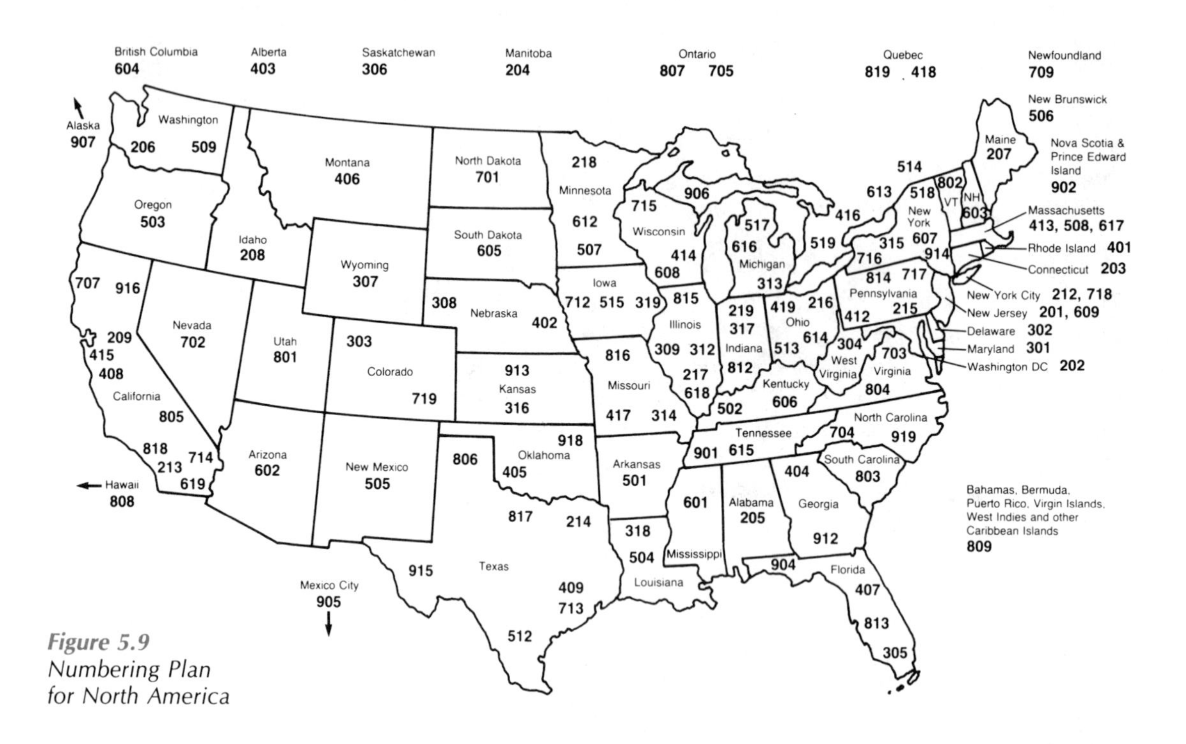

Figure 5.9
Numbering Plan for North America

The plan assumes "all number calling," that is, telephone numbers composed entirely of numbers, with no letters of the alphabet being used. Under the plan, a customer's international telephone number consists of a maximum of twelve digits, plus access code (if one is required). The first one, two, or three digits represent the code assigned to the country, and the remainder constitute the national telephone number, which is composed of the routing code (area code) plus the subscriber's local telephone number.

In order to assign country codes, the world was divided into nine geographic zones, sometimes referred to as World Zones, each of which was assigned a digit, as follows:

1 North America
2 Africa
3 Europe
4 Europe
5 South and Central America
6 South Pacific
7 Union of Soviet Socialist Republics
8 Far East
9 Middle East and Southeast Asia

All countries within a zone were assigned codes beginning with that zone's digit. Since the plan was formulated to meet the telephone requirements of each country for the year 2000, the number of digits in a country code is determined by the number of national telephones forecast for that year. Countries with the highest requirements were assigned a single-digit number, and those with the lowest requirements were assigned a three-digit country code.

Because the composition of national telephone numbers is determined by each individual country, there are many variations in national numbering plans. As discussed in the previous section, United States national telephone numbers are composed of ten digits. By prefixing the digit "1"—the digit assigned to North America—the national number becomes the American's unique international telephone number.

INTERNATIONAL DIALING

International direct distance dialing permits a customer to dial a telephone anywhere in the world without the assistance of an operator. This service is provided by the major long distance companies. To place an international call, the customer dials an *international access code,* followed by the *country code* plus a *routing code* (area code) plus the subscriber's telephone number. Thus, a call from the United States to Frankfort, Germany, would be completed by dialing 011-49-611-432-432. The 011 is the international access code, 49 is the country code, 611 is the area code, and 432-432 is the subscriber's telephone number.

TELEPHONE SYSTEMS

Telephone systems may be classified into three broad types: single-line systems, key systems, and private branch exchanges.

SINGLE-LINE SYSTEMS

Single-line telephone systems consist of an individual line and one or more telephone sets. Each telephone line is capable of accommodating several extensions. Single-line systems are available for both residential and business customers. However, the rates charged business customers are substantially higher than those charged residential customers for the same service. Single-line systems are used by residential customers and very small businesses. They comprise the largest market for telephone systems today.

Telephone service features available for single-line customers include the following:

- ☐ *Call Waiting ("Camp-on").* A person using a line can be signalled that another call is waiting.
- ☐ *Abbreviated Dialing.* Commonly used telephone numbers are replaced by two-digit numbers to shorten dialing time.

Figure 5.10
Key Telephone Set

(Courtesy of Southwestern Bell Freedom Phone)

- ☐ *Call Forwarding.* Incoming calls can be forwarded automaticallly to any telephone number.
- ☐ *Third-Party Add-On (Conferencing).* A user can dial another station, thereby setting up a three-party conference call.

The four previously-described telephone service features are provided by the local telephone company and are described as *custom-calling features.* They may be obtained individually or as a package. Discounts are generally available for ordering the package. Additionally, abbreviated dialing can also be provided on certain types of telephone sets; that is, the service is inherent in the telephone set itself.

KEY TELEPHONE SYSTEMS

Key telephone systems are an arrangement of key telephone sets (Figure 5.10) and associated circuitry located on a customer's premise that permit more than one telephone line to be terminated on one telephone instrument. Key systems are also called *multi-line systems.* Any one of the lines can be used by pressing the key associated with that particular line. Key systems are used by business customers who do not require a larger system.

Service features that are inherent in the system itself include: *call hold,* which allows the called party to place the call on hold while answering another incoming call; *call pickup,* which allows the called party to answer the call at any one of a number of telephones; *call line-status signal,* which indicates a line in use by a lighted lamp; and *interconnection among on-premises locations,* which allows any extension to be interconnected with any other extension.

Additionally, the custom-calling features previously described for single-line systems are also available for key systems. These features are not inherent in the system; they are provided by the telephone company through its central office equipment.

PRIVATE BRANCH EXCHANGES

A *private branch exchange (PBX)* is a switching system installed for the exclusive use of one organization and generally located on the customer's premises. The system can place calls to or receive calls from the public network, as well as handle calls within the organization. Calls between stations (telephones served by the PBX) are dialed directly. One directory number is listed for the entire organization. *Direct inward dialing (DID),* or calls direct to certain extensions without the use of a switchboard, can be provided by some PBXs.

PBX systems can be purchased, rented, or leased from the local telephone company or from any one of a number of other vendors. Justification for installing such a system depends on the number of telephones an organization needs and the ratio of internal telephone calls to external telephone calls (Figure 5.11).

Early PBXs were equipped with a switchboard, which required operators to switch calls by manuallly interconnecting one telephone line with another. When automatic PBXs first appeared, they were described as PABXs (*private automatic branch exchanges*) or CBXs (*computerized branch exchanges*). In today's computerized business environment, the distinction is no longer meaningful since virtually all PBXs are completely automatic.

PBXs are available from most vendors of telephone equipment. The heart of a modern PBX is typically a small computer with a solid-state switching network. Frequently used programs are stored in the computer's memory. New features can be added by updating the software. Each product's features differ somewhat from that of other companies, but the popular ones include the following:

- *Automatic call forwarding.* By dialing instructions to the computer, a person may direct the PBX to forward incoming calls to another location.
- *Automatic call back.* When a call is placed to a busy number, the caller may instruct the PBX to call back when the number is free.
- *Telephone number retention.* Users who move can retain the same telephone numbers when they move to a new station because the tables used by the PBX are easily changed; no rewiring is necessary.

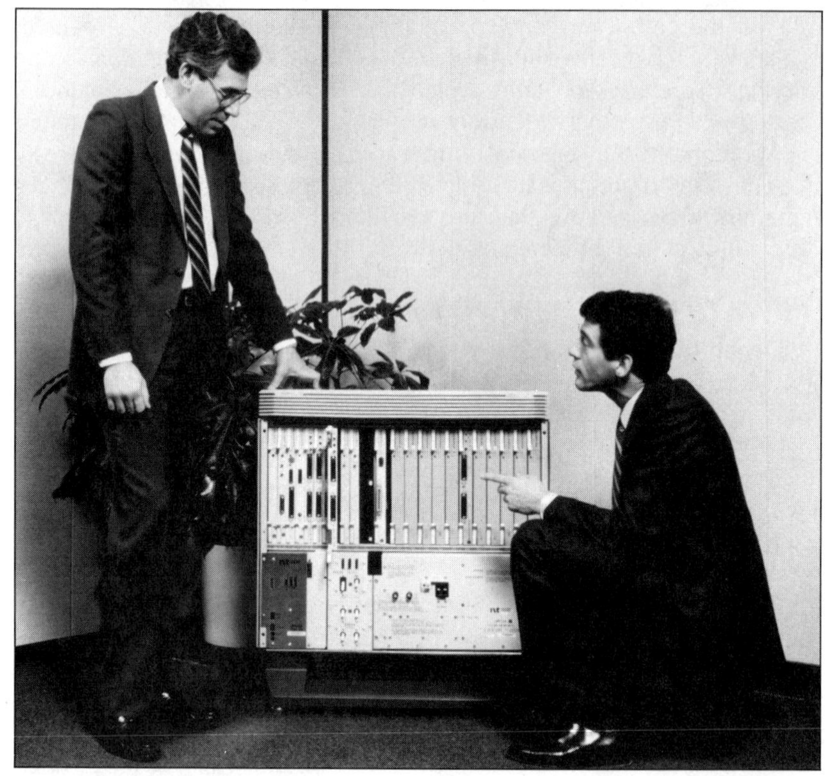

Figure 5.11
Meridian SL-1ST PBX. This PBX is designed for a 30-to 400-line business site.

(Courtesy of Northern Telecom)

- ☐ *Pushbutton station selection.* The attendant has a status light to indicate when the line is busy and a button for controlling each station.
- ☐ *Call waiting ("camp-on") signal.* A person using a line can be signaled to indicate that another call is waiting.
- ☐ *Outgoing call restriction.* Designated stations may be prevented from making outgoing calls.
- ☐ *Call holding.* A user can place a call on "hold" while dialing or talking to another station, and then return to the interrupted call.
- ☐ *Third party add-on.* A user can dial another station, thereby setting up a three-party conference call.
- ☐ *Charge listing by stations.* The computer can furnish a listing of the toll calls placed by each station, giving the number, time, duration, and cost of each call.
- ☐ *Pushbutton to dial-pulse conversion.* Signals can be converted so that pushbutton telephones can be used even when the local central office accepts only rotary-dial pulsing.
- ☐ *Abbreviated dialing.* Commonly used telephone numbers are replaced by two-digit numbers to shorten dialing time.

- ☐ *Automatic call transfer.* Incoming calls to a busy station can be automatically transferred to another designated station.
- ☐ *Distinctive ringing.* Different ringing tones are used to identify incoming calls so that the person called has some indication of the call's origin before answering. Ringing tones may distinguish among internal, external, secretary, and specific extension calls.
- ☐ *Automatic least-cost routing on corporate networks.* A call is routed from the corporate network by the least expensive route (for example, first choice, tie line; second choice, WATS line; third choice, direct distance dialing).

CENTREX

An alternative to the PBX is Centrex. *Centrex* is a service offered by the local telephone companies in which the switching equipment is located as part of a central office on the telephone company's premises. Centrex provides direct inward dialing to the called telephone. It also permits abbreviated dialing to other telephones within the organization and connections to outside telephones after dialing an access code (usually the digit "9"). Additionally, there are a wide variety of enhanced service features available from the various telephone companies.

TELEPHONE SETS

Many different kinds of telephones are now offered to customers and can be obtained from phone centers, department stores, and electronic stores. The telephones are equipped with a connector plug that fits into a standard jack or outlet, enabling the customer to attach them to a house or building's internal wiring without the assistance of a telephone installer. In addition, many telephone instruments are modular; they can easily be taken apart and reassembled. Modular construction facilitates the replacement of defective components in a telephone, reducing telephone maintenance costs.

Until the middle 1950s, telephones came in only one color — standard black — but today all telephone manufacturers offer a wide variety of decorator colors to harmonize with many decors. Telephones are also available in a variety of shapes and designs; replicas of antique telephones, reproductions of storybook characters, and novelty items of many types and descriptions.

TELEPHONE SETS AND DIALING

When the telephone receiver (handset) is lifted off the hook, an electrical current flows between the telephone instrument and the central office. Some telephone sets send the called telephone number to the central office by dial pulses; others send it by a pair of audio tones.

Figure 5.12
Single Line Telephone

(Courtesy of Southwestern Bell Freedom Phone)

■ ***DIAL PULSING*** Earlier model telephones are equipped with rotary dials that operate by interrupting the flow of line current a number of times, depending upon which digit is dialed. The interruptions cause a series of *dial pulses* that communicate the digit to the central office switching equipment and control the progress of the call through the switching system.

Dial pulses are generated at a 10-pulse-per-second rate. The number of dial pulses in a sequence equals the value of the digit, except for the digit zero, which is represented by ten pulses. The digits are represented by a relatively long off-hook interval, the length of which depends on how fast the caller can rewind and release the dial.

Rotary dial telephones are still in use today, but they are being replaced by pushbutton telephones incorporating a 12-button keypad. Some pushbutton telephones send the called number to the central office by dial pulses; others use audio tones. If the telephone set uses dial pulses, it must be equipped with a buffer to store digits that have been dialed until they can be out-pulsed. Storage is required because pushing a button on a keypad is faster than operating a rotary dial device.

■ ***DUAL TONE MULTIFREQUENCY (DTMF)*** A newer technique for communicating with the central office is dual tone multifrequency. DTMF service requires the use of a pushbutton telephone that generates tones (Figure 5.12). Pressing one of the keys causes an electronic circuit in the keypad to generate a pair of frequencies that produce a tone resembling a musical note. Each digit has its own distinctive tone that can be identified by the central office switching equipment. Bell companies call their DTMF service Touch-Tone.

■ ***SERVICE SELECTION*** DTMF has several advantages over dial pulse transmission:

1. *Speed.* Transmission rates are about 10 digits per second with the tone signaling method, and 1 digit per second with dial pulsing.
2. *Accuracy.* Impatient users of rotary-dial telephones frequently reach wrong numbers because they try to speed up the dialing process by forcing the dial mechanism back to its original position. Also, because of the shorter dialing time, the DTMF caller does not have to remember the number being called for as long a time, thereby improving dialing accuracy.
3. *Computer communication.* "Talking" to a computer or any system incorporating computer technology by a telephone usually requires DTMF.

DTMF is an optional telephone service provided by the local telephone company for an additional charge. Customers automatically receive dial pulse service unless they specify DTMF.

Since subscribers are responsible for providing their own telephones, they must select instruments that are compatible with the requested type of service.

CORDLESS TELEPHONES

Cordless telephones consist of a base station connected to the telephone line and a cordless handset that uses low-powered radio transmission to and from the base station (Figure 5.13). They have become very popular for use as portable telephones in the home and out in the yard. The units are designed with enough range to use them on an average residential lot. However, since they use radio transmission, anyone with a cordless telephone on the same frequency can listen to the conversation, thus making privacy a real issue. Some of the newer cordless telephones use simple scrambling techniques to alleviate this problem.

FEATURE TELEPHONES

Telephone sets that provide capabilites beyond the basic functions of placing and receiving calls are often described as *feature telephones* or *smart telephones*. These telephones can be further categorized as *convenience telephones* and *intelligent telephones,* according to the number and level of capabilities they provide. The classifications suggested herein will probably change as newer technology provides us with additional enhanced telephone capabilities. Today's smart telephones may well be the standard telephones of the future.

■ ***CONVENIENCE TELEPHONES*** Convenience telephones are also called *comfort telephones*. They are equipped with features that make the calling process more convenient and the caller's work easier, thereby saving time. Typical convenience features include: abbreviated dialing, where the telephone stores a list of frequently called numbers that can be dialed with

Figure 5.13
Cordless Telephone

(Courtesy of Southwestern Bell Freedom Phone)

one or two digits; last number redial, another form of abbreviated dialing; and hands-free operation, where a loudspeaker and microphone amplify the conversation, thus freeing the caller to move about while talking.

■ ***INTELLIGENT TELEPHONES*** An intelligent telephone has many more features than a convenience telephone, often with expanded capabilites, such as a larger memory for storing speed-dial numbers. Additionally, an intelligent telephone has features that might not traditionally be associated with a telephone. Such features include: a visual digital display of the calling telephone number, a time-of-day clock, an elapsed time for the current call, a built-in calculator to compute the cost of a toll call, automatic answering and message recording (Figure 5.14), a built-in facsimile machine, or a freeze-frame picture.

Figure 5.14 *Answering Maching Telephone*

(Courtesy of Southwestern Bell Freedom Phone)

SUMMARY

Telephony refers to the electrical transmission of speech over a distance. The telephone is the basic instrument of all communications technology. Words spoken into a telephone mouthpiece are converted into electromagnetic impulses and transmitted over telephone lines. At the receiving instrument, these impulses are reconstructed into speech so that even the voice characteristics of the speaker are recognizable. This phenomenon is basic not only to telephony but also to the entire field of telecommunications as well.

Telephone systems consist of four basic parts:

1. telephone instruments
2. telephone lines that connect the telephone instruments to a central office
3. the central office that interconnects the telephone lines
4. trunks that interconnect central offices so that any caller can reach any telephone

The process of transferring a connection from one telephone circuit to another by interconnecting the two circuits is known as switching. The central office serves as a switching center.

In order to connect any telephone to any other telephone, it is necessary to have a numbering plan that identifies each telephone as unique. In the United States, each telephone is assigned a seven-digit number, which is used for calls within a specified local area. Additionally, the country is divided into geographic areas, each of which is assigned a three-digit code called the area code, which precedes the local telephone number. International calls are dialed by prefixing an international access code and a country code to the national telephone number.

Telephone systems include:

1. single-line systems
2. key telephone systems
3. private branch exchanges

The largest category of telephone systems is single-line systems. They consist of an individual telephone line but may include several extensions. They are used by residential customers and very small businesses.

Key telephone systems (also called multi-line systems) permit more than one telephone line to be terminated on one telephone instrument. Any one of the lines can be used by pressing the key associated with that particular line. They are generally used by small businesses whose needs do not require a larger phone system.

A private branch exchange (PBX) is a switching system installed for the exclusive use of one organization. The system is generally located on the customer's premises. Although a PBX can place calls to and receive calls from a public network, its main function is to handle calls within the organization. Today's PBXs are completely automatic. They contain a small computer with solid-state switching capabilities, which enables them to provide a wide variety of service features. Centrex is an alternative to a PBX.

Modern telephones are modular, i.e., equipped with a connector plug that fits into a standard jack or outlet, enabling them to be unplugged and taken to a service center for testing in case of trouble.

Rotary dial telephones communicate with the central office by dial pulsing. Pushbutton telephones may use either dial pulsing or dual tone multifrequency. The main advantage of DTMF is that it permits communication with computers.

Cordless telephones use low-powered radio transmission to and from the base station.

Modern technology has turned plain old telephone sets into sophisticated telephone instruments that are part telephone and part computer, thus enabling them to provide a variety of telephone service features that make the work easier and increase an organization's productivity. These telephones are generally described as feature telephones, which means that they provide capabilities beyond the basic functions of placing and receiving telephone calls.

REVIEW QUESTIONS

1. What is telephony?
2. Describes how the telephone works.
3. What is the function of the central office?
4. What is meant by the term *switching?*
5. What is the function of a modem? Why is it necessary to use a modem?
6. What is a numbering plan area?
7. What is a key telephone system?
8. What is a private branch exchange?
9. In what ways are pushbutton telephones more efficient than rotary-dial telephones?
10. How do feature telephones differ from traditional telephones?

VOCABULARY

telephony
switchhook
central office
telephone exchange
subscriber loop
local loop
trunk
switching
analog signal
digital signal
Hertz
pulse code modulation (PCM)
numbering plan area (NPA)
international access code
country code
area code
key telephone system
private branch exchange (PBX)
step-by-step switching equipment
crossbar switching equipment
panel switching equipment
Strowger switch
Centrex
dial pulsing
dual tone multifrequency (DTMF)

CHAPTER

6

TELECOMMUNICATION NETWORKS

CHAPTER OBJECTIVES

After completing this chapter, the reader should be able:

- *To describe the major components of a telecommunication network.*
- *To distinguish between public and private telecommunication networks and describe the principal networks in each category.*
- *To describe the characteristics of the various types of transmission media.*
- *To describe the basic modes of transmission and discuss their uses.*
- *To discuss alternating current and distinguish between frequency and amplitude.*
- *To explain bandwidth and classify transmission channels according to their bandwidth.*
- *To explain multiplexing and describe the basic multiplexing techniques.*

The concept of networks is familiar to all of us. We use highway networks of country, intrastate, and interstate roads to take us from one location to another. Trains and airplanes greatly enhance their usefulness by being linked into a network. Ships sail in a network of navigable rivers and canals. The U.S. Postal Service operates a network to deliver mail to every state, city, and home in the nation and to link up with postal systems of other nations. Radio and television networks, such as the American Broadcasting Company (ABC), National Broadcasting Company (NBC), and Columbia Broadcasting System (CBS), use facilities of numerous local broadcasting stations.

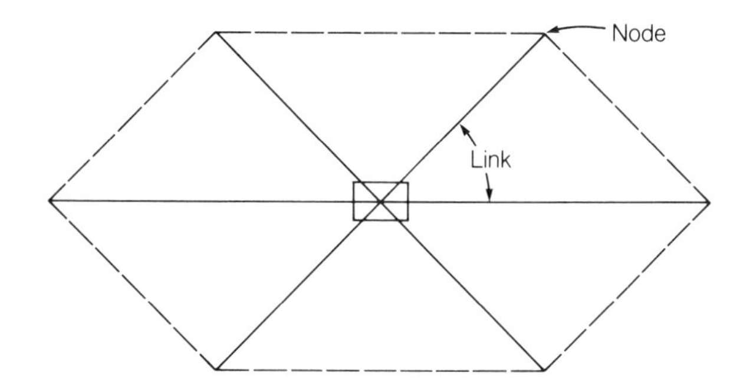

Figure 6.1
A Simple Local Network

Brenda Maddox, writing in *Beyond Babel,* describes the world's telephone networks as "technical miracles." She says, "The networks can make millions out of billions of possible connections among telephones thousands of miles apart in a matter of seconds. And they can just as swiftly unmake them and make millions of new paths in quite different directions."[1]

A telecommunication network interconnects a number of stations using telecommunication facilities (Figure 6.1). The network consists of transmission systems, switching systems, and station equipment. The physical circuit between two points is referred to as a *link.* A *node* is a point of junction of the links in a network. A basic principle of networking is that each of the stations in the network must be able to connect with any of the other stations in the network. The stations on a network intended for voice communications are, of course, telephones. Stations on data networks may be telephones or other devices, such as terminals, printers, facsimile units, or computers.

This chapter deals primarily with long distance networks since these intercity networks embody all of the fundamental principles of networking.

PUBLIC AND PRIVATE TELECOMMUNICATION NETWORKS

Telecommunication networks may be public or private. They may be designed for voice, data, message, or image transmission. They may be local or global — intercity, interstate, or intercontinent — in scope. They may use any of a number of transmission channels: open wire, paired cable, coaxial cable, radio, satellite, waveguide, or optical fibers. Regardless of their differences, the electronic principles that make these networks work are common to all types of telecommunication networks.

There are many telecommunication networks in operation, and they provide a wide variety of services. Some of these are public-switched networks (PSNs), private-line voice networks, audio program networks, video program networks, private-line data networks, packet-switching networks, and public-switched data networks.

Most networks fall into two main categories: the public-switched networks and private- or leased-line networks. Unfortunately, each of these two classifications is also known by a number of other names, a fact that causes some confusion. For example, the following network terms are synonymous: *public switched, public, switched, message toll service (MTS), long distance, direct-distance dialing (DDD),* and *inter-exchange facilities*. Similarly, the terms *private line, leased line, bypass, dedicated line, full-time circuit,* and *tie line* are used interchangeably.

PUBLIC-SWITCHED NETWORKS

Public-switched networks provide business and residential telephone service for voice and data transmission to the general public. Users share common switching equipment and channels; thus, callers wait their turn for service if all the facilities are in use. Fees paid for the use of the network are assessed on a per-call, per-minute, per-mile basis. Public-switched networks are by far the largest category of networks in terms of volume of traffic and revenues.

Until the late 1960s the Bell System was the only communications common carrier offering nationwide voice network services to the public. All long distance calls initiated by independent telephone company subscribers were switched to Bell System facilities for completion.

The MCI decision opened the way for other communications common carriers to establish point-to-point communication routes and combine them into networks. The decision also ordered the Bell System to provide interconnect privileges to the specialized common carriers, thereby allowing them to compete directly with AT&T for long distance business.

Initially, SCCs offered service only for private- or leased-line facilities. Later they expanded to include public-switched long distance services. Today, the public can choose from a number of SCCs, such as MCI, Western Union, and US Sprint, which offer private- or public-switched network services.

■ ***PACKET-SWITCHING NETWORKS*** Public packet-switching networks were originally designed to provide a more efficient method of transferring data over networks. They are still used primarily for this purpose; however, digitized voice may also be transmitted using packet-switching techniques.

In packet transmission a data message is divided into discrete units called *packets,* which are routed individually over the network. Since each of the various packets of the message can be routed over different transmission facilities, they may arrive out of sequence. Each packet contains control information that enables the message to be reassembled in proper sequence before it reaches its final destination. Packet switching is efficient because packets use the network only for the brief time they are in transit, in contrast to a circuit-switched message, which requires the use of the line for the duration of the message. The concept of packet switching is analagous to computer time sharing. In each case several users share the same facility in what appears to be the same time.

Packet-switching networks are described as value-added networks because the transmission lines are supplemented with computerized switches that control traffic routing and flow. A standard feature of packet switching is automatic error detection and correction of transmitted packets.

The first packet-switched network was ARPANET, the U.S. Defense Department's Advanced Research Projects Agency Network, which is used primarily by government and academic institutions. Several vendors currently offer packet-switching services for data communications, including Telenet, Graphnet, Tymnet, ITT, AT&T, MCI, and US Sprint.

INTEGRATED SERVICES DIGITAL NETWORK (ISDN)

The *Integrated Services Digital Network (ISDN)* is an international concept whose objective is a digital, public telephone network. It embodies a network architecture (system design) that represents a fundamental change in the way voice and data telephone traffic are managed. At the present time, ISDN is more an evolving set of interface standards than a network. The basic premise of the standards is that all forms of communications input will be reduced to a digital form prior to transmission.

An end result of such a network will be that modems and other data communications devices will become entirely obsolete. When in place, this totally digital network will extend from the user's terminal, through the local telephone company central office, to the user's destination, providing rapid transmission of voice, data, and video information.

ISDN is not proprietary to any particular vendor. Its interfaces and protocols are specified by the Consultative Committee for International Telephone and Telegraph, a division of the International Telecommunication Union of the United Nations. Over 150 nations are participating to provide this standardization.

ISDN's system architecture, called the Open System Interconnection (OSI) model, was developed by the International Standards Organization (ISO). The OSI architecture is a set of software programs distributed in seven layers or software modules. Each layer has a specific task to perform and requires its own unique protocol to enable it to function.

In today's environment, at least two local loops (each connecting the subscriber with the central office) are required for a computer interaction and a voice call to occur simultaneously. ISDN electronics will create three information channels on the "one pair" connection to the central office. Two of these channels are identical and can be used for transmission of digital voice, data, or video. The third channel provides a signaling path that transports instructions between the central office and the terminal equipment.

Figure 6.2
Foreign Exchange Service

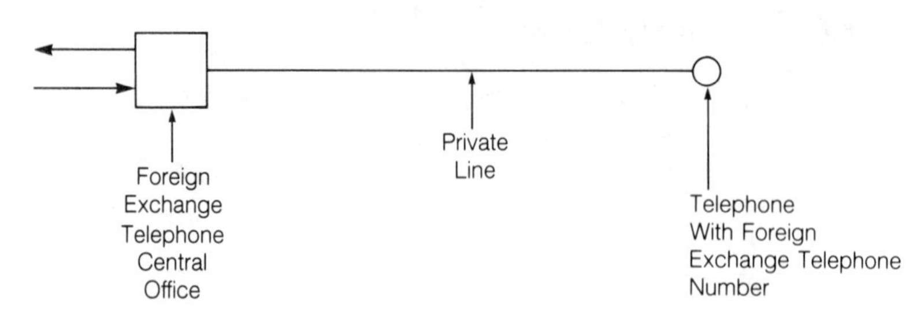

ISDN circuits integrate services that were completely separate in earlier architectures. The circuits can handle both voice and data at the same time. The net effect is that it no longer matters what kind of traffic is being transmitted over the network; the user interface and protocol are the same for all. Thus, a person with a home computer can have a telephone conversation at the same time that he or she receives or transmits data.

At present, not all facilities provided by the local telephone companies and interexchange carriers are capable of digital transmission. Before ISDN can be implemented totally, it will be necessary for the telephone companies to upgrade their central office switches, access loops, interoffice channels, and interexchange channels to digital facilities.

Several partial implementations of ISDN networks are already in place in the United States and in several European countries. The first implementation of a total ISDN environment was placed in operation in 1986 by the Illinois Bell Telephone Company. It provides ISDN, on a trial basis, to the McDonald's Corporation headquarters.

PRIVATE NETWORKS

Private or leased facilities are dedicated to the exclusive use of one subscriber; the line — or network of lines — is always available for the customer's use. The subscriber pays a flat monthly fee for the service and is entitled to use the lines on an unlimited bases.

Foreign Exchange (FX) Service (Figure 6.2) is service in a telephone exchange that is "foreign" to (outside) the one in which the user is located. This service is implemented by a private leased line that connects the subscriber's station (telephone, PBX, or modem) to a central office in another (foreign) exchange. The subscriber can then make an unlimited number of calls to any number associated with the foreign exchange area for just the cost of a local telephone call. The subscriber is given a directory listing in the foreign area directory. FX is a two-way service, so people from the foreign exchange area may also call the subscriber using what to them is a local telephone number. The call goes over the subscriber's FX line; the caller does not pay long distance charges.

Subscribers who have substantial call volume to a foreign exchange location may find it advantageous to obtain FX service. For example, a firm located in Lansing, Michigan, might use FX lines to Detroit if much of its business were there. Some large city department stores have FX lines to suburban locations to encourage business from suburbanites by permitting them to call the main store toll free.

Permanently connected private lines have several advantages. For one thing, the line—or network of lines—is always available; there are no busy signals. Another advantage is that private lines can be *conditioned* or specially treated to reduce distortion and improve transmission quality. Lines are conditioned by the addition of electronic components to the circuit. The customer can select the grade of transmission quality, as defined by the carrier's specifications, that best meets transmission requirements. Conditioning results in fewer errors during transmission, fewer echos, and less *crosstalk,* a condition in which one pair of wires picks up the transmission on adjacent wires and a conversation becomes faintly audible. Further, lines that have been conditioned are capable of higher transmission speed, thereby sending or receiving more information in a given amount of time. Line conditioning is a service offered by long distance carriers; it can be performed only on private voice-grade lines.

The cost effectiveness of leased lines depends upon the amount of time subscribers use the lines and the mileage covered by the circuit. If the lines are used only briefly each day, a public-switched network is probably more economical since leased lines are charged at a flat rate regardless of time usage. However, subscribers requiring considerable transmission time often realize substantial savings by using leased lines.

Because of their many advantages, large companies and government organizations frequently use leased lines. Private facilities range from point-to-point telephone lines to nationwide switched voice and data systems. The latter are private long distance networks that offer intercity circuits under a special pricing arrangement.

The implementation of services such as Common Control Switching Arrangements (CCSA) by AT&T in 1963, Enhanced Private Switched Communications Service (EPSCS) in 1976, and Electronic Tandem Networks (ETN) in 1979 provided customers with private network communication via dedicated full-period circuits connected to the telephone company switching centers.

Similar types of services were made available by most of the other common carriers. Large-volume customers benefited substantially from the economies of scale that resulted from the use of public-switched telephone facilities for establishing private-line networks.

Virtual private networks (VPNs) are one of the latest additions to the list of private network services that are being offered by common carriers.

VIRTUAL PRIVATE NETWORKS

A *virtual private network* is a mixture of public and private network circuits in a customized arrangement. Virtual networks use software within the interexchange carriers' systems to provide enhanced routing and network management capabilities combined with the sharing of facilities in the public-switched network. A network path is established on each call. Thus, users enjoy the pricing advantage of a private network with the efficiencies of the public network.

Virtual network service is offered by all the major long distance carriers. AT&T calls its service SDN (software defined network), MCI calls its service V-Net (virtual networks), and US Sprint names its service VPN (vertical private network).

■ ***LOCAL AREA NETWORKS*** The term *local area network (LAN)* describes a configuration of telecommunication facilities designed to provide internal communications within a limited geographic area. The network can be used to interconnect telephones, computers, terminals, word processors, facsimile machines, and other office machines within a building, a building complex, a campus, or a metropolitan area. An important advantage of using a LAN is its ability to share expensive devices like printers and disk storage.

LANs usually make use of the latest technologies to meet the needs of a data communications-intensive environment. They can also be designed to handle combinations of data, voice, and video transmission.

Local area network users often require access to external domains and to other LANs. This access can be accomplished through the use of a *gateway* or a *bridge*. A gateway serves to connect dissimilar networks, while a bridge is used to connect similar LANs.

TRANSMISSION MEDIA

Transmission systems are the links that interconnect the nodes of a telecommunication network. They may consist of any one of a variety of transmission media. For many years the principal media used for telecommunications transmission were wire conductors—open-wire pairs, wire-pair cables, and coaxial cables. Today, a number of other media such as microwave radio, satellites, waveguides, and optical fibers are in use.

OPEN-WIRE PAIRS

Early telephone and telegraph transmission lines consisted of open-wire pairs. The lines contained uninsulated wire suspended between poles with cross arms. The wires were either copper, copper-clad steel, or galvanized steel, and they required about 12 inches of separation to prevent momentary shorts.

Open-wire pairs served the nation's telecommunication needs for many years. However, they had several inherent disadvantages:

(Reproduced with permission of AT&T)

Figure 6.3
A City Street with Open-Wire Pairs

1. They were very susceptible to storm damage; thus, their maintenance cost was high.
2. They were subject to crosstalk.
3. They were unsightly; as more and more telephones were installed, the number of open-wire curcuits strung on poles in the large cities reached the saturation point (Figure 6.3).

Today, it would be impossible to meet telephone requirements with open-wire facilities. (See Figure 6.4.)

WIRE-PAIR CABLES

Wire-pair cable consists of copper conductors insulated by either wood pulp or plastic and twisted into pairs (Figure 6.5). Twisting minimizes the interference between pairs when they are packed into a large cable. Each of the two-wire circuits is capable of carrying one telephone channel. The cables may be either suspended from poles or buried underground. Wire-pair cables have replaced open-wire pairs almost entirely.

COAXIAL CABLES

A coaxial cable consists of one or more hollow copper cylinders with a single wire conductor running down the center. A single cable can carry a very large number of telephone calls. The name "coaxial" is derived from the fact that the cylinder and the center wire each have the same center axis (Figure 6.6). Because coaxial cables transmit at very high frequencies,

Figure 6.4
Open-Wire Pairs

(Reproduced with permission of AT&T)

they have little distortion or signal loss. Therefore, they are a better transmission medium than either open-wire pairs or wire-pair cables.

Figure 6.7 shows a cable composed of 12 coaxial units surrounding a core of conventional wire conductors. It can handle approximately 1,000 telephone conversations simultaneously.

Figure 6.8 shows a sample length of Co-ax 20, a 20-unit coaxial cable capable of carrying 18,740 telephone calls at once.

MICROWAVE RADIO

A microwave radio system sends signals through the atmosphere between towers usually spaced about 20 to 30 miles apart. The system amplifies the signals and retransmits them at each receiving station until they reach their destination. Microwave radio operates at the high-frequency end of the radio spectrum. The signals follow a line-of-sight (straight-line) path, and the relaying antennas must be within sight of one another. (See Figure 6.9.) One of the principal problems affecting microwave radio is the variation in signals caused by changes in atmospheric conditions. Moisture and temperature conditions can cause the radio beam to bend, resulting in fading.

The principal advantage of microwave radio is that it can carry thousands of voice channels without physically connected cables between

(Courtesy of The Western Electric)

Figure 6.5
A Twisted Wire-Pair Cable

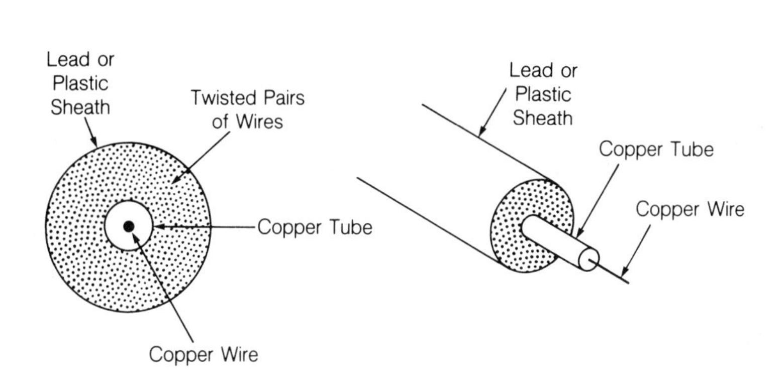

Figure 6.6
A Coaxial Cable

Figure 6.7
A 12-Unit Coaxial Cable

(Reproduced with permission of AT&T)

Figure 6.8
A 20-Unit Coaxial Cable

(Reproduced with permission of AT&T)

(Courtesy of MCI Telecommunications Corporation)

Figure 6.9
A Microwave Installation. A microwave relay tower located in the desert. All microwave antennas are on towers within sight of one another. Relay towers are usually spaced 20 to 30 miles apart.

points of communication, thus avoiding the need for continuous right-of-way between points. Further, radio is better able to span water, mountains, or heavily wooded terrain, which pose barriers to wire or cable installation. Most long distance links today are either coaxial cable, microwave radio, or optical fiber. (See Figures 6.10 and 6.11.)

COMMUNICATION SATELLITES

Communication satellites relay microwave transmissions. The satellite functions as a microwave tower located high in the sky. The *transponder* is equipment that receives a signal, changes its frequency so that the outgoing signal does not interfere with the incoming signal, and retransmits it. Satellites with solar-powered batteries orbit directly over the equator at a distance of 22,300 miles above the earth so that they travel at exactly the same speed as the rotation of the earth. Because of their very high altitudes, they can receive radio beams from any location in the country and reflect them back to a portion of the world (Figure 6.12). Their very high altitude also permits them to overcome obstacles that block line-of-sight transmission (microwave radio), such as mountains and the curvature of the earth. Satellites can handle very large volumes of voice and data transmission simultaneously.

Figure 6.10
A Microwave Installation. A microwave relay tower located on a mountain range. Since microwave transmission is line-of-sight, the towers are placed on elevated points when possible.

(Courtesy of MCI Telecommunications Corporation)

An efficient antenna system is a vital part of any radio system — microwave or satellite. The transmitting antenna radiates as much or the transmitter's energy as possible toward the receiver, and the receiving antenna collects as much of this energy as it can. A satellite antenna, or *earth station,* consists of a dish that points at the satellite in much the same way that a microwave relay tower points to the next tower in the relay chain. Probably most of us have noticed the satellite dishes appearing atop buildings or on the ground in ever-increasing numbers.

Satellite communication has several unique characteristics that make it very attractive from both performance and economic standpoints. These include the following:

1. The cost of a satellite is not dependent upon the distance between stations.
2. Natural barriers, such as mountains, oceans, or densely wooded terrain, do not impede placement of satellite facilities.
3. Signals are available at any point within the satellite's path.

Where traffic moves between a few nodes of a network over great distances, satellite communication is the most cost-effective medium because the cost of the link is not dependent upon distance. In fact, the greater the distance, the greater the cost advantage.

(Courtesy of Western Union)

Figure 6.11
A Microwave Installation. Satellite antennas at Western Union's Glenwood, New Jersey, earth station capture signals from satellites. The microwave tower is used to send messages to other microwave towers.

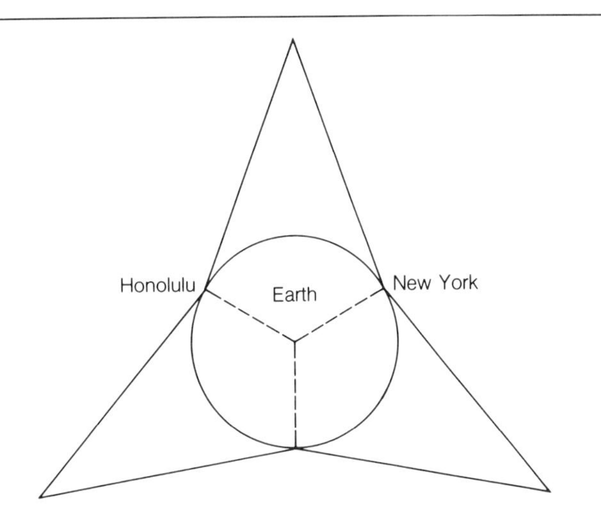

Figure 6.12
Satellite Coverage of Earth's Surface. A satellite orbiting 22,300 miles above the earth transmits to one-third of the world.

Figure 6.13
WESTAR IV Satellite

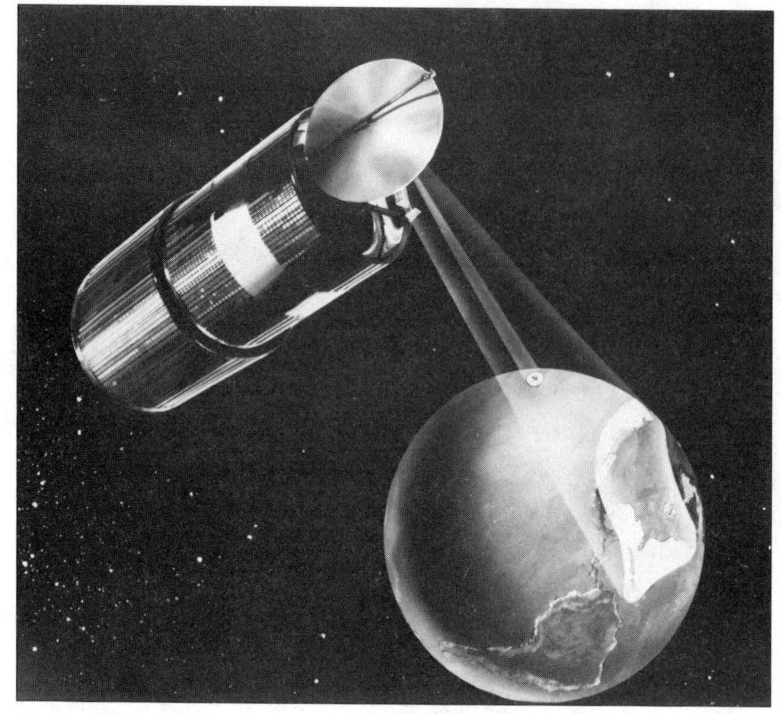

(Courtesy of Western Union)

Natural barriers make it difficult, and therefore expensive, to construct any terrestrial or submarine transmission system. Satellite systems effect substantial cost savings by avoiding these barriers.

Since a signal is available at any point within the satellite's path, the only additional cost for a new receiving node is the cost of the hardware. Thus, in the case of point-to-multipoint distribution, the satellite has a significant cost advantage. Furthermore, the addition of new receiving nodes greatly increases the cost effectiveness of the system.

A number of businesses provide voice and/or data transmission service by satellite (Figures 6.13 and 6.14). These companies either own their satellites or purchase capacity from another organization. Some companies offering satellite transmission services include: RCA Americom, Western Union, American Satellite Company, GTE Satellite Communication System, Hughes Communications, Inc., and Communication Satellite Corporation.

The distribution of printed text has been changed considerably by the use of satellites. *The New York Times, The Wall Street Journal, USA Today,* and other national newspapers use satellite communications to reduce

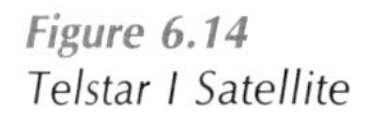

Figure 6.14
Telstar I Satellite

(Courtesy of Bell Laboratories)

their distribution costs. The copy is transmitted in facsimile form to a number of printing plants in various regional locations.

WAVEGUIDES

A *waveguide* (Figure 6.15) is a rectangular or circular metal tube down which very high frequency radio waves travel. The metal tube confines the radio waves and channels them to a point where they are released into the air to continue their travel over microwave transmission facilities. Waveguides can transmit greater amounts of power with less energy loss than coaxial cables.

Rectangular waveguides have been used for some time to connect microwave transmitting equipment to the microwave towers. Their use is generally limited to distances of less than a thousand feet. The newer and more efficient waveguide is the circular type, which consists of a precision-made pipe about two inches in diameter. This type of waveguide is capable of transmitting frequencies much higher than rectangular waveguides.

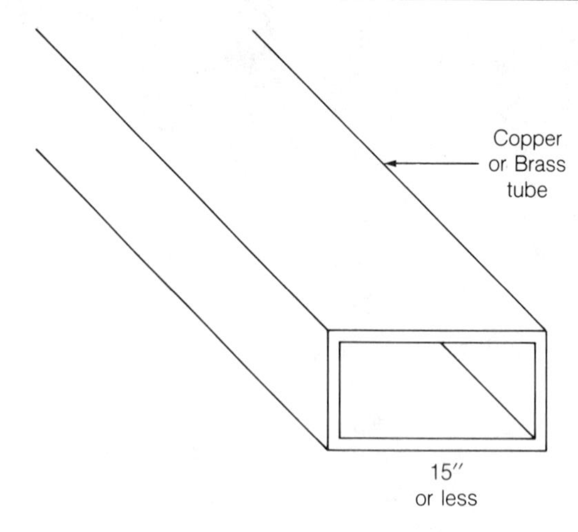

Figure 6.15
Waveguide

The waveguide is attractive because of its wide bandwidth and low-loss transmission characteristics. Deterrents to its use are its critical engineering requirements and very high cost.

OPTICAL FIBER SYSTEMS

The newest and most remarkable technological development in transmission media has been fiber optic waveguides. *Fiber optics* are hair-thin filaments of transparent glass or plastic that use light instead of electricity to transmit voice, video, or data signals. The fibers act as waveguides and have the potential of carrying an extememly high bandwidth with low signal attenuation. Covered to prevent light loss along the line, the fibers are bundled together into a flexible cable (see Figure 6.16). In most applications, optical fiber systems combine optical fibers with *laser* (an acronym for light amplification by stimulated emission of radiation) technology to generate very high frequency beams of light with tremendous information capacity. Some optical fiber systems use light-emitting diodes (LEDs) as a source of light.

In lightwave communications, the transmission sequence begins with an electrical signal. The signal is transformed into a light signal by a laser or other source that couples the light into a glass fiber (Figure 6.17) for transmission. The light signal is renewed along the way by a repeater unit. At its destination the light is sensed by a receiver and converted back into electricity.

Optical fiber systems have many advantages:

1. The hair-thin size of the fiber lightguides makes possible cables that are much smaller and lighter than their wire counterparts (Figure 6.18).
2. Lightwave systems cost much less to manufacture than wire systems of equivalent capacity.

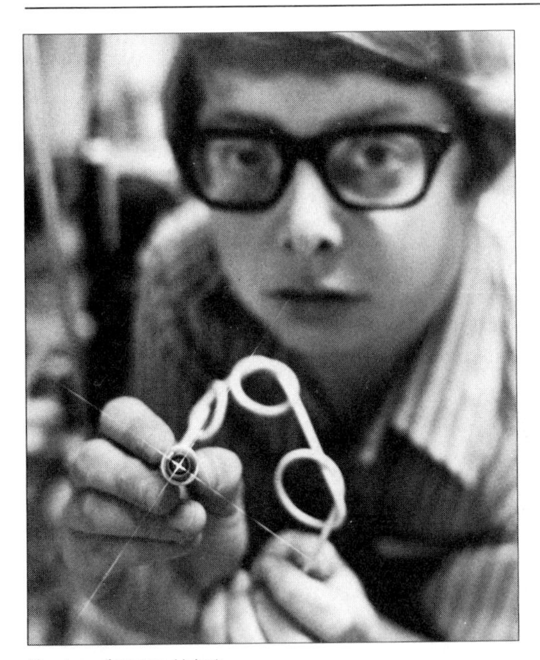

(Courtesy of Western Union)

Figure 6.16
Cross-Section of Lightguide Cable

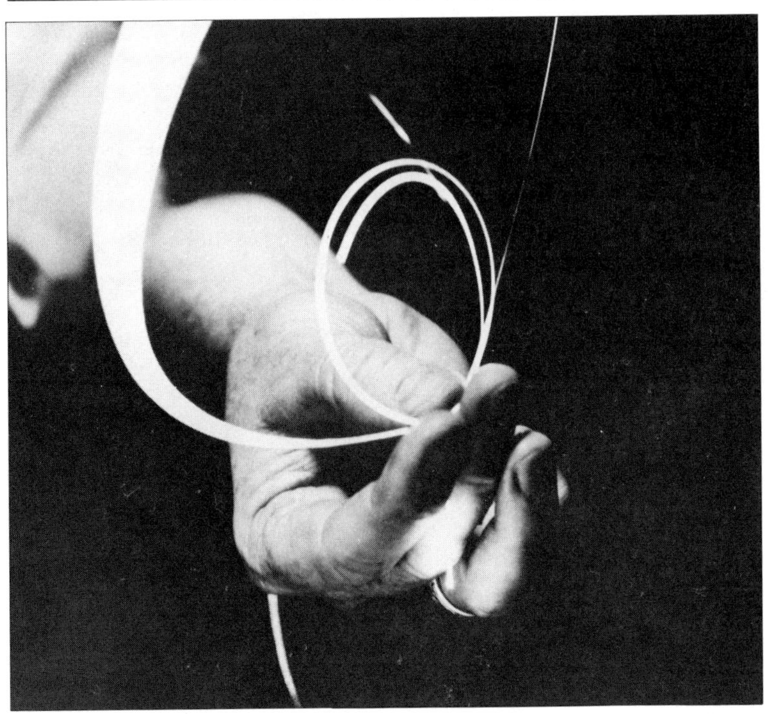

(Courtesy of Bell Laboratories)

Figure 6.17
Glass Fibers

Figure 6.18
Cables of Glass Fibers

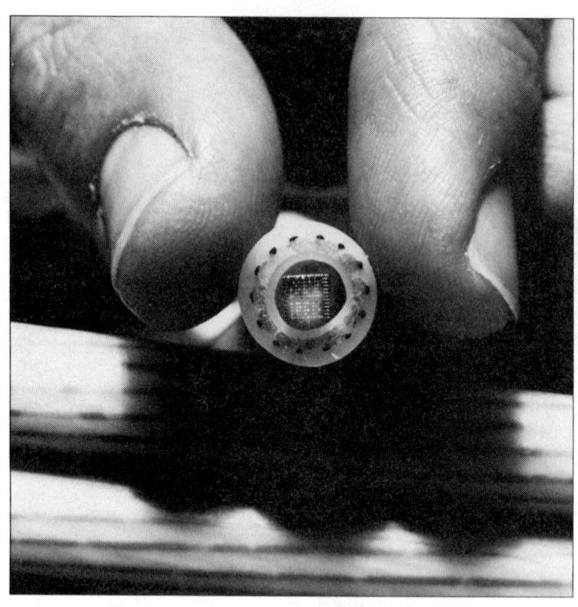

(Courtesy of Bell Laboratories)

3. The broader bandwidth of fiber optics provides greater communications capacity than traditional forms of transmission. A single fiber cable (two fibers) carries over 1,300 two-way conversations simultaneously; one twisted wire pair (two wires) carries only 24.
4. Fiber is capable of transmitting data at extremely high speeds. It easily handles data rates of 90M bits per second.
5. Transmission through lightguides is relatively immune to the effects of lightning and other types of electrical interference (Figure 6.19).

Fiber optic technology has revolutionized the long-haul telecommunications market, becoming the medium of choice for many applications. The first commercial application of fiber optics was a digital lightwave system called FT3, which transmitted 44.7 million bits of information per second. It began service in September 1980 and covered a 6.5 mile route of the Bell System, linking three central offices in Atlanta, Georgia. Soon all of the interexchange carriers began racing to expand their networks by the construction of fiber-optic facilities. By the end of the 1980s, all of the major cities in the United States were linked by fiber-optic networks.

In addition, a consortium of international telecommunications carriers combined their resources to lay fiber-optic cables across the Atlantic and Pacific Oceans. These transoceanic fiber-optic systems are now operational.

Figure 6.19
Lightguide Installation on the Golden Gate Bridge

(Courtesy of The Western Electric)

TRANSMISSION CHANNELS

Information travels from one point to another along a transmission link that carries an electrical signal. The link is called by any of a number of different names: *channel, trunk, path, circuit,* or *facility.*

In designing a voice or data network, it is necessary to decide which transmission medium will be used. Any of the transmission media discussed in the previous section can be used to move information. Also, two or more types of transmission media can be combined into a network. Practically every long distance telephone call travels over a variety of transmission media before it reaches its destination. Figure 6.20 illustrates the media that could be used to route an intercity telephone call.

Another important choice to make is the type and grade of circuit to be used. Here it is necessary to determine whether communication lines must transmit in one direction only or in both directions. If transmission will occur in both directions, the system's designer determines whether the transmission in both directions will be simultaneous. The principal criteria

Figure 6.20
Intercity Telephone Call Media

Route	Medium
Subscriber telephone to local central office	Wire-pair cable
Local central office to toll switching office	Wire-pair cable, coaxial cable, or fiber optic
Toll switching office to distant city toll office	Coaxial cable, microwave radio, satellite, or fiber optic
Distant city toll office to local central office	Wire-pair cable, coaxial cable, or fiber optic
Local central office to called party's telephone	Wire-pair cable

used in choosing the media and equipment are transmission requirements and economics.

ATTENUATION AND REPEATERS

As a signal travels along a transmission line, there is a natural loss of power; that is, the signal grows weaker. This loss in power is called *attenuation*. If the signal is not strengthened somehow, it will not reach its destination in a way that it can be understood. To compensate for this loss of power, amplification devices known as *repeaters* are inserted into the channel. A repeater is a device used to restore signals that have been distorted because of attenuation to their original shape and transmission level. Repeaters are also known as *amplifiers* since they amplify the signal while it still has enough magnitude to represent the original signal. Repeaters placed at equal distances throughout the transmission channel reinforce the signal strength so that when the signal is received at the terminating point it can be fully understood. Attenuation increases as frequency of the transmission bandwidth increases; it also increases as the diameter of the wire used in the circuit decreases.

ECHO AND ITS CONTROL

To provide high-quality speech transmission over long toll connections, a type of distortion known as *echo* becomes important. Echo is an electric wave that has been reflected back to the transmitter with sufficient magnitude and delay as to be perceived. Echo is a function of distance on the circuit; the greater the distance, the more time it takes for the echo to return to the person talking.

Echo is caused by a change in impedance (opposition in an electrical circuit) on the transmission line, causing the signal to be reflected at reduced amplitude. The resulting "hollow" sound can be objectionable; therefore, the telephone company installs *echo suppressors* to reduce it to a negligible level. Echo suppressors are devices that detect speech signals transmitted in either direction on a four-wire circuit. The suppressors introduce loss in the opposite direction of the speech transmission to suppress echos. In the public telephone network, echo suppressors are used typically in trunks longer than 1,850 miles.

Echo suppressors designed for voice cannot be used when data is transmitted over voice circuits, since the speech detector detects only speech, not other sounds. To transmit data over voice circuits, the echo suppressors must be disabled. This is accomplished by installing *echo suppressor disablers,* devices that transmit a tone that can be heard on the telephone as a high-pitched whistle. The tone puts the echo suppressor out of action until there has been no signal on the line for a period of approximately 50 milliseconds.

Echo suppressors have two main drawbacks:

1. They tend to clip the speech as they open and close the transmission path.
2. They have no capability for suppressing echo during two-way transmission, because both transmission directions cannot be opened simultaneously.

The newest technique to control echo is a device known as an *echo canceler.* Because these devices eliminate both of the above-cited problems, they are widely used in modern circuit construction.

MODES OF TRANSMISSION

There are three basic modes of transmission: *simplex, half duplex (HDX),* and *full duplex (FDX)* (Figure 6.21).

■ ***SIMPLEX*** The *simplex* mode of operation is used when transmission is in one direction only. Simplex circuits are two-wire circuits; however, transmission is in the same direction at all times. An analogy would be a radio announcer who talks to listeners who cannot reply.

■ ***HALF DUPLEX*** In *half-duplex* operation, transmission can be in either direction, but in only one direction at a time. HDX is often referred to as two-wire transmission. One station transmits information; and only when that transmission has been completed, can the other station respond. This mode could be compared to two debaters; each may speak, but only in turn and one at a time.

■ ***FULL DUPLEX*** *Full duplex* is used for transmission in both directions simultaneously. Full-duplex circuits are sometimes referred to simply as duplex circuits. There is no way that human communications can be compared with full-duplex transmission because when two people interrupt each other by talking at the same time, intelligent information cannot be conveyed.

Historically, the communication circuit had to be four-wire for full-duplex capability. However, two-wire circuits can be used in full-duplex mode if the amplifiers are designed to permit transmission in both direc-

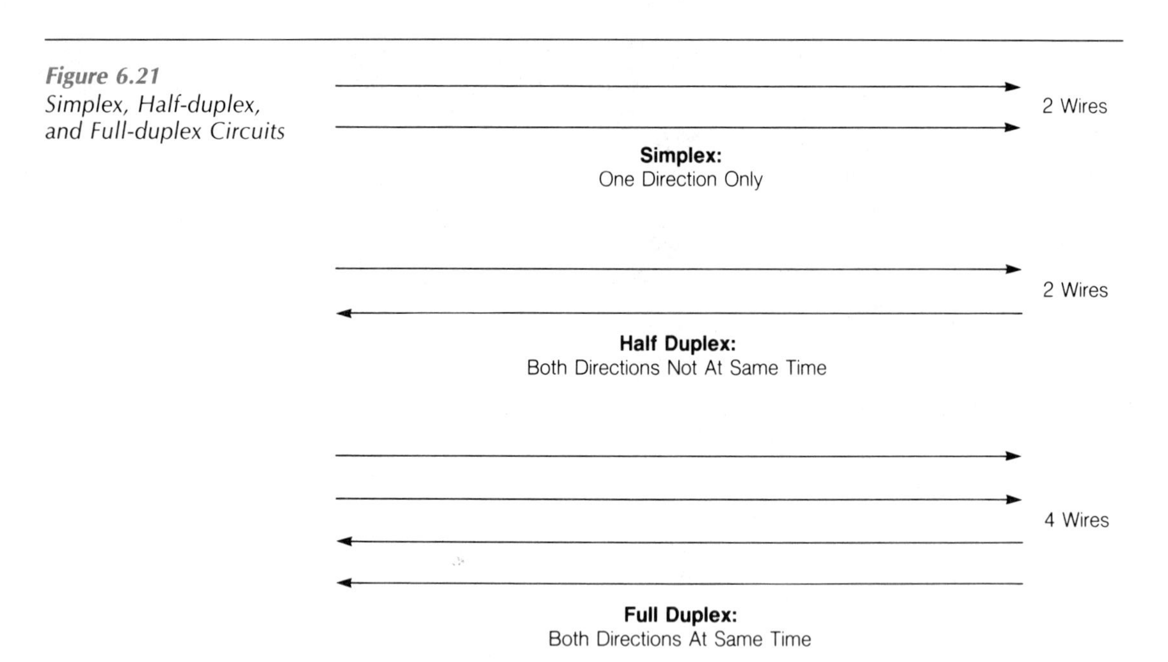

Figure 6.21
Simplex, Half-duplex, and Full-duplex Circuits

tions simultaneously or if a modem is used to split the band of frequencies into two parts.

■ ***USES OF EACH TRANSMISSION MODE*** Simplex circuits have limited use. Their principal use is in signaling, where they control the telephone switching process. They are not used in voice or data transmission. Either half-duplex or full-duplex circuits may be used for data transmission. With leased telephone lines, the user can choose between half-duplex and full-duplex circuits. Full-duplex circuits cost more than half-duplex circuits; however, when data communications require two-way simultaneous transmission, a full-duplex circuit is necessary.

ALTERNATING CURRENT

There are two types of electrical current: *direct current (DC)* and *alternating current (AC)*. Direct current travels in only one direction in a circuit (+ to −), while alternating current travels first in one direction (+ to −) and then in the other direction (− to +). Household current is alternating current—it reverses itself continuously. It is known as 60-cycle electric current because it reaches a maximum flow in one direction, then reverses itself to a maximum flow in the opposite direction; the reversal repeats back and forth, making 60 complete cycles in one second.

The number of cycles completed per second is the *frequency* of the current. It is expressed in Hertz (Hz). The higher the frequency, the more

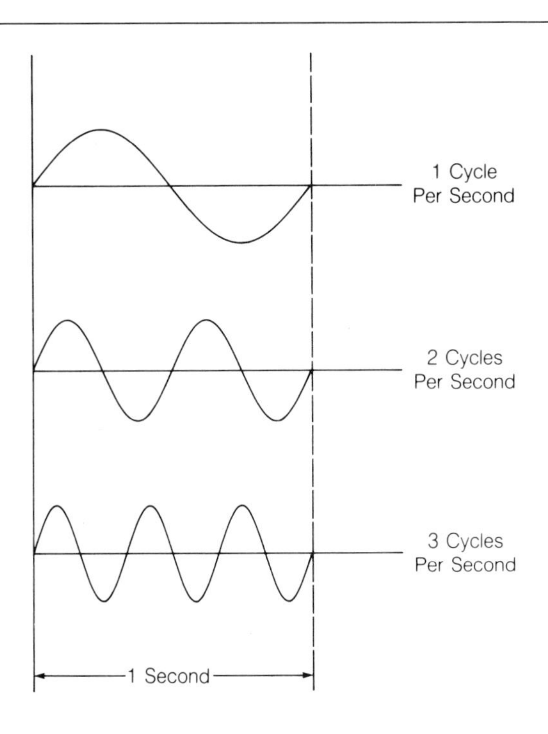

Figure 6.22
Sine Waves of Differing Periods

direction reversals each second. *Amplitude* is the amount of variation of an alternating quantity from its zero value. It is sometimes described as volume, intensity, or loudness. Frequency and amplitude are the two parameters of any electronic signal.

The frequency of oscillation of an alternating current can be represented as an undulating wave known to mathematicians as a *sine wave*. A sine wave can be drawn by plotting the power level of the alternating current against intervals of time (Figure 6.22). The alternating current starts out at zero power level, rises to a maximum at ¼ of a cycle, then drops back to zero after ½ of a cycle, continues to a maximum in the other direction at ¾ of a cycle, and goes back to zero power at one full cycle.

BANDWIDTH

Sounds that we hear consist of a mixture of electromagnetic waves that usually range from about 300 Hz to 3,300 Hz, although the range extends up to 20,000 Hz for the person who has exceptional hearing ability. However, since most voice characteristics are contained in sound waves in the range from 300 to 3,300 Hz, telephone lines are designed to transmit frequencies in this range. *Bandwidth* means the range of frequencies that are transmitted (Figure 6.23). It is the difference in Hz between the highest

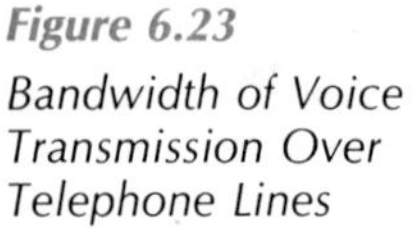

Figure 6.23
Bandwidth of Voice Transmission Over Telephone Lines

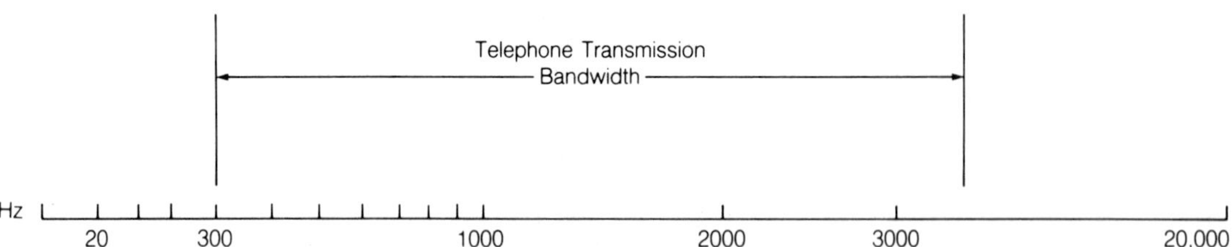

and lowest frequencies of a band. Thus, the *voiceband* is said to be 3,000 Hz, or the difference between 300 Hz and 3,300 Hz.

The chief difference between the grades of available transmission channels is their bandwidth. In classifying channels by bandwidth, there are three general categories:

1. narrowband (0 to 300 Hz)
2. voiceband (300 to 3,300 Hz)
3. broadband or wideband (over 3,300 Hz)

Narrowband channels are used for nonvoice service, such as teletypewriter and low-speed data transmission. Voiceband channels are used for voice transmission, foreign exchange (FX) service, and data communications. Voiceband lines are the most prevalent form of communication facilities. Wideband channels are used for high-speed data, facsimile, and video transmission. Their principal value, however, is for multiplexing many channels of various types onto a single bandwidth.

FREQUENCY

The *frequency* of an oscillating current — the number of complete oscillations the wave makes per second — can be controlled so that the current oscillates at a specific frequency. This frequency produces a given sine wave, which carries signals representing voice or other forms of communication.

Although telecommunication transmission systems are quite different from commercial radio systems, they both work on many of the same principles. A radio station broadcasts (transmits) at an assigned frequency. A listener can tune the radio to receive a particular station's transmission; the radio tuner selects the desired station by locating the frequency on which the station is operating. Each station is received on a different frequency. However, the sound is in the same bandwidth for all stations (300 – 3,300 Hz).

Similarly, a telecommunication channel is designed to operate at a particular frequency. This frequency serves as a carrier that transports specific communication bandwidths. Telecommunication systems differ from commercial radio systems in that the signals transmitted are confined to the medium (wire, cable, coaxial cable, microwave, satellite, or fiber optics) rather than broadcast to the air.

MODULATION

Modulation is the process of changing the form of a signal carrying intelligence (voice or data) to make it compatible with a different transmission medium. Modulation alters (modulates) a signal to enable it to be transmitted on a different frequency. Modulation improves the efficiency of a transmission by increasing the transmission capacity of a telecommunications channel through the use of higher frequencies.

The newer transmission media, particularly satellites and fiber optics, have wider bandwidths than earlier transmission media. In addition, newer and more efficient transmission techniques, such as pulse code modulation (PCM), have been developed. These factors have significantly increased the number of messages that can be carried by each channel.

In the transmission process, modulation is a function of imposing the signal that carries the intelligence of a message onto the carrier wave generated by the flow of alternating current. *Demodulation* changes the bandwidth back into the form of the original message signal.

There are two principal types of modulation: amplitude modulation and frequency modulation. *Amplitude modulation* (Figure 6.24) is the process whereby the intelligence of the signal is represented by variations in its amplitude or strength. *Frequency modulation* (Figure 6.25) is the process whereby the intelligence of the signal is represented by the variations in the frequency of the oscillation of the signal.

There are many other forms of modulation. Two forms that are used frequently in data transmission are phase modulation and pulse code modulation. *Phase modulation* is used in equipment operating at high speeds; it makes use of a frequency phase shift that occurs when a specific pattern of digital information is received. *Pulse code modulation* is one of the newest forms of transmission. This system uses a sampling process and requires specific types of modulation/demodulation equipment as well as special transmission facilities. These two types of modulation are described more fully in Chapter 8.

MULTIPLEXING

Multiplexing is the division of one transmission channel into two or more individual channels. It is accomplished by a *multiplexer* (Figure 6.26), a device that combines a number of low-speed channels into one higher speed channel at one end of a transmission and divides them into low-speed channels at the other. Multiplexers must be provided at both ends of the circuit. The process permits a voice-grade channel to carry two or more narrowband channels. Similarly, a wideband channel can carry several

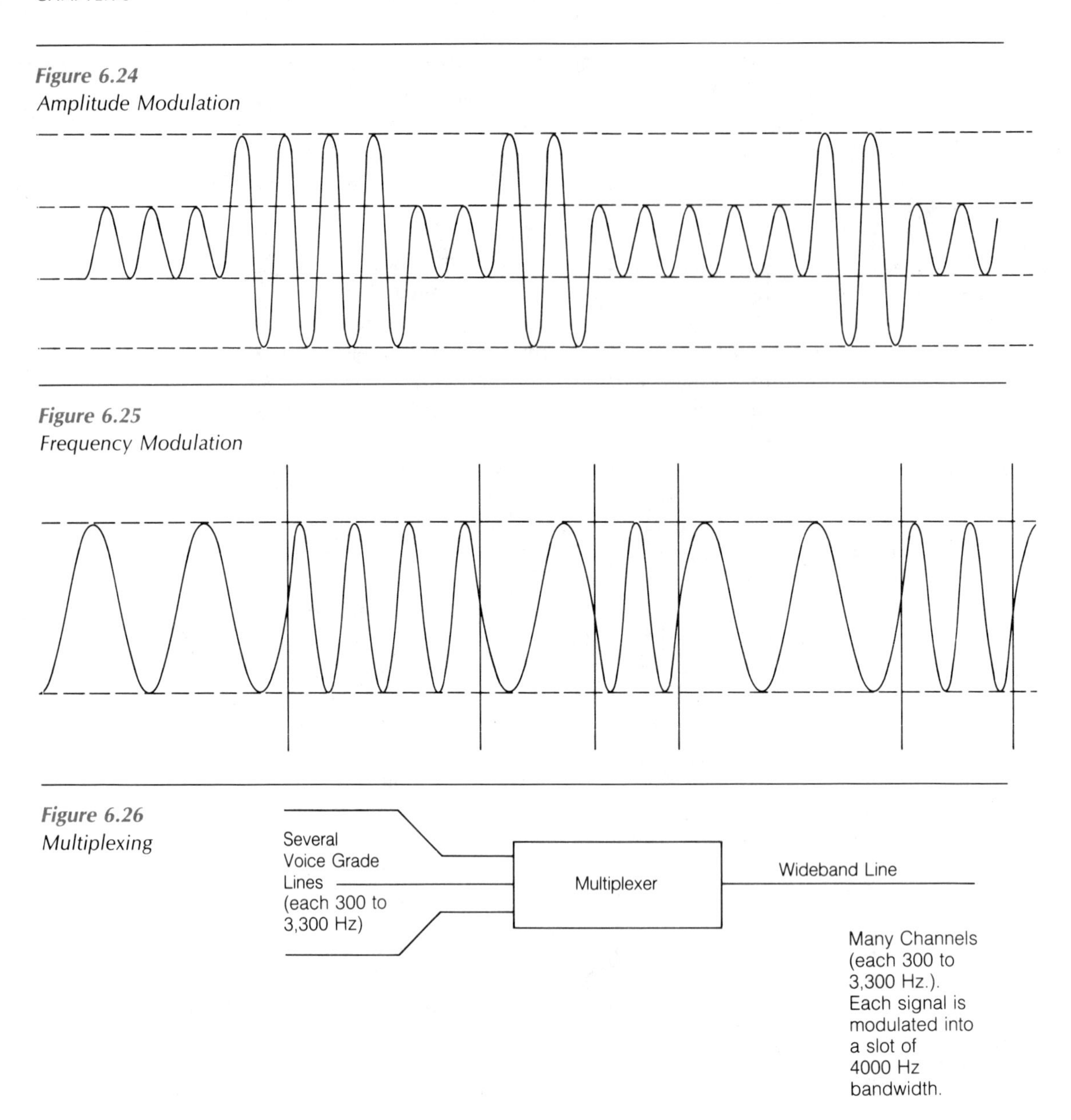

Figure 6.24
Amplitude Modulation

Figure 6.25
Frequency Modulation

Figure 6.26
Multiplexing

voiceband channels. Multiplexing reduces line costs by increasing the number of circuits that can be carried by a communications channel.

There are two basic multiplexing techniques: frequency-division multiplexing (FDM) and time-division multiplexing (TDM). *Frequency-division multiplexing* divides the channel frequency range into narrower frequency bands. *Time-division multiplexing* assigns a given channel successively to several different users at different times.

SUMMARY

Interconnecting a number of individual telecommunications lines into an integrated network makes it possible for any telephone to be connected to any other telephone anywhere in the world. Services such as distributed data processing and electronic mail depend on the existence of networks.

A telecommunication network consists of transmission systems, switching equipment, and station equipment. A basic principle of networking is that each of the stations in the network must have the capability of establishing a connection with any of the other stations in the network.

The stations on a voice network are telephone sets. Stations on a data network may be telephones, terminals, facsimile machines, or computers. Transmission channels may use paired cable, coaxial cable, radio, satellite, or optical fiber. The newer transmission media, such as satellites and optical fibers, make it practical to offer service to locations that would otherwise be inaccessible.

The two types of telecommunication networks are public and private. Public-switched networks provide telephone service to the general public. They are by far the largest category of networks in terms of volume and revenue.

Public packet-switching networks provide an efficient method of transferring data. In this technique, messages are divided into packets for transmission. The transmission channel is occupied only during the transmission of the packet; afterward it is available for the transmission of other packets. This contrasts with circuit switching, wherein entire messages are transmitted as a unit, and the transmission channel is occupied for the duration of the transmission.

The Integrated Services Digital Network (ISDN) is an international concept whose objective is a totally digital public network. At the present time, it is more an evolving set of standards than a network.

Private or leased networks are dedicated to the exclusive use of one subscriber. Both public and private networks are provided by AT&T, GTE, MCI, RCA, Western Union, US Sprint, and many other common carriers.

A virtual private network is a mix of public and private network circuits in a customized arrangement.

A local area network (LAN) is a private network designed to provide internal communications within a limited geographic area, such as a building, a building complex, a campus, or a metropolitan area. LANs incorporate state-of-the-art technologies to meet the needs of a data communications-intensive environment.

Transmission systems consist of a variety of transmission media: open-wire pairs, wire-pair cables, coaxial cables, microwave radio, satellites, waveguides, and optical fiber systems.

Information travels from one point to another along a transmission channel that carries an electrical signal. Attenuation is the loss of power that occurs when a signal travels along a transmission line. Repeaters or amplifiers are inserted into the channel to restore the signal.

In voice transmission, echo is an electric wave that is reflected back to the transmitter. It can be perceived as an objectionable hollow sound. Echo suppressors are used to reduce it. If data needs to be transmitted over a voice circuit, the echo suppressors must be disabled by using either echo suppressor disablers or echo cancelers.

There are three basic modes of transmission: simplex, half duplex, and full duplex. Voice transmission requires full-duplex circuits. The chief difference between the grades of available transmission channels is bandwidth. The three categories of bandwidth include: (1) narrowband, used for teletypewriter and low-speed data; (2) voiceband, used for voice and data transmission; and (3) wideband, used for high-speed data, facsimile, and video transmission.

The two characteristics of electrical current are amplitude and frequency. Amplitude measures the strength of the signal; frequency is the number of times a wave form repeats itself in a given period of time.

Modulation is the conversion of signals from pulse form to wave form. Demodulation is the reversal of the modulation process.

Multiplexing is the division of a communication channel into two or more narrower channels. This process is performed by an electronic device known as a multiplexer. Two multiplexers are required—one at each end of the circuit.

REVIEW QUESTIONS

1. What three elements comprise a telecommunication network?
2. What is the basic principle of networking?
3. Briefly describe packet switching. Why is it an extremely efficient method of transferring data?
4. What is ISDN?
5. Under what conditions would it be advantageous for a subscriber to obtain FX services?
6. Cite two advantages of private telephone lines.
7. What are some newer types of transmission media? Briefly describe their advantages.
8. What transmission medium would you use if you were planning a transmission link between New York and California? Why?
9. Briefly describe the three modes of transmission.
10. What are the three categories of bandwidth? Which one would be used for video transmission?
11. Describe the differences between amplitude and frequency modulation.
12. What is multiplexing and what purpose does it serve?

VOCABULARY

link
node
public switched network
message toll service
direct distance dialing (DDD)
dedicated line
tie line
Foreign Exchange Service (FX)
line conditioning
virtual networks
Integrated Services Digital Network (ISDN)
Local Area Network (LAN)
open-wire pairs
wire-pair cables
coaxial cables
microwave radio
communications satellite
waveguide
fiber optics
transponder
earth station
laser
light-emitting diode (LED)
channel
trunk
circuit
attenuation
repeaters
amplifiers
echo-suppressors
simplex
half duplex (HDX)
full duplex (FDX)
frequency
amplitude
bandwidth
narrowband
voiceband
broadband
modulation
amplitude modulation
frequency modulation
multiplexing

ENDNOTE

1 Brenda Maddox, *Beyond Babel* (New York: Simon & Schuster, 1972), 202.

CHAPTER

7

DATA COMMUNICATIONS

CHAPTER OBJECTIVES

After completing this chapter, the reader should be able:

- *To discuss the importance of being able to use the existing telephone network for data transmission.*
- *To discuss the developmenmt and importance of microprocessor technology.*
- *To discuss the development and uses of microcomputers.*
- *To define data communications systems and explain their purpose.*
- *To distinguish between centralized processing systems and distributed processing systems.*
- *To describe local area networks.*
- *To distinguish between online processing, realtime processing, and batch processing.*
- *To discuss the applications of realtime systems as contrasted with batch processing systems.*

The first union of computing and communications occurred in 1940 when Dr. George Stibitz demonstrated the use of an electrically operated digital computer developed by Bell Laboratories. Stibitz, a mathematician, took a teletypewriter to a meeting of the American Mathematical Society at Dartmouth College in Hanover, New Hampshire. The teletypewriter in New Hampshire was connected to the calculator in New York via telephone lines. Stibitz then invited conferees to type in problems; the machine sent back the answers in minutes.

In its broadest sense, data communications describes any method of moving data from one location to another. Data can be communicated by physical transportation — messengers, trucks, or airplanes — or by electrical means — input and output devices, electrical transmission links, and associated switching equipment. In many cases, the best way to get information from one place to another is by mail; this method is usually the most economical. However, when rapid transmission of information is important, electrical methods are required.

One of the first business applications of data communications was the Sabre airline reservation system, developed jointly by IBM and American Airlines. After six years of development, the system began operation in 1962. The system used telephone lines to link terminals placed throughout the United States with a central computer located in Tarrytown, New York.

Business and industry are increasingly recognizing the advantages of data networks. Although only about one percent of the computers sold in 1965 were linked to a data communication system, virtually all computers sold or leased in the United States today have communications capabilities.

EVOLUTION OF DATA COMMUNICATIONS

The earliest form of data communications was the *teletypewriter (TTY),* a printing telegraph instrument having a signal-actuated mechanism for automatically receiving printed messages. (*Teleprinter* describes a receive-only unit that has no keyboard.)

Teletypewriter systems, which transmit data at the rate of ten characters per second, have been in use for many years and are still useful for certain applications. However, the communication lines that carry the teletype signals are capable of transmitting information at much faster speeds than the teletype machines are capable of sending and receiving. This condition, coupled with the development of electronic computers with their high-speed processing capabilities, focused attention on the need for faster methods of data communications.

DATA TRANSMISSION RESEARCH

In an effort to solve this problem, the Bell System investigated the possible use of the telephone-switched network for data transmission. Since the telephone-switched networks had been constructed for analog voice transmission and since computers used digital transmission, extensive testing was required to determine the feasibility of using the telephone networks to transmit data.

The results of this research were published by the Bell System in 1960 in a report entitled *Capabilities of the Telephone Network for Data Transmission.* It showed that the telephone-switched network could be used satisfactorily for data transmission. This was a most important finding, for had the telephone network proved unsatisfactory, it would have been necessary to build a new national network to accommodate data communications. The time and expense involved in such a gigantic undertaking would have been quite a deterrent to the growth of data communications. Instead, this discovery began an era of rapid growth in data communications.

INDUSTRY GROWTH

Today, data communications comprises a significant portion of the traffic on the telephone network, and its share is increasing. A number of factors have influenced the rapid growth of data communications, including the following:

1. Advances in technology coupled with the dramatic drop in computer costs. Applications that formerly required an expensive mainframe, minicomputer, or standalone word processor can now be processed on microcomputers. As the office work force increasingly uses microcomputers to aid in their work, the demand for shared data bases that are accessed over telecommunications facilities escalates.
2. The advent of competition in the communications marketplace. Deregulation in the telecommunications industry has resulted in improved communications facilities, a wider variety of common-carrier services, and competitive pricing.
3. The trend of business organizations toward decentralization, mergers, and conglomerates. As companies grow larger and become more geographically dispersed, they turn more and more to data communications to solve their problems of internal communications.

TECHNOLOGICAL DEVELOPMENTS

Advances in technology have revolutionized the data processing and data communication industries.

MICROELECTRONICS

Microelectronics is a branch of electronics that deals with the miniaturization of electronic circuits and components. Electronics hardware has gone through three stages of development — from vacuum tubes to transistors to integrated circuits. (See Figure 7.1.) Each of these stages represented significant technological and practical advances, including the following:

1. a tremendous decrease in the size of equipment units
2. the capability to perform additional computing and communication functions
3. a substantial decrease in production costs
4. reduction of problems caused by excessive heat
5. lowered power requirements
6. increased reliability
7. lower maintenance costs

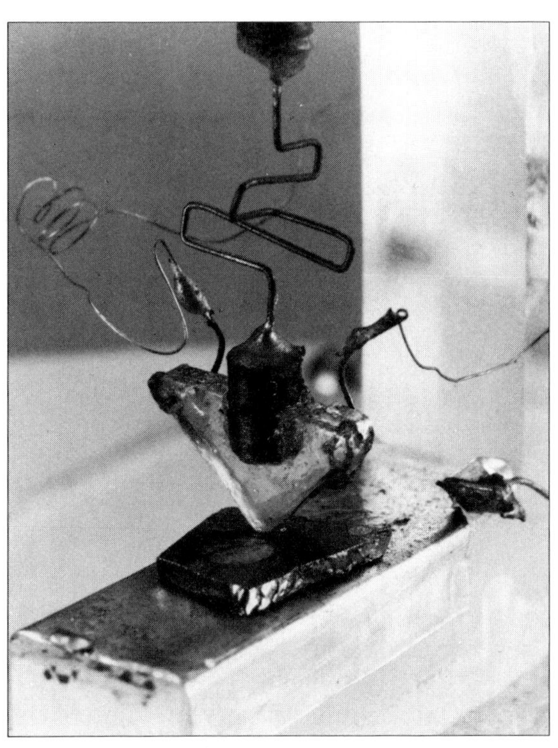

(Courtesy of Bell Laboratories)

Figure 7.1
Point Contact Transistor

With advances in electronic technology, computer components became smaller and smaller, reducing the amount of space required for circuit wiring and making it possible to combine several circuits on one circuit board. Later, many individual circuits were integrated into a single, complete electrical circuit on a small silicon chip. As the miniaturization trend continued, many thousands of circuits were integrated onto a single tiny chip, a process known as *large scale integration (LSI).*

Microelectronic circuits are produced by a process similar to photolithography. The circuits are designed according to specifications, photographed, and reduced in size. They are then etched on a thin silicon wafer, a process that enables them to be produced inexpensively. When the electronic circuitry for an entire central processing unit is placed on a single, very small silicon chip, the result is a *microprocessor* (Figure 7.2). A microprocessor consists of a central processing unit, memory circuits, and input/output devices; it contains all the circuits necessary for a true functioning computer.

Recent advances in microelectronic technology have increased the number of components that can be placed on a chip [*very large scale integration (VLSI)*] and their speed and reliability (Figures 7.3 and 7.4).

Figure 7.2
The One-Chip Computer: Offspring of the Transistor

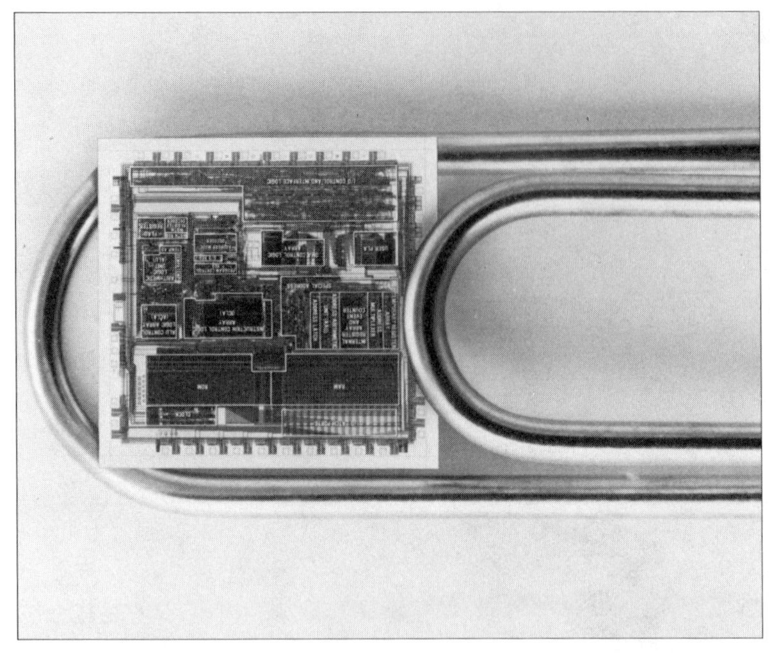

(Courtesy of Bell Laboratories)

Figure 7.3
Digital Signal Processor
The digital signal processor chip is used in such applications as speech synthesis, voice recognition, filtering, tone detection, and line balancing in digital communication systems.

(Courtesy of Bell Laboratories)

(Courtesy of Bell Laboratories)

Figure 7.4
Microprocessor Chip.
This microprocessor chip contains nearly 600 tiny, fast switches that operate at temperatures hundreds of degrees below zero. It is a step toward the day when high-speed computers, only about the size of a baseball, will handle the same amount of information now handled by a room-size computer, and do it faster.

Further, the cost of microprocessor chips has decreased substantially. Chips that cost hundreds of dollars apiece a few years ago now cost less than a dollar today. Sometimes described as "a computer on a chip," microprocessors have become the foundation for a whole new industry. They can be found in everything from automobiles to fountain pens.

THE MICROCOMPUTER

The microprocessor is the processing component of the small, but powerful *microcomputer*. The fact that microcomputers are relatively inexpensive has made it possible for individuals and small businesses to own them. Microcomputers, also called *personal computers (PCs)* or *micros*, may be used as standalone computers or linked to a network via telecommunications lines.

The smallest of all computers, microcomputers are now more numerous than all other computers. They are being used for every kind of task imaginable — from keeping records to preparing mathematical models of automobiles, from playing games to plotting the geographic levels of the earth for oil exploration. Their future applications are probably beyond our wildest imaginations.

■ ***MICROCOMPUTERS AS PERSONAL COMPUTERS*** A microcomputer at home can be used for many purposes. Many types of office work can be performed with microcomputers. By using the appropriate software package, you can perform such functions as writing letters and reports, maintaining appointment calendars, performing calculations, analyzing data, maintaining personal financial records, and managing your investment portfolio.

When microcomputers are connected to telecommunications facilities (thereby incorporating data communications), they link the home to the outside world. This permits users to access data bases, such as Compuserve and The Source, and to subscribe to services that allow them to shop at home, make plane or hotel reservations, order travel and entertainment tickets, and send messages electronically. By incorporating a FAX board in the PC, users can send facsimile messages to anyone.

■ ***MULTIUSER MICROS*** Until recently microcomputers were primarily personal computers—for individual use only. However, recent technological developments have increased the processing power of micros to the extent that it is now possible for several users to be on the system at once. These multiuser micros, often located in the same office, share the computer's power and its peripheral devices.

DEFINITIONS AND BASIC CONCEPTS

Data communications evolved from the union of communications technology and computer technology. The integration of these two makes it possible to transmit data to computers from remote locations.

Data communications is the movement of encoded data from one point to another by means of electrical transmission systems, including radio and optics. The term *data* refers to any representation, such as numbers, letters, or facts, to which meaning can be ascribed. In today's world, data implies facts, statistics, and other information that is *digitally coded* and intelligible to a variety of machines, including computers, terminals, electronic typewriters, and microcomputers.

Data communications may be distinguished from telegraphy chiefly by the fact that some form of processing is involved either prior to or after transmission.

Data communications is an integral part of a data processing system and, like voice communications, is a subsystem of the field of telecommunications. In fact, data communications is sometimes defined as "the portion of a telecommunications system concerned with transferring data."

DATA COMMUNICATION SYSTEMS

In the data processing cycle, data must be collected and moved to the processing unit before it can be processed. And before processed data can be used, it must be delivered to the user.

In the early days of the computer, a user had to be physically present at the computer site to access computer power, and data had to be physically transported from one site to another. However, telecommunication channels have made it possible for data to be transmitted electronically, thus servicing users at geographically dispersed locations.

Data communication systems, or networks, are designed to transmit data from one location to another electronically. The objective of data transmission systems is to provide faster information flow by reducing the time spent in collecting and distributing data. Data communication networks facilitate more efficient use of central computers by providing *message switching* capabilities. *Message switching* is the routing of messages among three or more locations using either circuit switching or store-and-forward techniques. If a telecommunication line is available, message switching is accomplished by instantaneous circuit switching. If all telecommunication lines are in use, store-and-forward procedures are used. Messages are accepted and stored in the computer memory until a telecommunication line becomes available, then forwarded to the next location.

DATA PROCESSING CONFIGURATIONS

All computer systems contain hardware for data input, central processing, and data output. They all use stored programs to perform basic operations. Their main differences are in size, storage capacity, processing speed, cost, and in the manner in which they are used. *Microcomputers* are the smallest general-purpose computers. But they have the same components as mainframe computer systems: input, output, storage, and processing. As one might expect, their input/output components are much slower, and the storage component has a smaller capacity than larger systems. *Minicomputers* are a medium-sized class of computers. They bridge the gap between micros and mainframes. But the way in which they are used makes them more characteristic of mainframes than of micros. Large computers, known as *mainframe computers,* are the backbone of the computer industry. They are capable of processing large amounts of data with very fast processing speeds. The major difference between minicomputers and medium-to-large size mainframes is in the number of remote workstations that can be serviced. *Supercomputers,* the fastest and most powerful of the mainframes, can provide service to thousands of remote workstations.

The earliest data processing configuration was *centralized processing,* wherein the processing for several divisions, functional units, or departments is centralized on a single computer. In this configuration, the input/output devices are located in the same area as the computer. The data input and output are in the form of physical media, such as punched cards and printed reports. Early computers were designed to handle one task at a time — an expensive mode of operation — and were centralized in a company because computers were very expensive and their operation required a special environment and operators with special technical skills. Thus, it was not economically justifiable to have computers in several locations within the organization.

The next development was the *data communications system,* wherein the components of the computer system are geographically separated but joined into a system by telecommunications channels. Data communications provides a means of performing data processing from a remote location. Hence, when first introduced, it was known as "teleprocessing." Three activities are involved: data capture on the terminal, data transmission over the communications facility, and data processing at the computer.

The most advanced electronic data processing configuration is *distribted processing,* which allows both computers and the data required for processing to be distributed throughout the organization. The data communication network is incorporated as an integral part of the system. However, the essential characteristic of distributed processing is that more than one computer processes the data. Distributed processing systems have the capability to process data at multiple points within a network.

A distributed system may contain microcomputers, minicomputers, or even larger general-purpose computers, referred to as host computers, located in the department where processing is required. These machines perform local processing tasks and transfer data between computers and terminals rapidly and efficiently.

Distributed processing has the following advantages:

1. By using smaller computers to perform some of the processing tasks, less complex (and less costly) equipment is required at the centrally located computer. This factor generally reduces costs for system design and programming.
2. Dividing a complex computer activity into simple components increases system reliability, because failure of any component will not render the entire system inoperative.
3. When processing is moved toward the user, each branch has greater control over its own data processing functions.
4. There is less traffic on communication lines, resulting in decreased communication costs.
5. Since data is entered into the system at a local site rather than a remote site, transactions may be processed almost as soon as they are entered, providing system users with more timely information.

Because of the many advantages of distributed processing, there is a growing trend toward decentralizing data processing activities.

LOCAL AREA NETWORKS (LANs)

A *local area network* is a system of hardware, software, and communication channels that connects devices such as mainframe computers, personal computers, word processors, input/output devices, printers, and

PBXs. It is usually confined to a building or a campus housing a group of buildings. LANs seldom extend for more than a few miles and are self-contained. However, they are not limited to communications within their defined area, but provide easy access to the outside world through a *gateway* (a bridge interface that enables a local area network to communicate with a long-haul network).

Most LANs are capable of carrying voice as well as data. They generally have a capability for high-speed transmission. This is a most important factor because the public-switched telephone network currently provides low-speed transmission.

The LAN concept began with the development of distributed processing in the 1970s. Its use has increased with the popularity of desktop microcomputers and other devices that require computer power wherever work operations are being performed.

Since LANs are private networks, it is not necessary to use a communications common carrier to transmit data among the components of the network. An organization can install its own communications channels using twisted pairs, coaxial cable, or optical fibers. It can also make use of a PBX to control the interconnection of the various devices. LANs can, however, be connected to a common carrier to include transmission to a remote location.

DATA TRANSMISSION SYSTEMS

Data transmission systems are designed to serve a variety of applications; thus, they differ in the way they function. The transmission system required for a given application depends on response time (the interval between data input and the system's response to the input). The timing of these responses is often critical. Some applications need a quick response; others require a response instantaneously; and still others need answers or processing at a later time.

TYPES OF PROCESSING

Three types of processing are available to meet a user's time-frame requirements: online processing, realtime processing, and batch processing.

■ ***ONLINE PROCESSING*** A type of processing giving immediate feedback is called *online processing. Online* means that the input device is directly connected to a computer for asking questions, doing calculations, and receiving responses. Online processing systems are *interactive* because they allow the user to communicate directly with the computer during processing.

An example on online processing is the automatic teller machine (ATM). The user of the ATM does not want to wait long for an answer to his or her request for cash, account balance, or acknowledgment of a deposit. This response need not be instantaneous; however, it should come within a matter of seconds.

Figure 7.5
Airline Reservation Center

(Courtesy of United Airlines)

REALTIME PROCESSING Another type of processing that is performed online is *realtime processing,* sometimes referred to as *online realtime processing (OLRT).* Realtime processing provides two-way communications between a terminal and a computer, processing transactions as they occur and providing instantaneous feedback to the user. The major difference between realtime processing and online procesing is the speed with which the data is handled, with realtime processing being the faster of the two.

An example of realtime processing is the airline reservation system. A reservation is confirmed immediately, an operation made possible by a network of computers that communicate with each other and perform the necessary data manipulations so that a reservation can be made (Figures 7.5 and 7.6).

Another example of a realtime processing system is computer-assisted instruction, wherein input data is processed instantaneously so that the response can affect subsequent input.

An essential part of realtime processing systems is their multiprogramming and multiprocessing capabilities. *Multiprogramming* means that two or more programs can be run simultaneously by interleaving their operations. A computer has the capacity to hold more than one program in storage at one time, but the *central processing unit (CPU)* can process only

(Courtesy of United Airlines)

Figure 7.6
Airline Reservation Center

one instruction at any one time. Processing functions are executed at high speeds, but input and output operations are relatively slow. Since the input and output operations cannot keep pace with the processing operation, the CPU is left idle much of the time. To reduce the time that the CPU is idle, realtime multiprogramming systems overlap input and output with processing. Thus, although the actual processing is performed sequentially, the system appears to be processing two or more programs at a given time in the same CPU. For example, in an interactive, overlapped terminal system, an operator keys in a record and the system processes and prints it all so rapidly that the functions appear to be performed at the same time. With multiprogramming, two or more independent jobs may be processed concurrently. Instead of allowing each job to run the complete cycle before it begins the next job, the computer switches back and forth between them, thus using the CPU to its fullest.

The concept of multiprogramming can be extended to users at distant sites using data communications channels. Multiprogramming is especially useful in these instances because a great deal of the time it takes to run a data communication program on a computer system is spent in data input and output operations. The use of multiprogramming allows a computer system to execute other application programs in the same time frame. Figure 7.7 illustrates how multiprogramming uses overlapped and nonoverlapped processing.

Figure 7.7 Overlapped and Nonoverlapped Processing

Nonoverlapped Processing

Input	-1-			-2-			-3-		
Process		-1-			-2-			-3-	
Output			-1-			-2-			-3-

Overlapped Processing

Input	-1-	-2-	-3-	-4-	-5-	-6-	-7-	-8-	-9-
Process		-1-	-2-	-3-	-4-	-5-	-6-	-7-	-8-
Output			-1-	-2-	-3-	-4-	-5-	-6-	-7-
Time Intervals	1	2	3	4	5	6	7	8	9

Another feature of realtime systems is *multiprocessing,* in which two or more CPUs are interconnected into a single system and one control program operates both processors. In addition to providing faster processing, multiprocessing systems can be set up to allow one processor to take over if another fails. However, the use of a second computer on a standby basis alone is not multiprocessing. Multiprocessing depends upon the interconnection of the computers and the control of both systems by a common operating program.

Multiprocessing systems are useful for applications in which time is a critical factor, such as airline reservations, intersive-care patient monitoring, and space flights.

Large computers in a multiprocessing system can also support multiprogramming. Each processor can function independently as a single computer and process two or more jobs nearly concurrently. Multiprogramming and multiprocessing increase the processing speed, power, and efficiency of the computer system. They are widely used in business today because their capabilities permit work to be done that could not be accomplished at all without them.

■ ***BATCH PROCESSING*** *Batch processing* is the processing of data accumulated over a period of time; each accumulation is processed in the same run. Batch processing is used when time is not critical. It differs from online and realtime processing in that an immediate response is not given to the user. Batch processing employs offline techniques, wherein the terminal is not directly connected to a computer.

The data for batch processing may be collected in their original form (time cards, invoices, report cards), accumulated over a period of time, and physically transported to the computer for processing. In this case, telecommunications would not be required. Or they could be transcribed onto an input medium (magnetic tape) that can be read by the computer, transported to the computer room in groups, or batches, and read into the computer storage. Or they could be entered via terminals, one record at a time, collected into a batch, and then transmitted to the computer over telecommunication lines as a unit. This method is called *remote job entry (RJE).*

Payroll processing generally employs batch techniques. Records for each employee are maintained for a period of time, then sent to the computer by any one of the above-described methods. When the computer has received all records, it processes the data in one run.

Another application that employs batch processing is student grade reports. As each instructor turns in grades, they are accumulated offline. When all instructors' grades have been received, the records are sent to the computer, and the data is then processed as a batch.

REALTIME—AN APPLICATION

The development and implementation of online realtime processing systems can make a substantial contribution toward improving services and reducing costs in many industries. (See Figure 7.8.) One such example is the telephone company's conversion of its directory assistance (formerly called "information") service to a realtime processing system.

For many years, directory assistance relied upon printed records. These records required daily updating as phones were installed and removed. The updating process was performed by company personnel using manual procedures. When a customer called directory assistance to request a telephone number, the operator searched for the number in a printed record (telephone directory). The operator had to flip pages constantly —a cumbersome, repetitive activity. Although the process was boring, it required a high degree of alertness and mental concentration to ensure speed and accuracy in output.

Now, using modern search techniques on a realtime basis, operators can search the computer files for number records by keyboarding a few identifying characters. The computer selects a few potential listings from which the operator can make a decision and relay the information to the customer who is waiting on the line.

For example, to search the computer files for the name ROBERT J. SMITH, 24708 HARRISON, DETROIT, MICHIGAN, the operator keys in several letters of the subscriber's last name, such as SMI; tabs over to another column and keys in one or two digits of the house or apartment number, such as 24; tabs over to still another column and types the first letter of the street name, H. This input produces all the listings for all the subscribers whose last name starts with SMI, whose house or apartment number begins with 24, and whose street name starts with H—perhaps ten or twelve in all. The operator then evaluates these listings in terms of

Figure 7.8
Computer Applications Using OLRT Systems

Application	Activity
Airlines	Flight schedules Reservations/cancellations
Artificial Intelligence	Computer games Linguistics (language translation) Robotics Simulating human dialogue
Banking	Account transactions Account inquiry Electronic funds transfer (EFT)
Education	Computer-Assisted Instruction (CAI) Student records Student registration Testing-scoring/distribution statistics
Health	Patient records/case histories Patient-care monitoring Diagnostic procedures Hospital information systems Computer-Assisted Research (CAR) Computer-Assisted Pharmacy Systems (CAPS) Psychoanalysis/therapy
Insurance	Policyholder information/records
Investments	Market quotations/transactions Market research Financial analysis/planning
Legal	Drivers' licenses and records National Crime Information Center Attorney business records Scheduling attorney time/cases/courtroom Legal research
Manufacturing	Computer-Assisted Design (CAD) Computer-Assisted Manufacture (CAM)
Politics	Computer voting systems Election return reports Fund raising Legislative administration
Retail Distribution	Point-of-sale (POS) systems Credit card authorization Inventory control Order processing Account records Electronic checkout
Science	Scientific research Scientific problem solving

the requested information and makes a decision. In a rare instance the operator might have to key in additional identifying information or even ask the customer for further information.

After deciding which one of the listings is the requested one, the operator identifies it on the screen (usually by pointing to it with a light wand) and presses a key. The computer then synthesizes a voice reply quoting the number to the customer. If the customer is not satisfied with this information and does not hang up, the computer signals the operator to return to the line to provide further assistance.

The use of realtime processing and voice synthesis for directory assistance has had two very desirable outcomes:

1. The average operator work time of a directory assistance call has been reduced from 36 seconds to less than 20 seconds, a reduction of 44 percent.
2. The operator's job is more interesting because the boredom of manual procedures has been eliminated.

Reducing the average time for directory assistance calls from 36 to 20 seconds may not seem like very much. However, given the very large volume of directory assistance calls the telephone companies handle, a reduction of 44 percent represents a significant savings in operating costs.

Job satisfaction is somewhat more difficult to measure but, nevertheless, very real. Using manual procedures, an operator is required to remain mentally alert while performing repetitive tasks at high speed and under the stress of time. Replacing manual procedures with computerized procedures makes the job more interesting and less tiring. These intangible benefits can have a positive effect upon employee morale and productivity.

COMBINED PROCESSING SYSTEMS

Many computer systems are capable of both realtime and batch-processing operations. The same data base and the same online terminals used for realtime processing can be used to supply input data for batch processing. Many applications use both realtime processing and batch processing at various stages in their processing cycle. Realtime processing is used when fast, two-way interaction is required; when these conditions are not required, batch processing is satisfactory and results in cost savings.

In the previous example of telephone company directory assistance calls, realtime processing is necessary to provide the information requested while the customer waits on the line. However, the same data base used to provide this information is also used for compiling and printing an annual telephone directory. In this case, the data base is read onto magnetic tape and processed using batch techniques. The tape is then sent to the printer, who uses it to automatically print the telephone directory.

Many business applications that use realtime systems also employ batch-processing techniques at some stage of their processing. The investment business provides another example of the use of combined systems. Brokerage firms use realtime systems for stock price quotations, research updates, trading transactions, and account information that must be up-to-the-minute. Batch-processing techniques are used for daily summaries, closing prices, customer statements, and various financial reports that are less dependent on time.

SUMMARY

Data communications is the movement of encoded data from one location to another by electrical means. Like voice communications, data communications is a subset of the field of telecommunications. Data transmission systems provide faster information flow by reducing the time spent in collecting and distributing data.

The rapid growth in data communications can be attributed to the fact that the telephone switched network can be used for data transmission. The digital signals used in computers may be converted to analog signals for transmission and reconverted to digital for computer processing.

Microprocessors have become the foundation for a whole new industry. The microprocessor is the processing component of the small but powerful microcomputer. The fact that microcomputers are relatively inexpensive has made it possible for individuals and small businesses to use them. The smallest of all computers, micros are now more numerous than all other computers.

When used as standalone computers, micros are ideal for maintaining all types of personal records. When connected to telephone lines, data communications is possible, thereby linking the home with the outside world. Their applications are many; one important application is their ability to access a wide variety of data bases.

The most advanced data processing system is distributed processing, wherein both computers and the data for processing are distributed throughout the organization. Distributed systems have the capability to process data at multiple points within a network.

A local area network is a system of hardware, software, and communication channels that connects such devices as computers, word processors, printers, and PBXs. It is usually confined to a building or a campus housing a group of buildings. Since LANs are private networks, it is not necessary to use a communications common carrier to transmit data among the components of the network.

Data transmission systems are designed to meet the user's needs with respect to response time. Some applications require a quick response; others an instantaneous response; and still others need answers at a later date. Three types of processing systems are available to meet these needs: online processing, realtime processing and batch processing. Realtime systems make possible many processing applications that could not be performed without the instantaneous response these systems provide. For example, reservations systems used by airlines, hotels, and car rental agencies require constant updating as events occur.

An essential part of realtime systems is their multiprogramming and multiprocessing capabilities. Multiprogramming means that two or more programs are executed simultaneously by interleaving their operations. In multiprocessing, two or more CPUs are interconnected into a single system, and one control program operates both processors. It is used when the computing power of a single CPU is not sufficient to process the jobs to be done within a given time.

REVIEW QUESTIONS

1. What is data communications?
2. What was the significance of the Bell System report entitled *Capabilities of the Telephone Network for Data Transmission?*
3. What are some of the factors that have contributed to the rapid growth of data communications?
4. What is a microprocessor?
5. Why has the use of microcomputers become so widespread?
6. What are distributed processing systems?
7. What is the purpose of local area networks?
8. What is the principal advantage of realtime processing systems?
9. When is it appropriate to use batch processing systems?
10. Distinguish between multiprogramming and multiprocessing.

VOCABULARY

teletypewriter
teleprinter
microelectronics
large scale integration (LSI)
very large scale integration (VLSI)
microprocessor
microcomputer
personal computer (PC)
data communications
data
circuit switching
store-and-forward switching
minicomputer
mainframe computer
centralized processing
distributed processing
local area network (LAN)
gateway
response time
online processing
realtime processing
batch processing
multiprogramming
multiprocessing

CHAPTER

8

DATA COMMUNICATION SYSTEMS

CHAPTER OBJECTIVES

After completing this chapter, the reader should be able:

- *To describe the components of a data communication system.*
- *To describe the principal types of data communication terminal equipment.*
- *To explain the function of a modem.*
- *To describe the data transmission controls.*
- *To describe the predominant information codes.*
- *To distinguish between phase modulation and pulse code modulation.*
- *To describe the principal types of line configurations.*

The increasing use of computers in business operations has created the need for rapid movement of data from one location to another. New computer hardware using the latest techonological developments requires a continuing evaluation of information movement procedures, in order to maximize the operational efficiency of the computer system. The manager has two options in moving data: physical transportation of the documents or electrical transmission of the information. If there is no immediate need for the data, physical transportation could be the best alternative. If speed is a factor, electronic data transmission is required. The data transmission system should provide the same degree of speed and efficiency as the computer system to realize the full system benefits. To use an analogy, a chain is no stronger than its weakest link. The manager is responsible for providing a data communication process that maximizes the information-processing system and is not the weak link.

Data communication systems are networks of components and devices organized to transmit data from one location to another—usually from one computer or computer terminal to another. The data is transmitted in coded form over electrical transmission facilities.

This chapter outlines the basic hardware, controls, and procedures of data communications and describes how they function together as an integrated data communication system.

COMPONENTS OF A DATA COMMUNICATION SYSTEM

The three essential components common to all data communication systems are: the *source,* the originator of the information; the *medium,* the path through which information flows; and the *sink,* the receiver of the information.

When data is being transmitted to a computer, the source in a data communication system is *remote terminals,* usually devices with typewriter-like keyboards used for entering data. Remote terminals might be in a different geographic location from the mainframe or CPU. The medium in a data communication system is the communication link, the facility that links remote terminals to the CPU. The medium could be wire, radio, coaxial cable, microwave, satellite, or light beams. Leased or public telephone lines are the most frequently used communication medium. The sink in a data communication system is the computer system that processes the data received.

To these basic components we could add the *modem* that transforms input into transmittable signals. Two modems are required—one to convert data from the source into a form that can be transmitted over telephone lines, and the other to reverse the process at the terminating end. In two-way communication the source and the sink could be constantly changing roles; that is, the terminal might alternate as both a source and a sink.

In a simple data communication system, an example of which is shown in Figure 8.1, the source might be a terminal keyboard; the medium, a telephone line; and the sink, a computer. If the system operates in a query-response mode, the terminal and the computer would change roles as the transaction progresses. Thus, the terminal, which was originally the source, would become the sink; and the computer, which was originally the sink, would become the source.

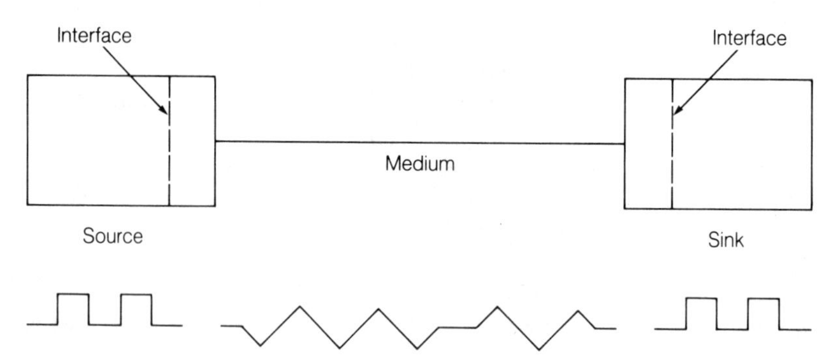

Figure 8.1
A Simple Data Communication System

TERMINAL EQUIPMENT

Originally, the word *terminal* meant the point at which data could enter and leave the communication network. In practice, however, it has taken on an additional meaning. The work *terminal* has become synonymous with *terminal equipment* and refers to any device capable of either input or output to the communication channel.

Terminals provide interfaces with computer systems so that people can insert or extract data. Terminals also provide a convenient way for people to exchange data directly. Terminals receive input data in coded form and convert it into electrical signal pulses for transmission to the computer. Similarly, at the receiving end they transform the electrical impulses into characters that can be read by humans. Thus, terminals function as translation devices for communication codes. Data can be entered into terminals either by human operators or by machines that collect data automatically from recording instruments.

Prior to 1968 the major terminal device was the teletypewriter, and the principal input medium was punched paper tape. Each keystroke on a transmitting teletypewriter (source) produces a sequence of electrical pulses determined by the coding representation for the keyed character. The electrical signals are sent over the communication channel (medium) to a receiving teletypewriter (sink), where they are reconverted to their original form.

The Carterfone decision lifted the ban on the attachment of customer-provided terminal equipment to the telecommunication network. This decision, along with phenomenal advancement in the computer industry, promoted competition among terminal suppliers, and the terminal industry grew rapidly. Today, there are many different kinds of data communication terminals offered by many manufacturers. Thus, system designers can be very selective in their choice of equipment. However, this wide variety can cause confusion and make selection difficult. An understanding of terminal characteristics and capabilities is helpful in the selection process.

There are undoubtedly many ways to classify terminals. These might include the following factors: operating speed, type of transmission (batch or online), memory capability, mode of operation (HDX, FDX), transmitting capability (send only, receive only, send/receive), method of input (keyboard, card reader, tape reader), output form (hard copy, soft copy), error control, type of applications, and, of course, price.

This chapter will discuss the following five broad categories of terminals:

- teleprinter terminals
- video terminals
- transaction terminals
- intelligent terminals
- specialized terminals

TELEPRINTER TERMINALS The teleprinter terminal resembles a standard alphanumeric keyboard with special function keys to provide transmission control capability, such as "line feed" or "bell signal." In the sending mode these terminals are controlled by an operator; in the receiving mode they are controlled by either the central processing unit of a computer or by another operator at a distant machine. Teleprinter terminals print one character at a time and produce hard-copy printout. They have no programming capability. (Intelligent terminals are a separate category.)

Teleprinter terminals are used on low-speed public or private telephone lines. These terminals are usually unbuffered; that is, they have no storage capacity. Each character is printed on paper at both the sending and receiving ends as soon as it is keyed. Teleprinter terminals are described as "dumb" terminals because of their limited capabilities.

VIDEO TERMINALS Video terminals consist of a keyboard and a *cathode ray tube (CRT),* a visual display device resembling a television screen. The keyboard is the input medium; the operator can enter both the data and the control commands that direct the operation of the computer. The CRT provides *soft copy,* a visual display with no permanent record. Display terminals may also be obtained with a printer attached, thus enabling them to print *hard copy.*

Most video display terminals use a standard typewriter keyboard, plus control and special function keys, such as "insert," "delete," and "repeat." As the operator types in a character, it appears on the screen. These terminals have a *cursor* (from the Latin *cursus,* meaning *place*), which is usually a blinking symbol that indicates current location on the CRT screen. The operator can move the cursor horizontally and vertically to any desired position. An important advantage of CRT display terminals is their capability to edit text. Errors detected on the screen can be corrected, usually by backspacing and striking over the error. Entire words, lines, or paragraphs can be deleted or repositioned through the use of special function keys. Changes in input copy are possible because some amount of

Figure 8.2
Video Display Workstation

(Courtesy of NCR Corporation)

the copy is held in a *buffer,* or temporary memory, until the user presses a special function key to transfer it to the main memory. Video display terminals with alphanumeric keyboards have become well known through their use in word processing machines.

A special type of video display terminal is the graphics terminal, which can display not only letters and numbers but also graphic images (charts, maps, and drawings). These terminals use matrix technology, in which many closely spaced dots are connected to draw lines and to plot data graphically. Graphic display terminals can accept input from a keyboard, an input tablet, and/or a *light pen*—an electronic drawing instrument equipped with a photoelectric cell at its end that allows the user to "draw" designs directly on the display screen. Some graphic display terminals are capable of displaying charts and drawings in different colors on the screen. The displayed material can also be reproduced as hard copy with a printer or plotter.

Scientists and engineers have long used graphics to represent information for review and analysis. Computer technology has extended the use of graphics to many other professions and industries as well. Computer graphics are currently being used in manufacturing to design automobiles, household appliances, and other products; in the health profession to assist in diagnostic procedures; in the transportation industry to design routes and schedules; and in the banking and investment industries to

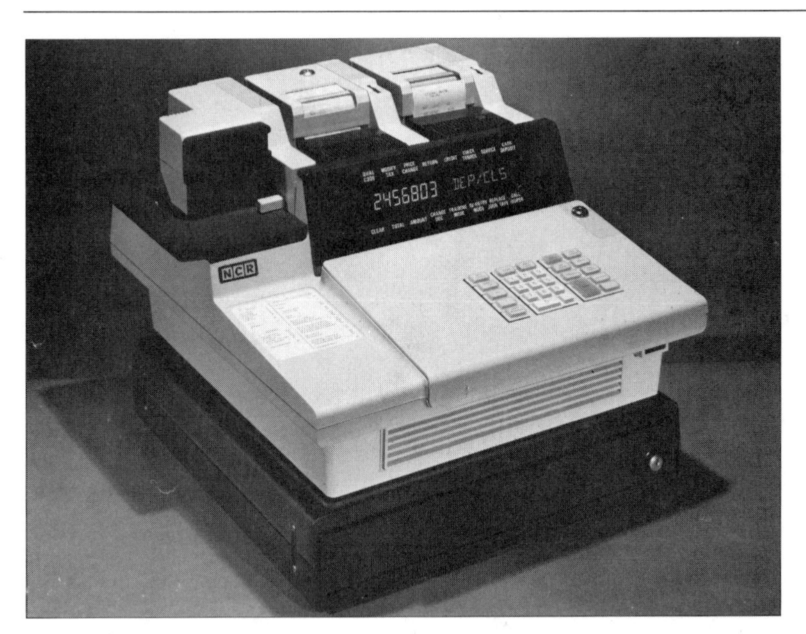

Figure 8.3
A Transaction Terminal

(Courtesy of NCR Corporation)

summarize and analyze financial data. They are also used to present business operating data that will help managers identify trends and relationships and make informed decisions. The adage "A picture is worth a thousand words" could be rephrased to "A graph or visual summary of data is worth many pages of computer printout."

Video displays are high-speed devices, since data output is not slowed down by being typed on paper. These terminals are especially useful when output from a distant location is needed quickly. They are used by airline reservation systems (Figure 8.2) to determine flight space availability, by hotels and motels to determine room availability, by brokerage firms to transmit stock market quotations, and by insurance companies to access and update policyholder records.

■ ***TRANSACTION TERMINALS*** Transaction terminals are designed for use in a particular industry application, such as banking, retail point-of-sale, or supermarket checkout.

In banking, transaction terminals are used to update both customers' passbooks and the bank's account records. Terminals are also used online for off-hours banking and for processing customer inquiries.

Retail point-of-sale terminals are used to record the details of the sale in machine readable form. Their functions generally include verifying credit, printing sales slips, maintaining a local record of transactions, and updating inventory control records. All of these functions, except credit verification, can be handled by offline terminals, generally by cash register-type

Figure 8.4
A Transaction Terminal

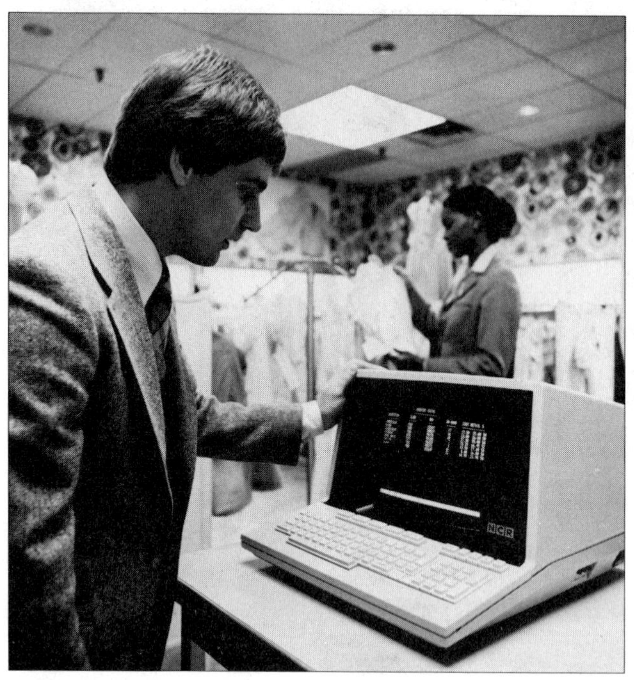

(Courtesy of NCR Corporation)

machines equipped with special keys to capture transaction data on paper or magnetic tape. Credit verification requires online access to storage files that may be built into the transaction terminal or may be in a central computer. (See Figures 8.3 and 8.4.)

Transaction terminals used in supermarket checkout lines can scan or read bar codes printed on the items being sold. (See Figure 8.5.) As the products go past the checkout point, the codes are read by a recording device or light pen that simultaneously prepares a cash register tape for the customer, records the sale, and updates the store's inventory.

Transaction terminals are easy to operate and may be used by persons with little technical knowledge. They have become an integral part of business operations because they help to increase productivity and control costs. (See Figure 8.6.)

■ ***INTELLIGENT TERMINALS*** An intelligent terminal is one that has a built-in microprocessor that can be programmed and that can execute stored programs. It could be a video terminal, a transaction terminal, or a specialized terminal. The most popular form of intelligent terminal is the microcomputer. A microcomputer can serve as a standalone computer or as an intelligent terminal linked to a mainframe.

Figure 8.5
NCR Retail Store System

(Courtesy of NCR Corporation)

Figure 8.6
NCR Lodging System

(Courtesy of NCR Corporation)

■ ***SPECIALIZED TERMINALS*** Two of the newer types of terminals are audio response units and pushbutton telephones. Audio response terminals are unique in that their output or response is verbal rather than printed or visual. The input device may be either a keyboard or a telephone. The computer has a built-in synthesizer, enabling the computer to assemble prerecorded sounds into meaningful words. Unlike some mechanical voices that have low-fidelity, robot-like characteristics, the voice quality of audio-response units is equivalent to that of the human voice. This synthesized response should not be confused with the response from an answering machine that transmits a recorded announcement.

The response from an audio terminal is designed for a specific type of message service. Messages are pieced together from sound fragments to produce a reply to a particular inquiry, such as a telephone number request from an information bureau, as described in Chapter 7. In this process an information operator finds the requested number and points it on the CRT screen with a light wand. The operator is then released from the call, and the audio response unit synthesizes the message (in this case, the requested telephone number) and transmits it to the customer waiting on the line. The combination of human operators and audio response units saves human effort, thereby improving productivity.

Rotary-dial telephones are not generally used for data transmission since the rotary dial cannot be used as an input terminal. The dial is used to establish a call using dial pulses that are digital in form. However, once the connection has become established, the dial pulses are ineffective for communicating with a computer.

The Touch-Tone telephone has been used for some time in voice communications, but its use as an input terminal for data transmissions is relatively new. Its widespread availability makes it particularly useful. A Touch-Tone telephone has a keyboard to replace the rotary dial. Pressing a key on the telephone set transmits a distinctive signal representing a number that the computer uses for the processing operation.

Many banking institutions offer a service that permits their customers to conduct certain banking transactions from their home or office using Touch-Tone telephones. For example, a customer can transfer funds from a savings account to a checking account and vice versa, or pay bills to a selected list of merchants and utilities by keying a sequence of numbers representing a code into the telephone set. An interesting feature of these transactions is that while the customer communicates with the computer by pressing the appropriate pushbuttons on the telephone, the computer communicates with the customer by using an audio response unit.

■ ***OTHER INPUT/OUTPUT DEVICES*** There are many types of input and output devices. Figure 8.7 summarizes the principal ones. Frequently, telecommunication systems use a combination of these devices to perform a specific function.

Figure 8.8 summarizes the important characteristics of the five categories of data communication terminals.

Document-Input Devices	Human-Input Devices	Input/Output Devices
Paper tape reader	Microcomputer	Typewriter
Magnetic tape reader	Typewriter keyboard	Printer
Punched card reader	Matrix keyboard	CRT screen
Magnetic card reader	Special keyboard	Facsimile machine
Optical character reader (OCR)	Touch-Tone telephone	Magnetic tape or disk
Magnetic ink character reader (MICR)	Teleprinter	Passbook printer
Mark sense reader	Light pen	Transaction printer
Microfilm	Light wand	Audio response unit
Facsimile machine	Stylus	Light display
Tape cassette	Voice instruction	Microfilm/fiche
Magnetic disk		Videotex terminal
		Word processor
		Telephone (computerized voice)

Figure 8.7
Input/Output Devices

Summary of the important characteristics of the five categories of data communication terminals.

- **Typewriter or keyboard terminals**
  - Hard copy output
  - Used on low-speed public or private lines
  - Both send and receive
  - Online or dial-up
  - Usually unbuffered
- **Video display terminals**
  - Feature a cathode ray tube (CRT)
  - Have editing capability
  - Buffer memory
  - Graphics display capability
  - Used on public or private lines
  - High-speed operation
- **Transaction terminals**
  - Used in business transactions
  - Designed for a specific application
  - Easy to operate
  - Usually buffered
  - Used on private lines
- **Intelligent terminals**
  - Feature processing capability
  - High-speed operation
  - Buffered
  - Used on public or private lines
  - Capable of operating independently of host CPU
- **Specialized terminals**
  - Audio response units—synthesized verbal output
  - Touch-Tone telephones — pushbuttons provide input to computer

Figure 8.8
Categories of Data Communication Terminals

TERMINAL SELECTION

The application dictates the type of terminal device required. Because of the wide variety of terminals available, the system's designer should understand the capabilities of the various types of terminals to match the capabilities with the applications.

Some of the questions that might be considered in the selection process include:

1. Will the equipment perform the required function or functions satisfactorily?
2. Is the terminal response time consistent with efficient operation?
3. Is the capability for error correction built in, or will excessive operator attention be required?
4. Are the technical specifications (such as transmission code, operational mode, synchronization techniques) compatible with other units in the system?
5. Could the operation under consideration best be performed at the centralized processor or at a terminal?
6. What degree of intelligence is required? Is processing capability required? Is programmability required?
7. Can the terminal be upgraded to meet future needs?
8. Will the advantages of improved operations justify the cost of the equipment?

MODEMS

A basic problem of data communications has been sending digital signals over transmission facilities designed originally for analog voice transmission. When the communications line carries analog signals, an interface device is required to convert to digital. This function is performed by a *modem,* a combination of modulator and demodulator.

The modem and the communications line can be connected directly (hard wired) or indirectly (acoustic or inductive coupling). Acoustically coupled modems are portable; they can be used with any available telephone. These modems can transmit and receive when the telephone handset is coupled, or cradled, into the proper position in the modem. With acoustic coupling, the DC signals are converted to audible sounds, which are then picked up by the transmitter in an ordinary telephone handset (Figure 8.9). The audible signal is converted to electrical signals and transmitted over the telephone network. The process is reversed at the receiving end. Because they involve an extra conversion step (digital to audible to electrical) that can introduce noise and distortions, acoustic couplers generally are not as reliable as direct electrically connected modems. A direct connection to the communications line is preferable.

Modems come in a variety of shapes, sizes, and forms. They are often classified by speed; however, there is no consensus on the range of speeds in the various groups. Low-speed modems are sometimes defined as those

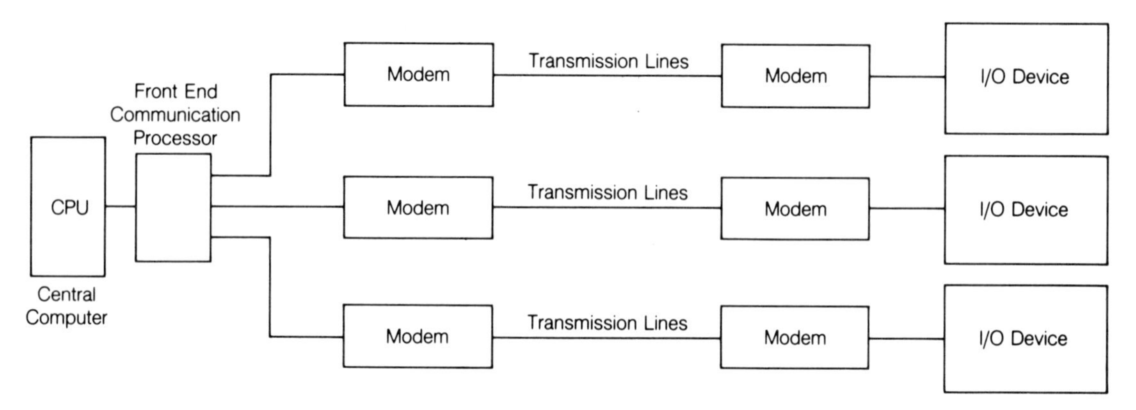

Figure 8.9
Components of a Data Communication System

that operate up to 1200 or1800 bps (bits per second), medium-speed modems up to 4800 bps, and high-speed modems up to 19,200 bps.

Not all transmission facilities are analog. Many of the newer facilities are digital because of their ability to provide good, economical voice communications, as well as their ability to support data communications in digital form. Of course, if the transmission facility carries digital signals, no modem is required.

Most major long distance carriers offer digital data service, a dedicated point-to-point data transmission service. Local access to and from these digital networks is usually over a transmission path known as a T-1 carrier. T-1 carriers are widely used because they provide 24 digital transmission paths over a single pair of twisted copper wires for a relatively low cost. They can be used for either data or voice transmission. If data is to be transmitted, no modem (or conversion to digital) is required. If voice is to be transmitted, the voice analog signal must be converted to a digital signal prior to its transmission over the digital T-1 circuit.

FRONT-END PROCESSORS

As terminals are added to a data processing system, the number and complexity of the operations necessary to handle them grow, and the demands on the CPU increase tremendously. To relieve some of these demands and to control the flow of data and ensure compatibility within the system, a *communications control unit (CCU)* or *front-end processor* is employed. This processor "front ends" a mainframe central processor by functioning as an auxiliary computer system that performs network control operations. This releases the central computer system to do data processing. Most front-end processors are minicomputers. The functional components of the front-end processor may be freestanding pieces of equipment or combinations of components integrated in one or more equipment units. Some of the functions performed by the front-end processors include:

1. Line access — connecting communication lines to the main computer
2. Line protocol — monitoring line control procedures
3. Code translation — translating the internal code of computer systems into communication codes
4. Synchronization — ensuring that the incoming signals are compatible with the requirements of the computer.
5. Polling — polling terminals to inquire whether they are ready to receive a message or whether they have a message to send.
6. Error control — checking accuracy of data received using parity techniques.
7. Path routing — choosing an alternate path to avoid heavy traffic or excessive error rate.
8. Flow control — controlling the flow of the signal from the processing unit to its destination.

CENTRAL PROCESSING UNIT

The intelligence of the receiving unit in a data communication system is generally a computer whose principal component is a central processing unit. The CPU is the heart of the computer system. It controls the input, data transfer, calculations, logic, and output operations of the system.

TRANSMISSION LINKS

The components of a data-communication system are connected by *transmission links.* These links can be wire, radio, coaxial cable, microwave, satellite, or light beams. Transmission links are discussed in more detail in Chapter 6.

DEVELOPMENT OF DATA CODES

Data is transmitted over telecommunication facilities and entered into computers in a code based on binary digits.

THE LANGUAGE OF DATA

The binary numbering system is the language of data. The word *binary* (from the Latin *bini,* meaning *two-by-two*) indicates that two digits make up the system. A binary digit can be represented mathematically by a + or a − or a 1 or a 0. The + (1) represents the flow of electricity; the − (0) represents the absence of the flow of electricity.

In data communications, the smallest element of information is the *bit,* derived from the contraction of binary digit. Thus, each of the binary digits, 0 or 1, is referred to as a bit. In the same way that a bit represents a unit of information, a *baud* represents a unit of signaling speed. Each is characterized by the presence or absence of a pulse of electricity in a trans-

A	B	C	D	E	F	G	H	I	J	K	L
1	1	0	1	1	1	0	0	0	1	1	0
1	0	1	0	0	0	1	0	1	1	1	1
0	0	1	0	0	1	0	1	1	0	1	0
0	1	1	1	0	1	1	0	0	1	1	0
0	1	0	0	0	0	1	1	0	0	0	1

Magnetic Tape

Figure 8.10
A Five-Channel Code

mission channel. In a transmission system using binary representation, the term baud is synonymous with bits per second. However, in transmission systems using other than binary representation, baud and bits per second are not synonymous.

Some transmission systems combine groups of bits in *dibits,* a transmission concept wherein two bits are transmitted at a time, thereby doubling the transmission speed. Groups of bits may also be combined into other configurations, thereby departing from the binary concept. In these configurations, baud and bits per second are not synonymous.

BASIC CONCEPTS OF CODES

The American National Standards Institute (ANSI) defines a *code* as "any system of communication in which arbitrary groups of symbols represent units of plain text of varying length." Similarly, a *code set* is "the complete set of representations defined by a code," and *coding* is "the process of converting information into a form suitable for communications." For example, the representation of the character *r* by a group of bits, such as 01010, is an example of coding.

One frequently used medium for encoded data is magnetic tape. The presence of one type of magnetic mark represents the binary 1; the presence of another type of magnetic mark represents the binary 0. Magnetic marks are combined in various ways to construct codes.

Figure 8.10 illustrates a 5-channel code recorded on magnetic tape. The channels run the length of the tape, and each coded character occupies a column across the width of the tape.

To expand the number of characters that can be represented, 6–, 7–, and 8–channel codes have been developed. These tapes follow the same general pattern as the 5–channel tapes, except that a parity bit is often added in the last channel for error detection.

VOICE TRANSMISSION OF DATA

Voice transmission over telecommunication facilities takes place in a conversational mode similar to face-to-face conversation. The words

Figure 8.11
A Simple Telegraph

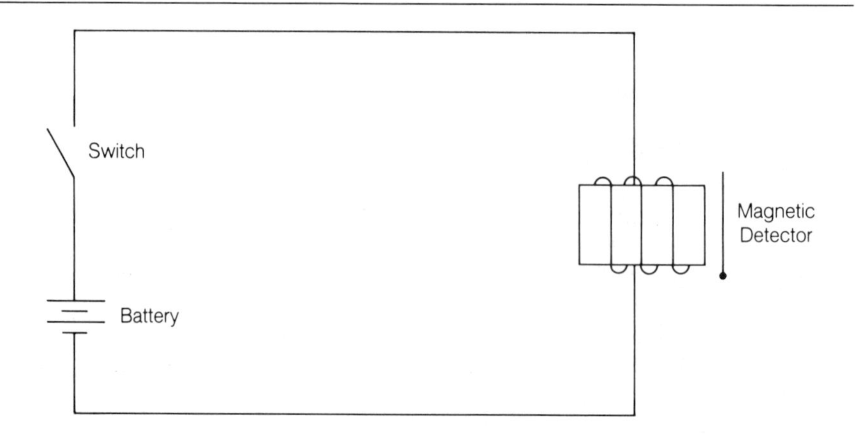

spoken into a telephone travel over an electronic connection to another telephone, where they are reproduced in the same form and with the same characteristics as when they entered the communication system.

Data can be transmitted by voice over a telephone, but the speed and accuracy of the transmission is not satisfactory for most purposes. For example, sales representatives could verbally give hourly reports of sales volume to a central location from all sales locations. However, this method would be a very slow process. In addition, the collected data would be unreliable unless painstaking verification procedures were used. A more efficient method of transmitting data is to encode the message into a form that can be sent rapidly and accurately over telecommunication lines. To enable to data message is be understood, both the sending and receiving persons or machines must have the same understanding of coding details.

ORIGIN OF CODES

The development of codes used in data communications began with the invention of the telegraph by Samuel F. B. Morse in 1844. The clicks of the telegraph were unintelligible until a code was developed. The telegraph consisted of sending and receiving stations connected by an electrical circuit. The sending station contained a source of power and a switch that could be opened and closed to send pulses of electricity timed as dots and dashes (Figure 8.11). The receiving station contained a sensing unit that could detect the opening and closing of the switch. An electromagnet functioned as the sensing unit and detected the presence of a dot or a dash as transmitted by the sender.

The telegraph operator could send short or longer bursts of electric current by pressing down the switch for short or longer periods of time. A short burst of current represented a dot, and the longer burst (equal to three dots run together with no space between) represented a dash. Thus, the receiver could interpret each signal in sequential manner.

Step	Computer		Terminals
1	Term #1, Any Msg?	---------------------→	
2		←---------------------	Nothing now
3	Term # 2, Any Msg?	---------------------→	
4		←---------------------	Yes
		←---------------------	Here is Msg
		←---------------------	Here is Parity Check
5	Parity is OK	---------------------→	
6		←---------------------	Here is More Msg
		←---------------------	Here is Parity Check
7	Parity is OK	---------------------→	
8		←---------------------	End of Msg
9	Term #3, Any Msg?	---------------------→	
10		←---------------------	Nothing Now

Figure 8.12
Line-Polling Protocol

DATA TRANSMISSION CONTROLS

Data communication systems generally use controls for three reasons:

1. to specify the rules to be followed in transmitting data
2. to detect errors in data transmission
3. to ensure system compatibility

PROTOCOLS

As the options and intelligence available in data communication terminals increased significantly, technology mandated new rules and procedures for efficient operation of the system. These communication controls are known as *protocols;* they are formulated by the equipment suppliers. They may govern communication lines, types of service, modes of operation, circuit compatibility, or total networks. This system is analogous to the use of traffic rules (highway ordinances) to control the efficient flow of traffic over city streets and highway networks.

Early protocols were referred to as *handshaking.* However, current usage of this term generally refers to the fact that a connection has been established and that the communication line is ready for the message. In present-day usage, protocol includes both handshaking and *line discipline,* a term that denotes the sequence of operations involving the actual transmitting and receiving of data. Many vendors use *line discipline* synonymously with *protocol.* Figure 8.12 illustrates line-polling protocol. Line polling is similar to a two-way conversation between a computer and a terminal wherein each confirms to the other the status of a message.

ERROR DETECTION

Morse code is used primarily for the transmission of information from one person to another. If the receiving operator fails to understand a signal or if

Figure 8.13
Odd and Even Parity

Letter **M**	1	0	1	1	0	0	1	0	Even Parity Code
Letter **J**	0	1	0	1	0	0	1	0	Odd Parity Code

The eighth bit is the parity code.

a sending error has been made, it is easily discerned by either operator. It can be corrected by a retransmission of the questionable part. However, when data is sent between electronic devices, there is a need for a systematic way to detect whether the message is valid or if something has gone wrong.

A basic problem in using voice facilities for data transmission is the presence of noise and distortion. The resulting errors in transmission require some error-control mechanism.

There are a number of methods of detecting errors. However, the most commonly used methods add redundance to the message to detect when a character is in error. A classical method of error detection is *parity checking,* which involves the use of a single bit, known as a *parity bit,* for the detection of errors. A description of parity checking is helpful in understanding how error detection can be built into a coding structure.

Parity describes a condition wherein the total number of 1 bits in each character is always even or always odd, depending upon the parity system being used. When a parity system is being used, the transmitting equipment automatically adds one noninformation-carrying bit, called a *parity bit* or *check bit,* to the characters being transmitted (Figure 8.13). This enables the computer to run its own check on every character it processes.

There are two levels of parity error detection: character checking and block checking. The simplest and least expensive method of error checking is character checking odd or even parity. The parity bit that is added will be either a 0 bit or a 1 bit, whichever is required to make the total number of 1 bits even or odd, depending upon the parity system being used. In a system using even parity, the total number of 1 bits, including the parity bit, should be even. In a system using odd parity, the total number of 1 bits, including the parity bit, should be odd. In a 7-bit code with even parity, the parity bit would be the eighth bit. If the character is represented by an even number of 1's, the parity bit added will be 0. If the character is represented by an odd number of 1's, the parity bit added will be 1. Checking consists of determining whether the data received conforms to the parity system being used. Character parity checking cannot detect errors involving the loss or addition of 2 bits in a character or compensating errors. For these types of errors, a longitudinal system known as *block character checking* or *longitudinal redundancy checking* (Figure 8.14) is employed.

Vertical Redundancy Checking

	Character				
	1	**2**	**3**	**4**	**5**
1	0	1	0	0	1
2	1	0	0	0	0
3	0	0	1	1	0
4	0	1	1	1	1
5	0	0	0	0	1
6	0	0	0	0	0
7	1	1	1	1	1
Odd Parity	1	0	0	0	1

Longitudinal Redundancy Checking

	Character					Block Parity
	1	**2**	**3**	**4**	**5**	
1	0	1	0	0	1	1
2	1	0	0	0	0	0
3	0	0	1	1	0	1
4	0	1	1	1	1	1
5	0	0	0	0	1	0
6	0	0	0	0	0	1
7	1	1	1	1	1	0
Odd Parity	1	0	0	0	1	1

Figure 8.14
Vertical and Longitudinal Redundancy Checking

With longitudinal redundancy checking, an additional 7 bits are added at the end of a specific block or message. Parity bits are added to all of the 1 bits, all of the 2 bits, all of the 3 bits, and so on until all of the bits have a bit at the longitudinal end, providing either even or odd parity at the end of the message. This checking significantly increases the probability of error detection.

SYNCHRONIZATION

Data signals, or bits, are transmitted in a specific code and transmission sequence. It is important that the coding and sequencing be sent properly so that devices that receive the data signals will be able to interpret them as intended. The transmitting and receiving devices in a data transmission system must operate in step with each other to allow communication between them. Determining and maintaining the correct timing for transmitting and receiving information is known as *synchronization*. Special equipment must be provided to accomplish synchronization.

Most data transmissions are serial in nature; that is, one bit is sent behind the other. Both control bits and message data are contained in the same message stream. There are two main forms of transmission: asynchronous and synchronous.

Figure 8.15
Asynchronous and Synchronous Transmissions

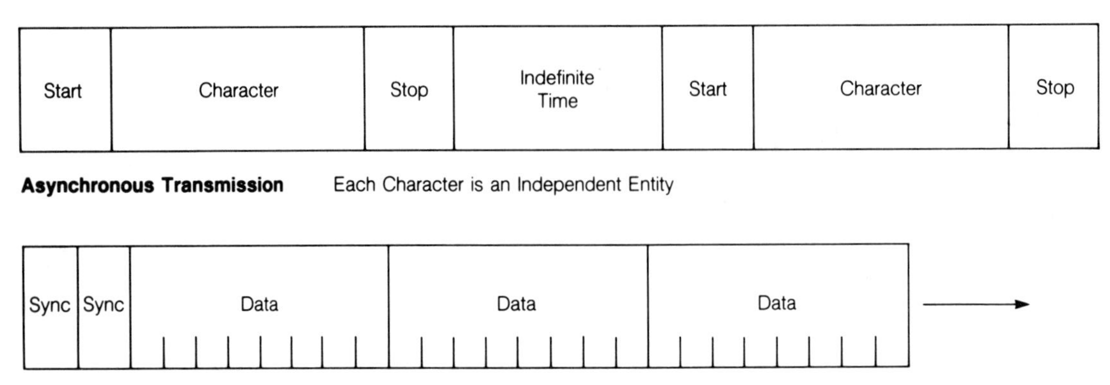

Asynchronous transmission is often referred to as "start/stop" transmission because additional start and stop bits are transmitted with each character to identify the beginning and the end of the group of data bits. With asynchronous transmission, characters can be sent at irregular intervals. Synchronization is accomplished on a per-character basis, and synchronization of the terminals is reestablished as each character is received.

In *synchronous transmission,* characters are sent in a continuous stream without framing bits between characters. The sending and receiving devices operate continuously at essentially the same frequency and are synchronized, or kept in step, by electronic clocking instruments. (See Figure 8.15.)

The two types of transmission may be differentiated by the fact that in asynchronous transmission each character is transmitted as an independent entity with start and stop bits to indicate to the receiving device that the character is beginning and ending. In synchronous transmission, whole blocks of data are transmitted in units.

Asynchronous transmission is useful when transmission is irregular. It is less expensive than synchronous transmission because it requires less sophisticated circuitry. Synchronous transmission makes more efficient use of the transmission medium; it permits higher transmission speeds because of the elimination of the start and stop bits.

PREDOMINANT INFORMATION CODES

Codes for representing information vary in two respects: the number of bits used to define a character and the arrangement of bit patterns in each particular character.

There are many transmission codes in use today. Some of the more commonly used codes are the International Telegraph Alphabet (ITA) No. 2, the American Standard Code for Information Interchange (ASCII), and the Extended Binary Coded Decimal Interchange Code (EBCDIC).

INTERNATIONAL TELEGRAPH ALPHABET (ITA) NO. 2

The International Telegraph Alphabet (ITA) No. 2 code replaces, and has many of the same characteristics as, the Baudot code. It is used primarily in international telex transmission. A description of the Baudot code illustrates the principles of code construction and provides a basis for understanding more extensive coding systems.

The *Baudot code* is a 5-bit code used by older teletype machines. (Figure 8.10 illustrates a 5-bit code.) The code is sequential; that is, a particular *control character* defines the subsequent series of characters until a new control character appears. There are 32 different combinations in any 5-bit code ($2^5 = 32$). The two control characters, letters-shift (LTRS) and figure-shift (FIGS), change the meanings of subsequent characters in much the same way as the shift lock on a typewriter does. Using letters and shift characters increases the number of available code configurations to 64 (64 − 2 shift characters). However, since 3 codes plus blank are the same in either shift, there are really only 58 different characters.

The Baudot code had several limitations, namely:

1. It was very limited in punctuation and special character codes.
2. It had no parity bit or inherent method for validating transmission accuracy.
3. The sequential nature of the code meant that if a control character was missing, an entire portion of the message would be unintelligible.

The early Baudot code was replaced by ITA No. 2, an asynchronous code commonly—but incorrectly—referred to as the Baudot code.

STANDARDIZATION OF CODING SYSTEMS

During the 1960s a great many codes were developed, presenting the systems designer with a wide array of choices. It soon became apparent that some standardization would be desirable so that different computers and terminals could communicate. As a result, the American National Standards Institute developed the American Standard Code for Information Interchange (ASCII). This code has been adopted by the government and the military as their standard.

ASCII

ASCII is a 7-bit code, plus one bit for parity per character. The 7 bits provide 128 characters ($2^7 = 128$). Transmission may be either asynchronous or synchronous. ASCII is used more extensively than any other code in the United States because it is a standard in the communications industry. Its alphanumeric codes are shown in Figure 8.16.

Figure 8.16
The ASCII Configuration of Alphanumeric Characters

	Bit 7654321		Bit 7654321		Bit 7654321		Bit 7654321
A	1000001	**K**	1001101	**U**	1010101	**1**	1000110
B	0100001	**L**	0011011	**V**	0110101	**2**	0100110
C	1100001	**M**	1011001	**W**	1110101	**3**	1100110
D	0010001	**N**	0111001	**X**	0001101	**4**	0010110
E	1010001	**O**	1111001	**Y**	1001101	**5**	1010110
F	0110001	**P**	0000101	**Z**	0101101	**6**	0110110
G	1110001	**Q**	1000101			**7**	1110110
H	0001001	**R**	0100101			**8**	0001110
I	1001001	**S**	1100101			**9**	1001110
J	0101001	**T**	0010101			**0**	0000110

EBCDIC

The Extended Binary Coded Decimal Interexchange Code *(EBCDIC)* is IBM's System 360/370 code. It is an 8-bit code with 256 characters ($2^8 = 256$). It is generally transmitted in synchronous systems and has no provision for parity bits, although some users have modified it to provide for them. Because of the many possible ways to make the modification, users often end up with an incompatible interface even though they are using the same basic coding system. Its alphanumeric codes are shown in Figure 8.17.

SPECIALIZED DATA TRANSMISSION TECHNIQUES

Chapter 7 discusses amplitude modulation and frequency modulation. In addition to these basic types of modulation, there are two specialized types: phase modulation and pulse code modulation.

PHASE MODULATION

As used in data transmission, the word *phase* refers to the relative timing of an alternating signal. Two signals of the same frequency differ in phase if one signal is behind the other by any amount that is not an exact multiple of the frequency. A sine wave starts at what is known as the *baseline,* rises to its peak at its 90-degree point, returns to the baseline at its 180-degree point, continues to its lowest point in the negative direction at its 270-degree point, and returns to the baseline at its 360-degree point. Thus, it completes one cycle. It then starts a new cycle and repeats the process indefinitely unless it is interrupted. (See Chapter 6, Figure 6.22 for an illustration of sine waves.)

	Bit 87654321		Bit 87654321		Bit 87654321		Bit 87654321
a	10000001	**u**	00100101	**M**	00101011	**1**	10001111
b	01000001	**v**	10100101	**N**	10101011	**2**	01001111
c	11000001	**w**	01100101	**O**	01101011	**3**	11001111
d	00100001	**x**	11100101	**P**	11101011	**4**	00101111
e	10100001	**y**	00010101	**Q**	00011011	**5**	10101111
f	01100001	**z**	10010101	**R**	10011011	**6**	01101111
g	11100001	**A**	10000011	**S**	01000111	**7**	11101111
h	00010001	**B**	01000011	**T**	11000111	**8**	00011111
i	10010001	**C**	11000011	**U**	00100111	**9**	10011111
j	10001001	**D**	00100011	**V**	10100111	**0**	00001111
k	01001001	**E**	10100011	**W**	01100111		
l	11001001	**F**	01100011	**X**	11100111		
m	00101001	**G**	11100011	**Y**	00010111		
n	10101001	**H**	00010011	**Z**	10010111		
o	01101001	**I**	10010011				
p	11101001	**J**	10001011				
q	00011001	**K**	01001011				
r	10011001	**L**	11001011				
s	01000101						
t	11000101						

Figure 8.17
The EBCDIC Configuration of Alphanumeric Characters

In *phase modulation,* the phase of the signal is shifted to respond to the pattern of the bits being transmitted . In shifting the phase, the sine wave rises to its peak, then returns to the baseline at its 180-degree point. Then, instead of continuing in the negative direction, it starts upward toward another peak; that is, toward a new 90-degree point. Where no shift occurs, the signal is represented by alternating bits; that is, 0,1,0,1,0,1. To shift the phase, the signal is represented by either two 1's or two 0's adjacent to each other; i.e., 1,0,1,1,0,1,1,0 (Figure 8.18).

When very high transmission speeds are required, phase modulation is preferred over other systems of modulation. Phase modulation is less affected by extraneous noise and is less error prone than frequency modulation.

Two-phase modems employ dibits, groups of two binary bits. These permit twice as many bits to be sent over the line in the same bandwidth in the same amount of time, thus increasing the speed. Although phase modulation is more costly than amplitude or frequency modulation, its speed and efficiency make it very valuable for data transmission.

PULSE CODE MODULATION

A frequently used technique in data transmission is *pulse code modulation (PCM).* PCM converts analog signals into a series of coded digital pulses for transmission. At the receiving end the pulses are converted back into analog signals (Figure 8.19). PCM uses a fast sampling technique in which

Figure 8.18
Phase Modulation

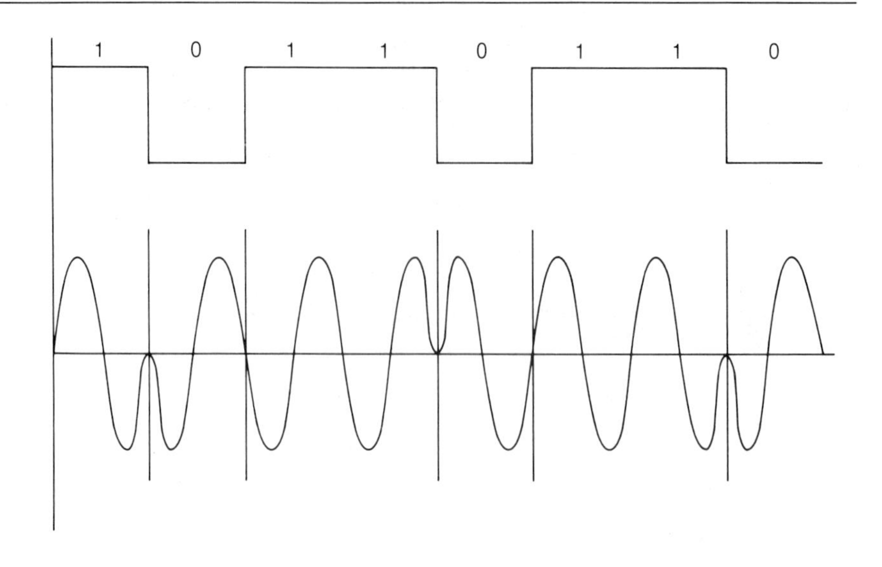

Figure 8.19
Pulse Code Modulation

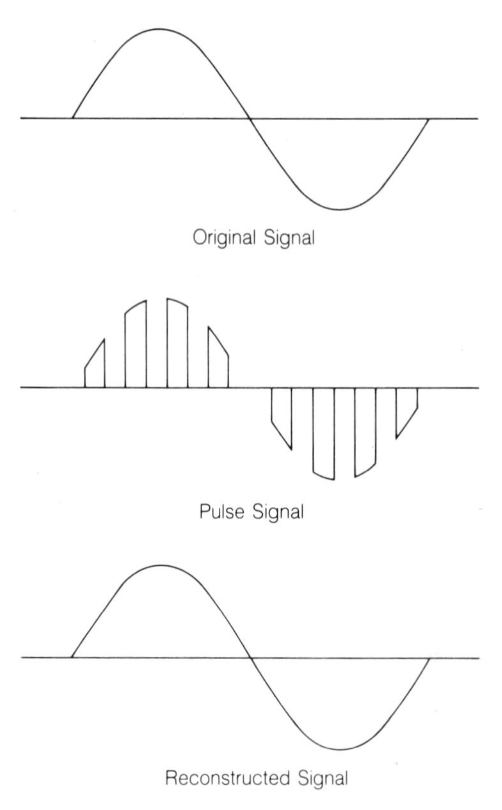

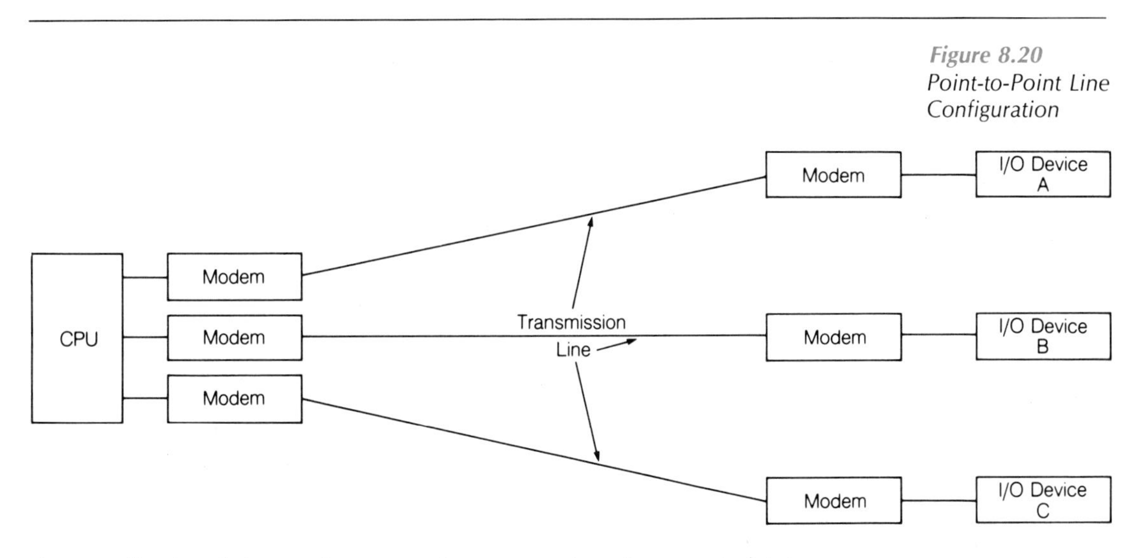

Figure 8.20
Point-to-Point Line Configuration

the amplitude of the analog wave is measured and converted into a numeric value. The numeric value, in binary form, is transmitted. Many signals are sampled with no interference. Pulse code modulation results in a highly reliable, noise-free digital signal. This technique offers a very cost-effective way to increase the capacity of transmission lines and to improve the efficiency of data transmission.

LINE CONFIGURATIONS

This chapter has explored the many facets of a data communication system. Regardless of the physical components of the system and the electronic techniques employed, the terminals and computer systems must be arranged in some type of line configuration. There are two principal types of line configurations: point-to-point lines and multidrop or multipoint lines.

POINT-TO-POINT LINES

The simplest network consists of one circuit between two points and is called *point-to-point.* As the name implies, point-to-point lines directly connect two points in a data communication network (Figure 8.20). These lines are relatively expensive because each terminal uses a different line into the computer system.

MULTIDROP OR MULTIPOINT LINES

If additional terminals are added to the point-to-point line, it becomes a *multipoint* network. A multipoint or multidrop line has more than one terminal connected to the computer system. However, only one terminal can transmit to the computer at a time. (See Figure 8.21.)

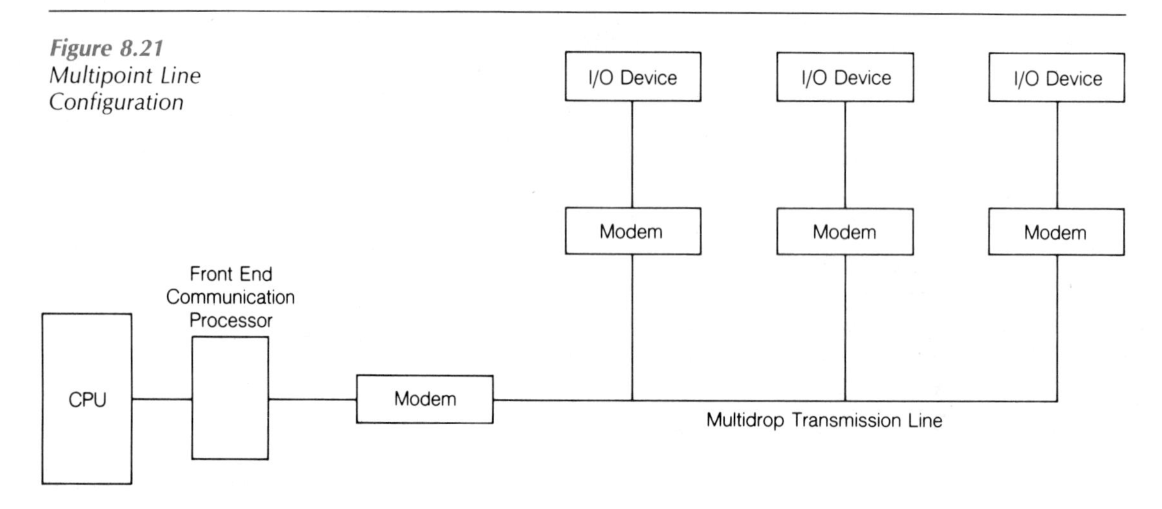

Figure 8.21
Multipoint Line Configuration

The number of terminals that can be served by one line depends on how much they are used. Multipoint operation is substantially less expensive than point-to-point operation because the per-line terminal cost drops appreciably when the line is used by a number of terminals.

SUMMARY

Data communication systems are designed to transmit data from one location to another—usually from one computer or terminal to another. The data is transmitted in coded form over electrical transmission facilities.

The three essential components of a data communication system are the source (remote terminals), the medium (transmission links), and the sink (the computer).

Five broad categories of terminals include: typewriter or keyboard terminals, video terminals, transaction terminals, intelligent terminals, and specialized terminals—audio response units and pushbutton telephones.

Telecommunication lines were originally built to carry voice communications consisting of analog signals. Since terminal devices transmit digital signals, an interface device or modem is required to convert the data to analog form so that it can be transmitted over telecommunication lines. If the transmission lines carries digital signals, no modem is required. There are two types of modems: hard-wired modems, which are permanently connected; and acoustic couplers, which are portable.

The CPU is the heart of the computer system. To relieve some of the demands on the CPU and to control the flow of data and ensure compatibility, a front-end processor is used.

Data is transmitted over telecommunications facilities and entered into computers in a code based on binary digits.

Data transmission controls are used to: (1) specify the rules to be followed in transmitting data (protocols), (2) detect errors in data transmission (parity checking), and (3) ensure system compatibility. Two levels of parity error detection are character checking and block or longitudinal redundancy checking.

Coding and sequencing should be sent properly so that the devices receiving the data signals can interpret them as intended. The transmitting and receiving devices must operate in step with each other, or in synchronization. Two forms of transmission are asynchronous and synchronous. In asynchronous transmission, each character is transmitted as an independent entity with start and stop bits to indicate to the receiving device that the character is beginning and ending. In synchronous transmission, whole blocks of data are transmitted in units. Synchronous transmission makes more efficient use of the medium; it permits higher data transmission speeds because of the elimination of the start and stop bits.

The ITA No. 2 code is commonly — but incorrectly — referred to as the Baudot code. This code is not used for data transmission; however, it provides a basis for understanding more extensive coding systems. The predominant transmission code today is ASCII; it is a standard in the communications industry.

Two specialized types of modulation are phase modulation and pulse code modulation. In phase modulation, the phase of the signal is shifted to respond to the pattern of the bits being transmitted. When high transmission speeds are required, phase modulation is preferable. It is less affected by extraneous noise and is less error-prone than other systems.

Pulse code modulation uses a fast sampling technique whereby the binary signals entered into the transmission system are representative of the original analog signal. It results in a highly reliable, noise-free digital signal. This technique offers a very cost-effective way to increase the capacity of transmission lines and to improve the efficiency of data transmission.

The two principal types of line configurations are point-to-point lines and multidrop or multipoint lines. Point-to-point lines are relatively expensive because each terminal uses a different line into the computer. Multipoint operation is less expensive than point-to-point operation because the per-line terminal cost drops appreciably when the line is used by a number of terminals.

REVIEW QUESTIONS

1. Name and identify the three essential components of a data communication system.
2. Name several different categories of terminals and describe them briefly.
3. What are some of the factors to be considered in terminal selection?
4. Name and describe briefly the two principal types of modems.
5. What are protocols and what is their purpose?

6. What is the purpose of parity checking?
7. Briefly describe the two forms of synchronization.
8. What is the predominant transmission code today? Why is this code predominant?
9. What is the purpose of phase modulation, and how is it accomplished?
10. What is the purpose of pulse code modulation, and how is it accomplished?
11. Name and briefly describe two types of line configurations.

VOCABULARY

source

medium

sink

remote terminals

modem

terminal equipment

buffer

soft copy

hard copy

transaction terminal

intelligent terminal

T-1 carrier

front-end processor

communications control unit

transmission link

binary

bit

dibit

American National Standards Institute (ANSI)

code

code set

protocols

handshaking

line discipline

error detection

parity checking

block character checking

longitudinal redundancy checking

asynchronous transmission

synchronous transmission

International Telegraph Alphabet (ITA)

American Standard Code for Information Interchange (ASCII)

Extended Binary Coded Decimal Interexchange Code (EBCDIC)

phase modulation

pulse code modulation (PCM)

point-to-point lines

multidrop or multipoint lines

CHAPTER

9

TELECOMMUNICATIONS SERVICES

CHAPTER OBJECTIVES

After completing this chapter, the reader should be able:

- *To define local exchange telephone service* and discuss the basic methods of assessing rates for this service.
- *To describe the various types of long distance telephone service.*
- *To describe the various types of specialized telephone service.*
- *To describe the various types of service features available on private branch exchanges.*
- *To define electronic mail* and discuss its advantages.
- *To describe the various types of electronic mail.*
- *To describe the various types of image services.*

Over the years, telephone service developed from two basic types—residential and business—to offerings that gave customers more options for communicating over a distance. There were shared (party) lines, services offering bulk volume discounts, services dedicated to the exclusive use of one organization, services for people on the move, services dedicated to the public welfare and safety, and auxiliary services to facilitate calling. Advances in technology, along with innovative pricing and marketing techniques, provided the impetus for these offerings.

Advances in technology are also responsible for the development of equipment that allows users to perform part of the service, thereby reducing the need for telephone operators and technicians. Examples include direct-dialing, which permits the customer to dial most long distance calls directly, and the modular jack, which permits the customer to install basic telecommunications devices.

In today's competitive telecommunications marketplace, new services are constantly emerging. This chapter examines the wide variety of telecommunications services available over and above Plain Old Telephone Service (POTS).

TELEPHONE SERVICES

Each of the four subsets of telecommunications — voice, data, message, and image — has its own complement of service offerings. Although some of these categories overlap, we will discuss them separately in order to help put them into perspective.

LOCAL EXCHANGE SERVICE

An *exchange* provides service for a specific geographical area. It consists of one or more telephone offices and the physical plant and equipment necessary to provide communication services to the area.

Local exchange service is public telephone service to points within the designated geographical area, known as the *exchange area* or *local service area*. The exchange area for a central office is defined in the telephone directory. Typically, the area includes customers served by other nearby central offices.

A call made to another point in the exchange area is called a *local call*. Figure 9.1 illustrates a telephone exchange map of the areas that can be called locally from a given telephone location.

Local calls are not billed individually. Instead, rates for local exchange service are usually assessed in one of two ways:

1. *flat rate service,* which entitles the user to an unlimited number of calls within a specified local service area for a fixed monthly rate
2. *measured rate service,* under which the user is charged according to a measured amount of usage each month

In most metropolitan areas, calls to exchanges that are contiguous to the local exchange are classified as *zone calls*. Zone calls are billed in *message units*. The length of the call and the distance involved determine the number of message units per call. For example, four message units might mean four separate calls of short duration to nearby locations. Or they might mean one or two calls of either longer duration or to a more distant location, or a combination of both time and distance.

The number of local calls, as well as the duration and distance of each, is considered under a newer form of rate assessment, known as *measured local service*.

Individual service is one telephone line that serves one subscriber. It is available to both residential and business customers. *Party-line service* is one telephone line serving more than one subscriber. It is generally available only to residential customers.

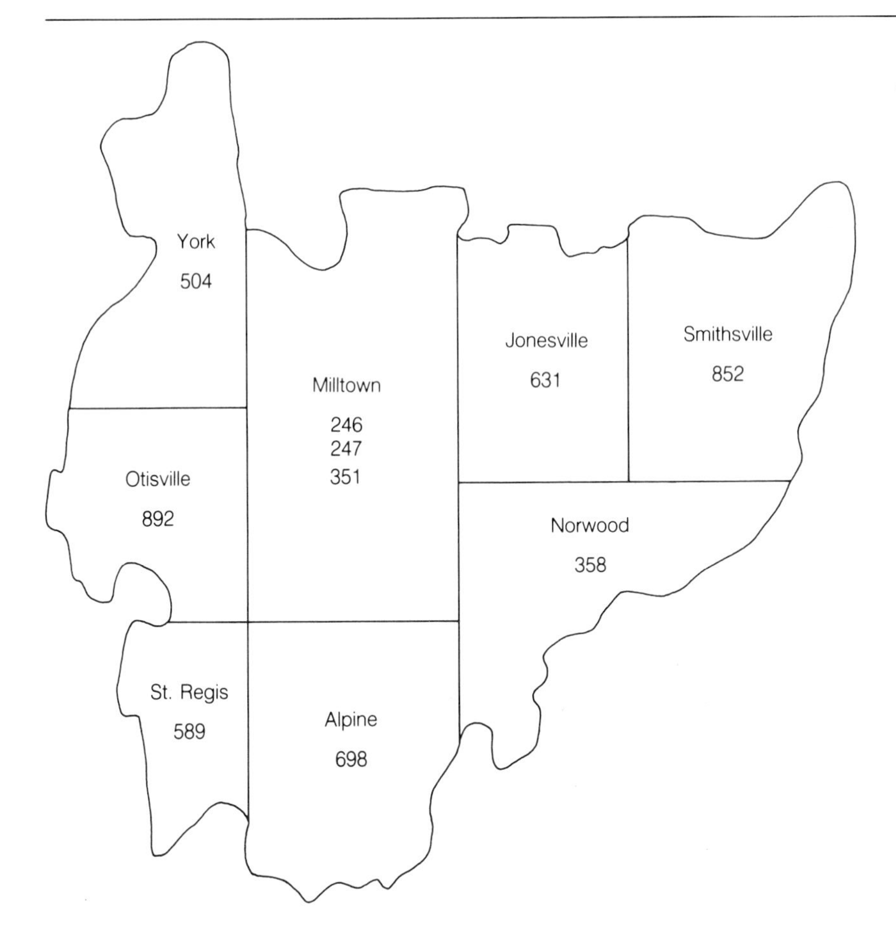

Figure 9.1
A Telephone Exchange Map

Extended area service (EAS) may be purchased in many localities. With EAS, a subscriber can make calls to a designated area beyond the local exchange area, at local exchange rates rather than long distance rates. A subscriber in a city who makes a great many calls to a suburb beyond the subscriber's local exchange area can save money by purchasing extended area service rather than paying toll rates on the calls.

■ ***LIFELINE SERVICE*** The FCC sponsors two programs to assist low-income families obtain basic telephone service, thus furthering the national goal of universal service. One program provides a federal subsidy of $2 per month credit per family meeting the eligibility requirements for Lifeline Service. The payment is made to any telephone company that offers tariffed Lifeline Service. The other program, called Link Up America, provides a 50 percent discount on telephone installation charges to families ordering Lifeline Service. Both programs are subsidized by revenues obtained from access charges levied on long distance companies.

Many states' regulatory commissions have approved tariffs for Lifeline Service that provide additional subsidies of up to $2 more per month and have permitted the local telephone companies to raise other rates to cover these subsidies.

LONG DISTANCE SERVICES

Long distance services are called intercity services or toll services. They include any calls beyond the local service area and are charged under a tariff separate from local exchange tariffs.

In today's deregulated telecommunications environment, long distance telecommunications service is provided by a number of carriers on a highly competitive basis. The services described hereafter are available from most long distance providers.

■ ***DIRECT-DISTANCE DIALING (DDD)*** DDD telephone service permits users to dial telephones beyond their local service area without the aid of an operator.

■ ***OPERATOR-ASSISTED CALLS*** Operator-assisted calls require an operator at a switchboard to complete the call. They cost substantially more than dialing direct. They can be classified as follows:

- ☐ person-to-person calls, in which the caller wishes to reach a particular person or extension number
- ☐ collect calls, in which the person or firm being called agrees to pay the charges for the call
- ☐ calling-card service (formerly known as credit-card service), in which callers who have a telephone calling card can have long distance calls charged to their regular monthly bill
- ☐ third-number calls, in which long distance calls are billed to an authorized third telephone number (a telephone number different from that of the calling or called telephone)

Although callling-card service is classified as operator-assisted service, many pushbutton telephones now have a mechanized calling-card service in which the caller dials these calls without the assistance of an operator. Also, most public telephones are now equipped for handling calling-card calls.

Another service available on all types of operator-assisted long distance calls is *time and charges.* A caller can obtain this service by asking the operator for the length and cost of any call. There is an additional charge for this service.

■ ***WIDE AREA TELECOMMUNICATIONS SERVICE (WATS)*** Wide Area Telecommunications Service is a pricing mechanism that permits customers to make long distance voice or data calls and have them billed at a bulk rate instead of on an individual-call basis. WATS differs from other discounted long distance services in that WATS is provided over a dedicated line to the long distance company. The service is economically attractive. For example, hotels and motels can resell the service to guests, thereby offering another avenue of profit.

WATS can be provided for either intrastate or interstate calls. Intrastate WATS may be either intra-LATA or inter-LATA. Intra-LATA calls are handled by the local telephone company; inter-LATA calls are routed to the designated long distance company. Interstate WATS is always provided by a long distance company.

Formerly WATS was known as a "banded service" because it provided service to established bands extending outward from the point of origin, and the customer could choose which band or bands to purchase. Additionally, at one time WATS was offered on a flat-rate basis. WATS is now a nonbanded service whose rates are determined by distance and length of conversation. The usual time-of-day, day-of-week, and volume discounts apply.

■ ***800 SERVICE*** Formerly known as Inward WATS, 800 service permits a caller to reach a telephone number outside the local calling area on a toll-free basis. Charges for the call are borne by the subscriber. Although used primarily by businesses, this service is available to both residential and business customers. The toll-free aspect of the service encourages callers to place calls to the number. All major long distance carriers offer 800 service.

This service is billed as a measured service. The cost of each call is determined by the distance, time of call, and duration of call. Calls are generally bulk-billed; however, individual billing with call details is available for an additional fee.

■ ***900 SERVICE*** The 900 service permits a business organization to receive telephone calls dedicated to a specific purpose. Calls are routed to a particular telephone number, enabling the caller to accomplish a specific purpose, such as registering a vote or an opinion or speaking with a celebrity. The 900 subscriber pays the cost of the basic service; the caller pays the cost of the call.

■ ***PRIVATE-LINE SERVICE*** With private-line service, the customer has the exclusive use of a leased circuit between two specific points. The circuit is not connected with the public telephone network. Private lines may be used for any type of transmission — voice, data, teletypewriter, video. (Private-line service should not be confused with individual service, a public network service in which one telephone line serves one subscriber.)

Private networks are a configuration of private lines and related switching facilities provided for the exclusive use of one customer.

■ ***FOREIGN EXCHANGE SERVICE (FX)*** FX is service to a telephone exchange outside of the one in which the user is located. A leased line connects the subscriber's telephone to a central office in the foreign exchange area. The subscriber can be listed in the foreign directory. Being a two-way service, FX permits the subscriber to call any number in the foreign exchange area and people from the foreign exchange area to call the subscriber. This service allows users to avoid long distance charges to the foreign exchange. (For a more complete description of FX service, see Chapter 6.)

SPECIALIZED TELEPHONE SERVICES

Telephone subscribers are generally familiar with most of the traditional residential or business services because they have used them or at least are aware of their existence. Another group of telephone services, however, is specialized services that are designed to meet a special need. They have one thing in common — people use them while away from their home or place of business. The two principal types of specialized services are coin telephone services and mobile telephone services.

■ ***COIN TELEPHONES*** Coin telephones fill a need that all of us have at one time or another — the need to make a call while away from our own telephone.

Public coin telephones provide service on the public network to persons away from their residence or place of business. They are installed in public areas, such as airports, hotel lobbies, stores, and outdoor locations. The telephone numbers for public coin telephones are not listed in telephone directories.

Semipublic coin telephones are installed where there is a combination of general public and individual customer need for the service, such as in a gasoline station. With these telephones the subscriber receives a listing in the telephone directory and guarantees a specific monthly revenue from the telephone. The revenue is offset by the coins collected in the coin box. If the revenue from outgoing calls is less than the guaranteed amount, the subscriber is responsible for the difference.

A person who places a long distance call from a coin telephone elects either to pay for it immediately or to use a billing option, such as calling-card service, third-number call service, or collect-call service.

Most coin telephones now provide *dial-tone-first service,* which permits customers to reach the operator and to dial certain calls, such as directory assistance or 911 (emergency), without depositing a coin. The dial-tone-first feature gives the caller some assurance that the telephone is working before coins are deposited. It also allows callers to use the telephone in an emergency without looking for coins first.

Throughout the history of the telephone industry, coin telephones were provided — and owned — by the local telephone companies. In today's deregulated telecommunications environment, coin telephones may also be owned and operated by an individual or an organization. This type of phone is commonly known as a *customer-owned coin-operated telephone (COCOT).* COCOTs are connected to the public network through a line provided by the local telephone company. Since these telephones do not come under the jurisdiction of any regulatory body, the owner can resell telephone service to the public with no restrictions on charges. The owner's only obligation is to pay the provider(s) of the service (the local telephone company, plus the long distance carrier if long distance service is used).

(Photo Courtesy of Motorola, Inc.)

Figure 9.2 Portable Cellular Telephone

Regardless of whether the coin telephone is owned by the local telephone company or a private individual or organization, the owner must choose a long distance carrier.

■ ***MOBILE TELEPHONE SERVICES*** Mobile telephone services use radio transmission. They include: cellular radio service, radio paging service, air-ground serice, VHF maritime service, coastal harbor service, high-seas maritime radio-telephone service, and high-speed train service.

Cellular radio service is an advanced form of mobile telephone service. It combines radio and computer technology to provide telephone service to vehicles on the move. Mobile telephones enable businesspersons to be more productive because they can be in touch with their offices or clients even when driving to and from work or business meetings (Figure 9.2).

Conventional mobile service uses two-way radio to connect the vehicle to the telephone system. The radio circuit establishes a talking path between the vehicle and an antenna connected to the telephone network (Figure 9.3). A single antenna serves an average-sized city, and communications can be maintained as long as the vehicle stays within range of

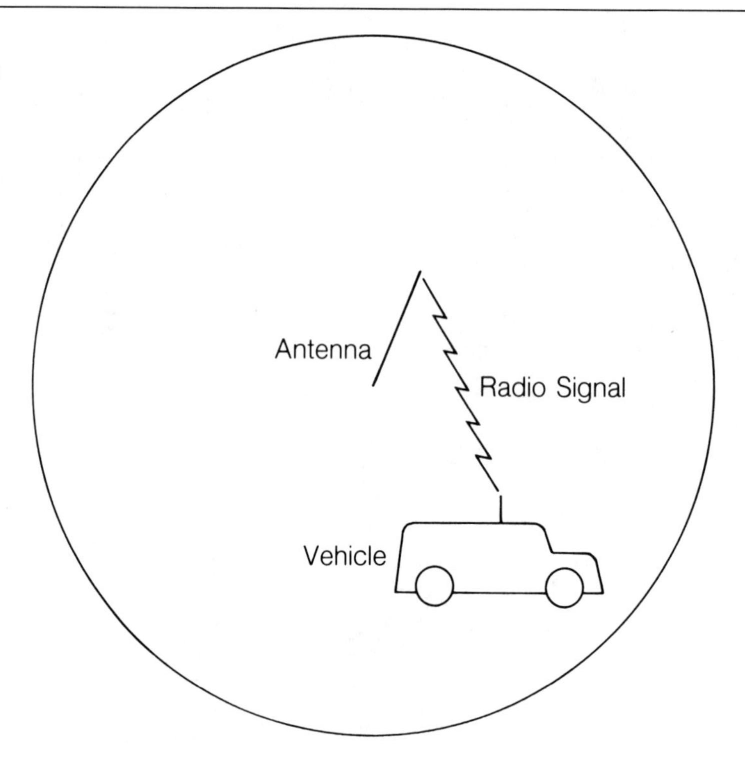

Figure 9.3
A Conventional Mobile Service Area

the antenna. However, conventional mobile service has several drawbacks, including:

1. the limited number of radio frequencies available to provide service
2. inferior transmission characteristics provided by a single, distant antenna
3. interference from other vehicles on the same channel
4. high service cost

Despite these disadvantages, mobile telephone service has been very popular. In fact, in many areas there has been a long wait to obtain this service.

Cellular radio is a newer concept designed to improve mobile telephone service and to make it available to more customers. The basic principle of cellular radio service is the division of one mobile telephone service area into a number of smaller areas called *cells*. (Figure 9.4). Each cell contains a low-powered radio communicating system that is connected to the local telephone network. The individual systems replace the single high-powered system that formerly served an entire area and permit one radio frequency to be reused many times. The chief advantages of cellular systems are their improved transmission and the ability to serve more customers.

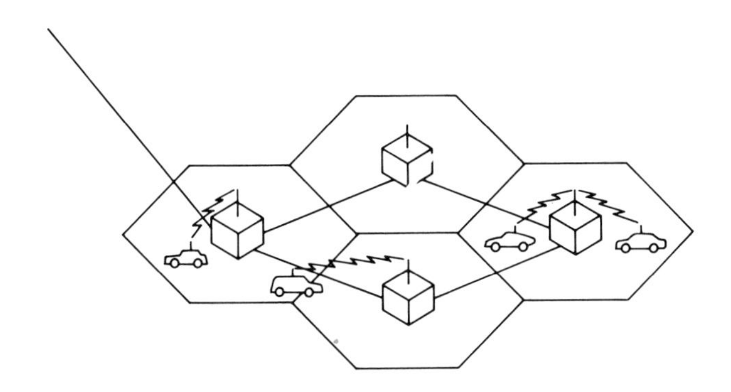

Figure 9.4
Cellular Radio Broadcast Areas

The system works like this: a call placed from a cellular telephone is transmitted via radio to an antenna system located on a local building. The antenna system relays the call through ordinary telephone lines to a mobile telephone switching office. From there it is transported to a telephone company central office to gain access to the public switched network. A vehicle located in a cell area is connected to the telephone network through this low-powered system. When the vehicle leaves the cell area, radio contact is automatically transferred to the radio system of the adjoining cell as the vehicle enters.

Cellular service is not solely for car phones. Recent improvements in batteries have made it possible to use small, hand-held cellular phones in almost any location — on the street, in an airplane, on a boat, or on the golf course.

According to Thomas Adams, Ameritech Mobile regional general manager, "The growth of cellular has exceeded everyone's wildest dreams, and we're still in our infancy."[1] In a little more than five years, the number of cellular phones in the United States went from 0 to 1.75 million, with projections of 4 million in use by 1990. Today, cellular service ranks as one of the nation's fastest growing industries.

Radio paging service is another form of telephone service made possible by radio technology. Like mobile telephone service, radio pagers provide a way to communicate with someone on the move. The principal difference between mobile telephone service and radio paging is that cellular telephone service provides two-way communications while radio paging is limited to one-way service.

Historically, radio pagers were used primarily by doctors and service personnel, who carried small units called *beepers* attached to their clothing. More recently their use has spread to businesspersons who carry pagers to enable them to be in constant touch with their offices. When the office needs to contact a person carrying a pager, someone dials a telephone number assigned to the paging device, and a radio signal causes the

Figure 9.5
Motorola's BRAVO Pager. Available in Numeric Display or Tone and Visual Alert Models

(Photo Courtesy of Motorola, Inc.)

Figure 9.6
Motorola's PMR 2000 Personal Message Receiver

(Photo Courtesy of Motorola, Inc.)

device to emit a signal, such as a beep or vibration. To receive the message, the user goes to a telephone and calls a predetermined number, usually his or her office. Because the signal transmission time is very short, many beeper units can be served by one transmitter.

Recent technological developments have expanded the usefulness of radio pagers. Now, in addition to *tone pagers* that supply only a beep, there are *tone-and-vibration pagers* that supply both a beep and vibration (the beep can be silenced in situations where sound could be distracting), *tone-and-voice pagers* that let callers send a spoken message along with the beep, *numeric-display pagers* that display only phone numbers, and *alphanumeric pagers* that can display numbers and letters (Figures 9.5 & 9.6).

Alphanumeric pagers (generally called simply "alpha pagers") represent the peak of pager technology. Although the earlier types of pagers currently own the largest share of the market, the ability of alphanumeric models to act as electronic mailboxes gives them the potential to eventually dominate the market.

Pagers have been shrinking in size ever since they were introduced, but alpha pagers are larger than their simpler cousins and weigh 3½ to 5 ounces. Alpha units can store 416 to 1,984 characters, which is more than a page of text.

Most paging devices are simple to use. For a tone-only page, callers dial the service's phone number, listen for the beep, and hang up. For a tone-and-voice page, callers listen for the beep, speak for six to ten seconds, and hang up after the cutoff tone. Sending a numeric page involves listening for the beep and using the Touch-Tone telephone keypad to enter the number. Alpha pagers generally require callers to enter the message on a computer terminal or relay the message through the service's operator.

Regardless of the type of pager, telephone lines deliver the message to the paging company's terminal. After the terminal identifies the pager number as being in service, it places the page in queue and transmits the message to one or more antennas that broadcast the information on the FM radio band. A special address, or cap code, in the message alerts the particular pager to receive the message.

A relatively new development is nationwide paging service; however, coverage consists of service "islands" that leave many gaps. The service requires the use of a more sophisticated device than is required for local paging. To use this service, the caller dials an 800 telephone number to obtain access to a satellite system providing nationwide coverage, waits for the beep, then dials the 6-digit code assigned to the subscriber's pager. As in all other paging services, the use of a Touch-Tone telephone is required.

Another form of mobile telephone service is *air/ground service,* which allows two-way telephone communication between aircraft in flight and parties on the public network. The service is provided by radio base stations connected to control terminals and mobile service switchboards. Major airlines offer this service.

Marine radio telephone services are used by ships at sea for two-way telephone service. There are three types of service, which differ basically in the distance range in which they operate:

1. VHF maritime service, which provides reliable communication in the very high-frequency band up to 50 miles offshore
2. coastal harbor service, which provides communications up to 1,000 miles offshore
3. high-seas service, intended for ships engaged in high-seas operations and transoceanic passages

High-speed train service provides two-way telephone communication between trains en route to a destination and parties on the public telephone network. Like air/ground service, this service is provided by radio base stations connected to control terminals and mobile service switchboards. This service is offered on Amtrak trains.

TELEPHONE SUPPORT SERVICES

Local telephone companies provide numerous support services, including business office services, community services, telephone directories, directory assistance, and intercept services.

■ ***BUSINESS OFFICE SERVICES*** The telephone business office takes orders for new services or changes in service, answers billing inquiries, receives bill payments, and completes the link from the customer to the rest of the telephone company's working forces. Service representatives coordinate these functions with customers.

■ ***COMMUNITY SERVICES*** Some telephone company services are of a civic nature; that is, they are provided primarily as a public service to the community. The program to make 911 the emergency reporting number is one example. Dialing 911 in many areas of the United States puts the caller in direct contact with an emergency center, which handles calls for the local police, fire department, or emergency medical service.

With the latest state-of-the-art technology, an emergency center can receive a display of the telephone number and location of the telephone that originated the emergency call. This is known as "enhanced 911 service."

In addition, many communities now provide direct lines to emergency centers. This feature eliminates the need to dial the telephone number 911. The user merely picks up the phone and is immediately connected to an emergency center. Such phones are installed in public areas that are readily accessible.

The cost of 911 service is borne by the providing community; the user pays no charge for the call. Some communities have enacted ordinances requiring the telephone company to add a surcharge to each telephone subscriber's bill. Thus, the cost of the service is borne equally by all members of the community.

■ ***TELEPHONE DIRECTORIES*** The White Pages contain an alphabetical listing giving each subscriber's name, address, and telephone number. For an additional fee, users select special directory listings, such as "If no answer, call . . . ," "After 5 o'clock call . . . ," or boldface type. An additional service is the withholding of a customer's listing from the directory (nonlisted number). This service offers more privacy, but because it involves special handling, telephone companies charge an additional fee.

The Yellow Pages contain an alphabetical listing of business subscribers by category of business. Subscribers can also purchase advertisements in the Yellow Pages.

In some areas, telephone directories also contain Blue Pages, which list the numbers for frequently called organizations such as government offices, media, and emergency agencies.

■ ***DIRECTORY ASSISTANCE*** Directory assistance service is designed to provide telephone numbers not included in the local telephone directory. Its use has grown to such proportions that many telephone companies now charge for the service. Directory assistance for foreign codes can be reached by dialing the area code and 555 – 1212. Although long distance directory assistance was traditionally a free service, present carriers now charge for it.

■ ***INTERCEPT SERVICE*** Intercept service informs callers of any changes regarding the telephone number they have dialed. Calls to a disconnected number are automatically routed to an intercept operator or to a computer. Either the operator or a recorded announcement informs the customer of the status of the number called and provides a new number if one is available. Most telephone companies today use computers in such instances, and this service is called automatic intercept service (AIS).

TELEPHONE SERVICE FEATURES

The telephone services described thus far are provided by the local telephone company, with the user frequently providing some part of the service. The following services are inherent in the telephone equipment— the switching system or the telephone instrument. The equipment may be provided either by the telephone company or by in interconnect vendor.

■ ***CUSTOM CALLLING FEATURES*** The four features that are generally classified as custom calling features are three-way calling, speed calling, call forwarding, and call waiting. All of these features can be provided by electronic switching systems. In addition, speed calling can be provided by telephones with electronic components.

□ *Three-way calling* permits a customer to add a third party to an existing conversation for a telephone conference call. When the third party answers, a private, two-way conversation with that party can be held before bridging the connection for the three-way conference.

- *Speed calling* (also called *automatic dialing*) permits a caller to reach certain frequently called numbers by using abbreviated telephone codes in place of the conventional telephone number.
- *Call forwarding* permits a telephone to automatically forward calls to another telephone number. The equipment can generally be programmed so that the call is forwarded only after a predetermined number of rings. This feature can be used to transfer calls to either a local telephone or a telephone in a distance city. When a forwarded call is subject to a toll charge, the charge is billed to the forwarding telephone.
- *Call waiting* permits a call to a busy telephone to be held while an audible tone notifies the called party that a call is waiting. The tone is audible only to the called party, who can decide whether to interrupt the existing conversation to find out who is calling.

PBX SERVICE FEATURES There are many different service features available to subscribers with private internal telephone systems (PBXs). Because these features vary from one system to another, in selecting a telephone system, the customer must evaluate the organization's needs carefully and match these requirements with the features. Not all features are available from all telephone equipment suppliers, but those listed here are available from most vendors.

- *Automatic call back* permits the caller to "instruct" a busy station to call back as soon as the busy station is free. The instruction is given by dialing an extra digit that tells the computer to reestablish the connection when both telephones are available.
- *Line privacy* is a station control feature for telephone lines that require special privacy. This feature is useful for data processing lines where the data could be contaminated or destroyed by outside interference. To obtain line privacy, the user activates a control key that excludes transmission interference by another person or electronic device.
- *Lockout* is a station control feature that ensures the confidentiality of a call. Many newer telephone systems permit a switchboard attendant to break into a call in progress; the lockout feature allows the user to "lock out" interruptions until the call is completed.
- *Automatic route selection (ARS),* sometimes called *least-cost routing,* permits the automatic selection of the most efficient routing of a call originating in a corporate network; e.g., tie line (first choice), WATS line (second choice), and direct-distance dialing (third choice). Some users can be denied access to the latter category, so that the call is delayed until one of the less costly routes becomes available.
- *Trunk prioritization* enables a customer to use its WATS lines or other specialized facilities to their fullest by stacking up the calls of those users having lower priority. When desired lines are all occupied, this option records the number dialed and makes the connection when a line becomes available.

- ☐ *Remote access* permits authorized personnel to place calls from other locations and be connected to a business PBX system. Users access the system by dialing a private telephone number plus a 3-digit security code. Then the authorized caller can place any type of call that could be placed from a PBX station.
- ☐ A *message-waiting indicator* is a message light on a user's instrument that can be lighted when the receptionist or message center presses a button or transmits a dial code. A lighted indicator tells the user that a message is waiting. (This should not be confused with *call waiting,* wherein a caller is on the line trying to reach a telephone that is busy.)
- ☐ *Identified ringing* provides distinctive ringing tones for different categories of calls. For example, internal calls, calls from a secretary, or calls from a given extension can be recognized by their unique ringing style. Thus, before answering the telephone the called person is given some indication of where the call originated.
- ☐ The *call pick-up* feature enables a person receiving a telephone call to have access to the incoming call on any telephone station in the system by entering a code. Also, others in the called party's group can pick up calls to take messages. This feature enables a telephone system to operate effectively without having all lines appear on each telephone station.
- ☐ The *individual call transfer* feature enables a telephone system user to transfer a call to another station without going through an operator, thus saving time for both users and operators.
- ☐ *Paging access* allows an authorized station user to have direct access to paging equipment. This feature is activated either by dialing a special code or by pushing a paging button on the instrument.
- ☐ *Recorded telephone dictation* equipment permits the user to be connected with central dictation facilities. It may also permit the user to access central dictation facilities directly from the public telephone network. This feature is useful for persons who are out of town or working at home.
- ☐ A *local maintenance* feature enables a special console to be used to reprogram the telephone system to change system features or stations. Some PBX equipment has *remote maintenance* capability, through which a service person can dial the telephone system to gain access to the central processor of the system to test or modify system features or programs. (See Figure 9.7.)
- ☐ *Electronic telephones* contain electronic components or microprocessors that enable them to perform a variety of special services. These telephones are described as "smart" or "intelligent" because of their enhanced capabilities. They can perform such services as speed dialing, last number redial, hands-free operation, automatic answering, and call timing.
- ☐ *Audio teleconferencing* enables people at a number of geographically distant sites to carry on a discussion among themselves. An audio

Figure 9.7
Automated Repair Service Bureau

(Reproduced with permission of AT&T)

teleconference differs from a conference call in that far more people can be involved in the former. Voice-only teleconferencing requires a bridge, which is an electronic device that allows from three to hundreds of people to converse from different locations. The bridge can be located in a PBX on the user's premises or at the local telephone company's central office. Participants all call one number, which has been permanently assigned to the user organization for private teleconferences. The bridge connects the participants automatically. Audio teleconferencing service is offered by the local telephone companies and by other companies that provide conference capability and assistance.

Another PBX service feature that is growing in popularity is *voice mail.* Voice mail is also called *voice messaging* or *voice storage and retrieval (VSR).* It is a computer application that digitally records voice messages, stores them in a data base, and forwards the message to the designated recipients. Voice mail enables recipients to hear the messages, but as yet it cannot convert the spoken words to printed or displayed text.

Telephone answering by voice mail systems resembles the service provided by telephone answering machines. Unanswered incoming calls are forwarded to the system, which plays a recorded message and invites the caller to leave a message. Some systems use a light on the telephone or special dial tone to notify the recipient of a waiting message. The recipient can retrieve messages from any tone telephone by calling the system and keying in a user code.

Voice mail systems are much more than answering machines, however. They preserve the actual works spoken by recording the voice in a digitized format (rather than on magnetic tape). Once messages have been digitized, they can be routed and transformed in ways not possible with other technologies. Voice messaging differs from telephone answering in that the sender of the message has several options, including sending messages immediately or at some future time, sending the same message to several recipients, or adding to someone else's message and rerouting it.

ADMINISTRATIVE SERVICE FEATURES

The previously described service features may be considered convenience features. Their purpose is to save time and make calling and maintenance easier. There is another group of service features whose purpose is to simplify record keeping and control telephone usage.

■ ***CALL DETAIL RECORDING*** *Automatic Identification of Outward Dialing (AIOD)* and *Station Message Detail Recording (SMDR)* are services that record call details. AIOD is the name of the recording system provided by interconnect vendors; SMDR is the Bell System's product. This equipment consists of a tape recorder with minicomputer features. It is activated on long distance calls made from a PBX system extension. Details of the call (data and time, originating extension number, called number, start and end times of conversation, type of line used) are recorded on tape. The tape is processed periodicaly, either in-house or by a service bureau, and a report summarizing the information obtained. This report is compared with the telephone company's billing so that the cost of each call can be allocated to the extension making the call.

Some systems do not record the calling station automatically, but require the caller to dial an identification code before the desired long distance number. Caller identification codes can also be used to restrict usage by unauthorized callers or to restrict usage to local calls. A complete listing of charges incurred permits supervision of telephone costs, which, in turn, can do much toward reducing an organization's telephone expenses.

■ ***SERVICE RESTRICTIONS*** Another way of controlling telephone usage costs is through service restrictions. There are two types of restrictions that are frequently employed: class of service restriction and originating restriction.

A *class of service restriction* limits use of a telephone station to certain types of calls. For example, a telephone station might be limited to only internal calls. With this feature, each telephone in a system can be provided with different levels of access to outside lines and telephone services. A telephone station might be restricted to accessing only certain cities or area codes. Or it might be restricted from using special long distance services such as WATS or FX lines.

An *originating restriction* restricts the telephone station from being used to place outgoing telephone calls. The station can receive calls in the usual manner. This restriction is often placed on telephones in conference rooms or unoccupied offices. The restriction is applied automatically; a person attempting to place a call by dialing "9" to obtain an outside line receives a busy tone. Some telephone systems restrict outgoing calls by requiring them to be handled by an operator.

The restrictions are customized to serve the needs of the organization. These features pay for themselves by controlling telephone costs.

WRITTEN MESSAGE SERVICES

The fastest, most efficient, and often the least expensive method of communicating is by telephone. The telephone provides two-way conversation, allowing exchanges of ideas and information. When a written record of the message is required for reference or legal purposes, the telephone must be supplemented by a written document.

The business letter is undoubtedly the most widely used form of written messgae communication. However, its production and delivery is a costly, time-consuming process. A letter must be composed, typed, placed in an envelope, stamped, delivered to a mailbox, picked up and taken to a post office, sorted again, and physically delivered to its destination. At its destination it frequently requires additional sorting and delivery to the proper department and person.

The many operations involved in the manual delivery of written communications make it a slow process. Further, since these operations are highly labor-intensive, they are expensive.

ELECTRONIC MAIL

Electronic mail (E-mail) refers to the delivery of mail, at least in part, by electronic means. Transmission via electronic information channels is essential; but delivery may be end-to-end electronic service, or it may be supplemented by physical delivery systems.

Messages can be sent electronically from a computer over any telecommunication transmission facility, including telephone lines, microwave radio, waveguides, coaxial cable, satellites, and fiber optics. Messages are transmitted and received in the form of electronic signals that are translated into readable messages by the receiving device (usually another computer). To transmit information there must be a terminal, such as a computer terminal, a teletypewriter, a communicating word processor, a facsimile machine, or a telephone.

There are many advantages to using electronic mail systems. Two of the most important advantages are speed and the elimination of "telephone tag." The fact that communications can be instantaneous across continents provides businesses a tremendous operational advantage — orders can be processed quicker and decisions can be made faster.

The other major advantage of E-mail is its elimination of "telephone tag," the exasperating phenomenon of trying to reach a person by telephone. The game goes like this: When a caller calls and the called party is not available, a message is left. Upon receiving the message, the recipient calls back only to find that now the original caller is not in, and so on.

The following research shows the inefficiency of telephone calls:

- ☐ 75 percent of all business calls are not completed on the first try
- ☐ 55 percent of all business communications are one way
- ☐ 76 percent of all business communications are not time-sensitive
- ☐ 50 percent of a business call is not business-related
- ☐ 60 percent of all incoming calls are less important than the work they interrupt
- ☐ more than 90 percent of written phone messages are incomplete or garbled[2]

Electronic mail has technically been around since the 1844 introduction of the telegraph. Modern E-mail systems, however, include facsimile systems, teletypewriter systems, private computer-based message systems, carrier-based message systems, and voice mail.

A major deterrent to the use of electronic mail systems is their initial cost. Although the cost of transmitting messages electronically is low, the costs of procuring the required equipment can be substantial. However, as costs of office supplies increase and office personnel and postal workers receive higher salaries, the costs of physical mail delivery can be expected to rise. Costs of electronic mail, on the other hand, are expected to drop sharply as a result of continued development in microprocessor, satellite, laser, and fiber optic technologies.

■ ***FACSIMILE SYSTEMS*** One of the most dependable ways of transmitting information is by the time-proven technology of facsimile, used in offices for over fifty years. *Facsimile,* abbreviated fax, is a system for transmitting a copy of a document to a distant point over telephone lines. Facsimile machines are also known as *telecopiers.* Connected to each other via telephones, these machines can transmit and receive any form of documents in less than a minute per page. The process can be compared to putting a copy of a document into a copier at one location and having it come out at another location (Figure 9.8).

Fax machines often look like copiers. To use a facsimile machine, the operator places the original document on a transmitting tray and establishes phone contact by dialing the number of the receiving fax machine. When the receiving machine answers with a high-pitched tone, the operator depresses a key. This causes the two machines to connect and establish transmission speed. The transmitting fax then scans the entire page of the document using reflected light and converts the graphic image into electronic signals, using a process known as digitizing. These signals are then compressed and sent through regular telephone lines. The receiving fax converts the signal into printer commands and reproduces an exact fac-

Figure 9.8
Lanier Facsimile Copier 2200

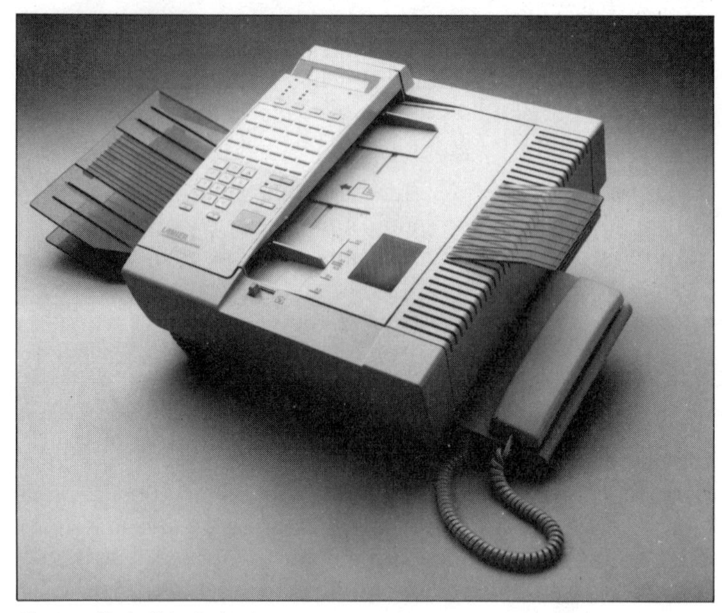

(Courtesy of Lanier Voice Products)

simile of the original document, usually on special thermal paper, although machines that use plain paper are available.

Fax has many advantages compared with other methods of communication. While nothing is as direct as the telephone, it is often worth more than a thousand words to have the picture, or the facts or figures, right in front of someone.

Fax requires no trained operator. With teletypewriter and other forms of E-mail, someone must retype the entire document. This takes time and can introduce errors. Additionally, teletypewriters cannot handle any form of graphic information.

Fax is usually not a cost-effective alternative to regular mail if time is not important. However, when time is important, fax can effect substantial savings in both time and money as compared to express mail. Today's fax machines can distribute a document to multiple locations around the country faster than an organization's interoffice mail can distribute a memo.

For the organization that does not have fax equipment, fax delivery services and public facsimile stations are available. Network Facsimile Service, a nationwide courier service, has fax machines installed in its service centers located in most major cities in the United States and abroad. The company offers a range of transmission and delivery services that combine fax transmission with courier pickup/delivery. Additionally, many copy centers and commercial mail centers, such as Mail Boxes, Etc., also offer public fax transmission.

During the last few years there has been an explosive growth in the use of fax. Some of the factors that have contributed to this growth include:

1. user friendliness — most fax machines are now about as easy to use as a copier
2. compatibility — the various vendors' products can be interconnected
3. advances in technology — fax machines are capable of providing faster transmission speeds, clearer copies, and feedback regarding receipt of transmission; they can transmit automatically with no one in attendance; they are smaller in size; and fax boards are available for use in personal computers
4. decreases in cost — both the price of fax machines and the cost of long distance telephone transmission have decreased substantially

Several directories are available for fax users: the *Official Facsimile Users Directory,* published by FDP Associates; *The Fax Phone Book,* published by D. A. F. Information Systems; and *International Public FAX,* published by Public FAX. The latter also contains other useful information, such as charges and hours of operation.

■ ***TELETYPEWRITER SYSTEMS*** *Teletypewriters* are electronically controlled typewriters that send and receive messages over communication lines (Figure 9.9). Teletypewriter machines can be obtained from Western Union Telegraph Company or from other vendors.

To communicate with each other, teletypewriter machines are interconnected to a network. The machines are referred to as "dumb" terminals; they usually have little or no intelligence because the intelligence for message delivery is in the networks. Organizations with a substantial volume of business often find it cost-effective to have their own (private) networks. The earliest public teletypewriter network was *Teletypewriter Exchange Service (TWX),* implemented by AT&T but later sold to Western Union. The best known and largest teletypewriter network is *Telex,* also owned by Western Union. A Western Union directory called *Infomaster* provides the names, locations, and identification of all subscribers.

Telex/TWX Combined Service (TCS) enables Telex and TWX customers to communicate with each other. Since Telex transmits on a 5-channel tape at 66 words a minute and TWX transmits on an 8-channel tape at 100 words a minute, the systems are basically incompatible. However, TCS service translates the form of signal carried on the transmission line so that these two types of message devices can communicate with each other.

In 1980 the International Telegraph and Telephone Consultative Committee adopted an international standard for a new text service called Teletex or "super Telex." Teletex is a high-speed service that can deliver messages at up to 30 times the speed of Telex. A version of the offering has been operating in West Germany and Sweden since 1981 and service between West Germany and the United States began in 1983. Believed by many to have commercial prospects far greater than either Telex or fac-

Figure 9.9
A Teletypewriter

(Reproduced with permission of AT&T)

simile, Teletex has been plagued by modem and software problems that have impeded its acceptance.

■ ***PRIVATE COMPUTER-BASED MESSAGE SYSTEMS*** Many general purpose computers can store messages and forward them between terminals having access to the CPU of the computer.

To send a message, the user establishes connection between the terminal and the computer, then keys in a user code or password, assigned to each person authorized to use the system. The user then transmits the address of the terminal for which the message is intended, followed by the message itself. This is done by keyboarding the message, editing it on the terminal screen, and sending the message to the memory of the computer, where it can be held until the person to whom it is sent is ready to receive it. Messages are sent over telephone lines, using either dial-up or direct-line service.

To notify the addressed online terminal that a message is waiting to be delivered, the computer flashes a signal on the terminal screen of the intended recipient. The person to whom the message is being sent accesses the computer, enters the proper password, and the message is transmitted onto the screen. There it can be read directly or printed as hard copy if a record of the message is desired. Messages can be sent at any time of day or night to either attended or unattended terminals. This permits low-priority messages to be sent during reduced telephone rate periods.

Some word processors are equipped with electronic components that enable them to send documents from one location to another over telephone lines or other telecommunication channels. When two communicating word processors are connected to each other, they are said to be *online* and can send and receive messages. With two-way communication capability, word processors can serve as message terminals for electronic mail service. They can be used as substitutes for conventional message equipment, such as teletypewriters.

■ ***CARRIER-BASED MESSAGE SYSTEMS*** Carrier-based message systems offer electronic mail service to the general public. Their service features vary from carrier to carrier. Some services are fully electronic end to end; others supplement electronic transmission by U.S. Postal Service or messenger delivery. Carrier-based systems include Western Union's telegrams, Mailgrams, and EasyLink INSTANT MAIL; MCI Mail; GTE Telemail; Digital's DECMail; and others.

Western Union offers two basic types of telegraph services: the *telegram,* which is a priority service, and the *overnight telegram,* which is a service with delivery the following morning. Delivery of either type of telegram is by telephone or teletypewriter unless the sender specifies messenger delivery. Physical delivery is available only in certain locations, as specified in the Western Union directory. Telegrams carry liability for mistakes or delays in delivery. Tariffs filed with the FCC set minimum delivery service standards for telegrams.

Mailgram is a fast communication service offered jointly by Western Union and the U.S. Postal Service. Mailgrams are transmitted electronically over Western Union's microwave network to the post office nearest the destination address. If filed by 7 p.m., they are delivered the next business day by regular mail carriers.

Messages may be entered directly into the Mailgram system by firms with teletypewriter service at a per message charge. Also, large volumes of messages can be entered directly into the Mailgram system from computer tapes prepared by either the customer or by Western Union.

Another Mailgram service is the Stored Mailgram, which permits the user to store frequently used information for later use. To access the information stored in the computer file, the customer dials a toll-free telephone number, identifies the account to be charged, and requests the computer to perform the desired service. The computer processes the messages and prepares them for Mailgram distribution.

A Western Union service that is available to customers who have electronic transmission equipment is *EasyLink INSTANT MAIL.* Message delivery is fully electronic if the recipient is also an EasyLink subscriber; otherwise, delivery is by either U.S. Postal Service or courier.

Another service for users with their own electronic machines (computers, communicating word processors, or Telex machines) is MCI Mail. Users of this service must keyboard the message themselves. Like EasyLink INSTANT MAIL, message delivery is fully electronic if the recipient is also an MCI subscriber; otherwise, delivery is by U.S. Postal Service or courier.

While these systems served early users well, they had at least one major drawback — the various systems had no interconnectivity. This problem was addressed by the CCITT, who formulated and subsequently adopted X.400 as its Message Handling System (MHS) standard.

Today, a new generation of electronic mail products are appearing in the marketplace. One of the first new-generation products is Digital's MAILbus transport system. This system, which carries messages between different electronic mail systems, consists of a central directory, a network management system, Digital's Message Router transport software, and a series of *gateways* (connections between two networks that use different protocols) among different popular E-mail systems, including Digital's ALL-IN-ONE, VAXMail, and ULTRIXMail; IBM's PROFS and DISOSS; MCIMail; and other systems that support the X.400 standard.

DATA SERVICES

Two basic services provided by common carriers enable data to be transported from one location to another electronically. They are digital-to-analog signal conversion and provision of digital facilities.

DIGITAL-TO-ANALOG SIGNAL CONVERSION

Conversion of digital signals to analog signals enables data to be carried over existing analog transmission facilities. The conversion is accomplished by modems at both the originating and receiving ends. Modems can be obtained from a number of suppliers and are available in a wide variety of transmission speeds. By using modems for signal conversion, customers can use either private lines or the public telephone network (including WATS and FX lines) for data communications.

DIGITAL DATA SYSTEM (DDS)

A digital data system uses interconnected digital transmission facilities that form a synchronous network for data communications. Since signaling is digital, no modems are required. Digital data systems are provided on a private-line basis between large metropolitan areas. They are available from many carriers.

IMAGE SERVICES

Image services communicate exact images such as pictures, blueprints, graphic displays, objects on picturephones, and freeze-frame television across distances. Our classification of telecommunication services as voice, message, data, and image is not absolute; some overlapping exists. Some written message services could also be classified as image because they produce both a written message and an image. Facsimile services and CRT displays of data fit this description.

Video services can be classified into five main categories: commercial television, cable television, satellite television, freeze-frame television, and conference-circuit television.

COMMERCIAL TELEVISION

Commercial television is the television that traditionally has come to us in our homes. It originates at a central location and is distributed to television broadcasting stations via a telecommunication network. The transmission facilities required to carry one commercial television program are equivalent to 600 voice-grade telephone lines. Television stations broadcast their signals through the air from their radio-like antennas. The signals are received on antennas and images are produced on television sets.

CABLE TELEVISION

Cable television signals are carried over coaxial cable distribution systems. The television signals are transported from a central source directly to receiving television sets. Cable television is a subscription service that requires a special line connecting the subscriber to the distributing point. There are many more channels available on cable television than on commercial television.

SATELLITE TELEVISION

Satellite television provides a picture signal that is transported directly from its source to an orbiting satellite. The satellite reflects the signal back to dish antennas connected to conventional television sets. Satellite transmission provides a greater number of channels and a wider variety of programming than either commercial television or cable television. Since the cost of a dish antenna is substantial, satellite television is used primarily by commercial establishments.

FREEZE-FRAME TELEVISION

Freeze-frame television is a technique for sending pictures over voice-grade telephone lines from one location to another. Pictures do not appear continuously. They are held for a period of time until the next picture is transmitted. Freeze-frame pictures are used as an adjunct to teleconferencing services.

■ ***VIDEOTELEPHONE SERVICE*** *Videotelephone service* adds the dynamic dimensions of visual connection to an ordinary telephone call. The service uses freeze-frame television technology to capture still, black-and-white "snapshot" images and send them over standard unconditioned telephone lines. No special equipment, installation, or communication circuits are required. The integrated monitor displays the caller's "live" image on the right for composing and transmission. The called party's image appears on the left as it is received (Figure 9.10).

Figure 9.10
Visual Telephone

(Reproduced with permission of Mitsubishi)

VIDEO TELECONFERENCING (CONFERENCE-CIRCUIT TELEVISION)

Video teleconferencing service enables a group of people in one room to see and hear people at a different location on a two-way basis. Camera switching is activated automatically by the speaker's voice. Some systems enable users to show slides, project graphics, produce hard copies, and record meetings on videocassette for later analysis. The full-motion videoconference most closely approximates the face-to-face conference.

Fixed locations are needed for the full-motion videoconferencing system since video transmission requires greater bandwidth capabilities than can be provided by the standard telephone hookup. Because videoconference transmission requires high-quality, wideband facilities, the costs are substantial. These costs must be weighed against the costs of time and travel required to bring both groups together in one location.

Telephone companies in the U.S. and abroad began offering commercial videoconferencing services in the 1970s. Technological improvements and competitive markets have since heightened in videoconferencing. Today, two basic options are available to businesses: the *public conference studio* and the *private conference studio.*

■ ***PUBLIC CONFERENCE STUDIOS*** Businesses not owning a videoconference room can use the facilities of public studios. AT&T has equipped public meeting rooms in many larger cities. Customers reserve time at each location and pay charges based on time and distance.

Major hotel chains such as Holiday Inn, Hilton, and Marriott have established public video meeting rooms. Hotel conference facilities usually feature one-way video reception with two-way audio interaction. Businesses that do not need in-house videoconferencing facilities but that require videoconferencing several times a year find these facilities particularly attractive. They have been proven to be cost-effective and time-efficient.

A number of companies specialize in videoconferencing and offer a wide range of services from consultation to "turnkey" (ready to go at the turn of a key) operations. Some public television stations have also entered the commercial videoconferencing market.

■ ***PRIVATE CONFERENCE STUDIOS*** For businesses wishing to set up their own conference studios, options are available for transmission services, terminal equipment, and conference rooms. The Picturephone Meeting Service offered by AT&T for public rooms can also be installed for private users on their own premises. Private users can connect with a public location or another private location. In addition to the capital outlay for room and equipment, transmission costs must also be considered.

SUMMARY

Organizations are constantly seeking ways to control costs. One way to do this is to get the user to do part of the job. The communications industry has incorporated new technology into their products, enabling the customer to perform many functions previously done by telephone company personnel. These functions range from installation and maintenance to direct-dialing of long distance calls. Each of these self-performed services results in costs savings for both the user and the telephone company.

Each of the four subsets of telecommunications — voice, data, message, and image — has its own complement of service offerings.

Voice services fall into four main categories — local exchange services, long-distance services, specialized telephone services, and telephone support services. Local exchange service is public telephone service to points within the local service area for a telephone. It typically includes customers served by other nearby central offices. Extended area service, which permits a subscriber to make calls to a designated area beyond the local exchange area without paying toll rates, is available in many areas.

Long distance services (also called inter-LATA services) include any calls that cross a LATA boundary. They may be obtained on either a direct-dial or operator-assisted basis. Business users with large volume may find it advantageous to use WATS, private-line, or foreign exchange service.

Specialized telephone services are designed to be used by persons away from their home or place of business. The principal types of specialized telephone services are coin telephones and mobile telephones.

Cellular radio is an advanced form of mobile telephone service. The basic principle of cellular radio is the division of one mobile telephone service area into a number of smaller areas called cells. Each cell contains a low-powered radio communicating system that is connected to the local telephone network. The individual systems replace the single high-powered system that formerly served an entire area and permit one radio frequency to be reused many times. A vehicle located in a cell area is connected to the telephone network through this low-powered system. When the vehicle leaves the cell area, radio contact is automatically transferred to the radio system of the adjoining cell as the vehicle enters. The chief advantages of cellular radio are improved transmission and the ability to serve more customers.

Another form of communication service for people on the move is paging service. The principal difference between mobile telephone service and radio paging service is that mobile telephone service provides two-way communication while radio paging is limited to one-way service.

Telephone companies also provide a number of support services to enhance the utility of their service offerings. These services include business office services, community services (direct contact with police, fire, and other emergency services), telephone directories, directory assistance, and intercept service.

Modern telephone equipment is also capable of providing a variety of enhanced features. These include custom calling (three-way calling, speed calling, call forwarding, and call waiting), automatic call back, line privacy, lockout, automatic route selection (least-cost routing), trunk prioritization, automatic identification of outward dialed calls (AIOD), station message detail recording (SMDR), and service restrictions.

Voice mail enables the recipients to hear their messages. Once these messages are digitized, they can be routed and transformed in many ways.

Electronic mail is the delivery of mail, at least in part, by electronic means. Its principal advantage is speed. The major categories of electronic mail systems are facsimile, teletypewriter, communicating word processors, carrier-based systems, and private, computer-based message systems.

There are two basic data transmission services provided by common carriers: digital-to-analog signal conversion, which uses modems so that data can move over analog facilities; and digital facilities, which are interconnected to form a synchronous network for data communications.

Image services communicate exact images (such as pictures, blueprints, graphic displays, and freeze-frame television) across distances. The principal types of video services are commercial television, cable television, satellite television, freeze-frame television, videotelephone and conference-circuit television (video teleconferencing).

Teleconferencing, or the use of telecommunications to conduct meetings, has evolved into an important business tool as technology and competitive pricing have increased the available options. Audio teleconferencing enables people at a number of geographically distant sites to carry on a discussion among themselves. They can hear, but not see, those with whom they are conversing. Audio teleconferencing differs from a conference telephone call in that far more people can be involved. Video teleconferencing, the most advanced — and most expensive — service, enables a group of people in one room to both see and hear people at a distant location on a two-way basis. Video teleconferencing most nearly approximates the face-to-face meeting.

REVIEW QUESTIONS

1. What is a telephone exchange area?
2. Describe extended area service. When would it be advantageous to subscribe to this service?
3. Name two types of savings effected by the use of direct-distance dialing.
4. What is the advantage of wide area telecommunications service (WATS)?
5. How does cellular mobile telephone service differ from conventional mobile telephone service?
6. What is the principal difference between mobile telephone service and radio paging service?
7. What is 800 telephone service?
8. What is the purpose of telephone support services? Name several types of telephone support services.
9. Describe two ways telephone custom-calling features can be provided.
10. What is the importance of PBX remote testing capability?
11. What is electronic mail? What is its chief advantage?
12. Describe the digital data system (DDS).
13. What is the difference between audio teleconferencing and video teleconferencing?

VOCABULARY

POTS

exchange

local service area

flat rate service

zone calls

message units

measured rate service

individual-line service

party-line telephone service

extended area service (EAS)

Wide Area Telecommunication Service (WATS)

800 service

foreign exchange service (FX)

coin telephones

customer-owned coin-operated telephones (COCOT)

cellular radio

mobile telephone service
paging service
radio pager
911 service
directory assistance
intercept service
automatic call back
line privacy
lockout
automatic route selection
identified ringing
call pick-up
audio teleconferencing
Automatic Identification of Outward Dialing (AIOD)
Station Message Detail Recording (SMDR)
class-of-service restriction
originating restriction
electronic mail
facsimile (FAX)
Teletypewriter Exchange Service (TWX)
Telex
computer-based message systems
carrier-based message systems
communicating word processors
gateways
voice mail
freeze-frame television
videotelephone
audio teleconferencing
video teleconferencing

ENDNOTES

1 Tom Andrew, "Phones on Wheels," *Michigan Bell Tie Lines* (February 1989), 7.

2 Michigan Bell Telephone Company, "Phone Lines Become 'Post Office' of the Future," *Tie Lines* (December 1988), 2.

CHAPTER

10

PRINCIPLES OF TRAFFIC ENGINEERING

CHAPTER OBJECTIVES

After completing this chapter, the reader should be able:

- *To discuss traffic as it relates to telecommunications.*
- *To distinguish between busy-hour traffic and peak traffic.*
- *To define traffic engineering and explain its purpose.*
- *To define traffic studies and explain their purpose.*
- *To describe how future telephone usage is predicted.*
- *To define grade of service and discuss its role in the provision of telephone service.*
- *To describe the two basic units of telephone usage measurement and explain their relationship to each other.*
- *To discuss traffic capacity tables and explain how they are used.*

In our mobile society the word *traffic* brings to mind automobiles and other channels of transportation. The term *traffic* is not limited to transportation, however. As used in communications, traffic denotes the flow of messages through a communication system. We are constantly reminded of transportation traffic because it is highly visible, especially when a traffic jam delays arrival at our destination. Because telephone traffic is invisible, we are not aware of it unless our calls are blocked, thereby inconveniencing us.

Both highway traffic and telephone traffic occasionally experience traffic jams that slow down or completely block traffic flow. From the user's point of view, the ideal situation would be never to encounter a traffic jam. However, this would require designing the highways or communication facilities so large that they would always meet peak demand. Obviously, this would not be feasible. From a practical point of view, user demand must be balanced with economic considerations.

The football fan en route to a big game might be understanding—although impatient—when experiencing a traffic jam caused by cars converging on the stadium. The same driver would not be as understanding of a similar delay twice a day, five days a week when driving to and from work. To provide good service and still be cost effective, highways must be engineered to accommodate busy-hour traffic, not occasional peak traffic.

Similarly, telephone networks experience *peak traffic* overloads (traffic jams) on certain holidays and times of unusual occurrences, such as storms, earthquakes, and other disasters. Telephone networks are engineered using the same basic principles as highway design. They must be able to accommodate normal busy-hour traffic rather than occasional peak loads.

TRAFFIC ENGINEERING PROCEDURES

Traffic engineering is the science of designing facilities to meet user requirements. The objective of telecommunications traffic engineering is to specify the quantities and arrangements of telephone trunks and switching equipment required to handle user traffic.

TRAFFIC STUDIES

To design a major telecommunication system to user needs, the needs must first be determined. This is done by conducting a *traffic study,* which is simply a count of calls classified by types, such as incoming calls, internal calls, local calls, WATS-line calls, data calls, and private-line calls. In early telephone systems, call counts were taken manually. Modern telephone systems incorporate call recording devices that produce call counts automatically.

Call data is usually summarized by half-hour internals. This data is to identify the *busy hour,* the two consecutive half-hour periods in which the largest number of calls occurs. The process is repeated for five days and the average computed. The resulting number is the *average busy-hour traffic count,* a very important factor in determining user requirements, since equipment and trunks are provided to handle busy-hour traffic. To be truly representative, it is important that the week chosen for the traffic study be a typical five-day week that does not include a holiday.

XYZ Company
Traffic Study

Period May 9–13, 1989					**Incoming and Outgoing Calls**		
	Mon	**Tue**	**Wed**	**Thur**	**Fri**	**5 Day Total**	**Daily Average**
9-9:30	70	68	63	71	73	345	69
9:30–10	131	126	122	134	137	650	130
10–10:30	139	129	126	143	153	690	138
10:30–11*	168	159	148	172	178	825	165**
11–11:30*	187	174	175	185	199	920	184**
11:30–12	162	156	152	163	172	805	161
12–12:30	136	132	137	140	145	690	138
12:30–1	94	90	87	93	96	460	92
1–1:30	132	122	136	135	145	670	134
1:30–2	148	136	137	138	151	710	142
2–2:30	147	142	140	146	155	730	146
2:30–3	141	138	135	146	150	710	142
3–3:30	165	146	149	158	172	790	158
3:30–4	168	165	152	165	175	825	165
4–4:30	142	136	129	150	153	710	142
4:30–5	129	125	138	141	148	680	136
Total Day	2,259	2,144	2,126	2,280	2,401	11,210	2,242

* Busy hour
** Average busy-hour traffic count

Figure 10.1
Sample Traffic Study of Incoming and Outgoing Calls

Figure 10.1 illustrates a traffic study summarization. The busy hour in this study is from 10:30 a.m. to 11:30 a.m. The average busy-hour count is 349 (165 + 184). Similar studies are prepared for other types of traffic, such as internal calling, WATS calling, OCC calling, and DDD calling. Figure 10.2 shows the hourly distribution of daily calls (from Figure 10.1). This graph demonstrates that approximately 15 percent of the total day's calls took place during the busy hour (10:30 a.m. to 11:30 a.m.). Figure 10.3 shows a record of daily average calls by months over several years — data used to predict daily average calls by months. This record is also used to predict daily average calls for future months. Similar records are prepared for other types of traffic, such as internal calling, WATS calling, OCC calling, and DDD calling.

The purpose of conducting a traffic study is to obtain data for use in predicting future requirements. As a rule, fairly accurate predictions can be made by taking a five-day count each month for several years. However, the more data available, the more accurate the prediction will be. Some businesses are also subject to seasonal fluctuations. Thus, it is important that each month of the year be represented. Monthly data listed for several years will identify *busy-season* traffic, just as half-hour data identifies busy-hour traffic.

Figure 10.2
Hourly Distribution of Daily Calls

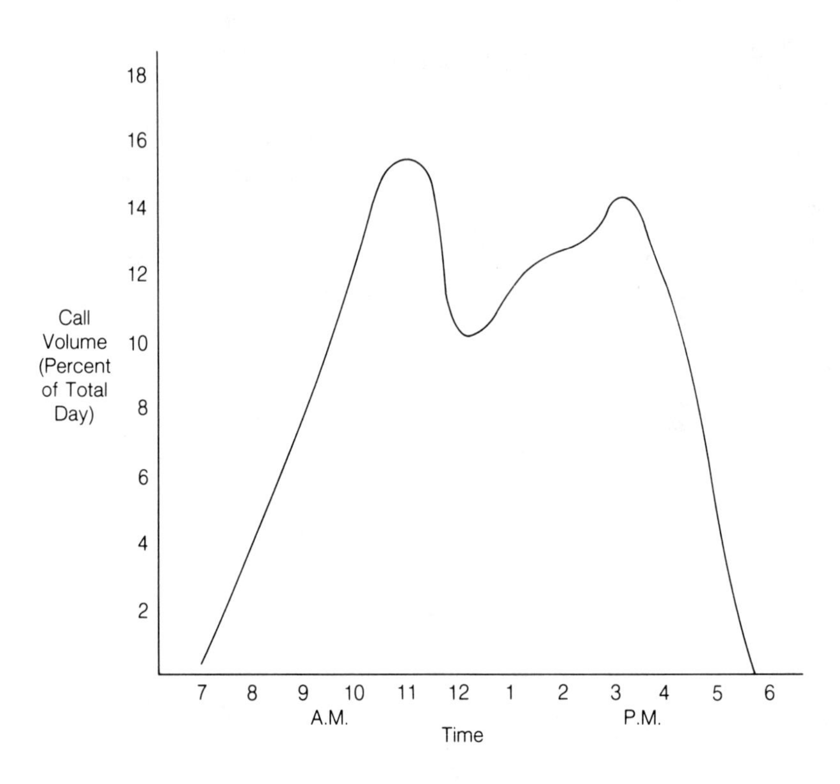

Figure 10.3
A Typical Record of Monthly Calls

XYZ Company
Record of Monthly Calls

Incoming and Outgoing Calls (Daily Average*)

	1986	**1987**	**1988**	**1989**	**1990**
January	1656	1754	1885	2210	
February	1675	1762	1912	2218	
March	1682	1776	1948	2232	
April	1690	1780	1976	2268	
May	1706	1796	1986	2242	
June	1710	1810	2022		
July	1650	1725	1850		
August	1625	1710	1810		
September	1750	1825	2129		
October	1726	1835	2142		
November	1781	1850	2168		
December	1795	1862	2185		

* 5 Day Average From Monthly Study

PREDICTING FUTURE TELEPHONE USAGE

To determine the amount of equipment required to provide good service, the traffic engineer must not only have data on the number of calls but also on the duration of the calls. This data is obtained by periodically conducting a *holding time study*. *Holding time* is the conversation time plus the time the equipment is engaged in establishing the connection. In other words, holding time is the total time that the telephone receiver is off the hook. The number of telephone calls multiplied by the average number of seconds per call (holding time) equals the number of seconds that the equipment is in use. For example, if the system handled 363 calls in an hour and the average call is 300 seconds, the equipment would be in use 108,900 seconds (363 × 300). This number is divided by 100 to determine the number of hundred call seconds (108,900 ÷ 100 = 1,089 CCS). *One hundred call seconds* is known as *1 CCS*. (In the abbreviation, the first C is the roman numeral for 100, the second C stands for calls, and the S stands for seconds.) Additionally, any combination of calls and seconds totaling 100 would constitute 1 CCS; that is, two 50-second calls, five 20-second calls, or ten 10-second calls. Telephone usage is measured in terms of CCS.

The traffic engineer is charged with the responsibility of providing the quantities and configuration of equipment that will be required to handle a predicted volume of telephone usage. The data obtained from traffic studies is used to predict the future volumes of traffic usage. Since traffic loads tend to grow from year to year at varying rates, the traffic engineer studies the traffic data and selects a rate of growth that reflects past performance. The future call volume can then be predicted by multiplying present call volume by the anticipated growth rate and adding this predicted increase to the present call volume. Figure 10.4 shows a graph of monthly incoming and outgoing telephone calls used to predict future call levels.

Telephone systems are engineered to provide sufficient equipment to handle busy-hour traffic. Therefore, the percentage of the total day's traffic that occurs during the busy hour must be calculated. The number of busy-hour calls can then be predicted by multiplying the predicted future call volume by the percentage of calls occurring in the busy hour. The following example illustrates these calculations:

Present number of calls per day: 2,200 calls
Predicted growth rate (based on history): 10%
Predicted calls (2,200 × 10%) + 2,200 = 2,420 calls
Percent of calls in busy hour: 15%
Predicted busy-hour calls: 15% × 2,420 = 363 calls

In older telephone systems, CCS were computed manually by measuring holding time with stopwatch observations and multiplying the holding time figure by the number of calls. The resultant figure was changed to CCS by dividing by 100. To illustrate, if we counted 100 calls averaging 300

Figure 10.4
Average Daily Telephone Calls by Month

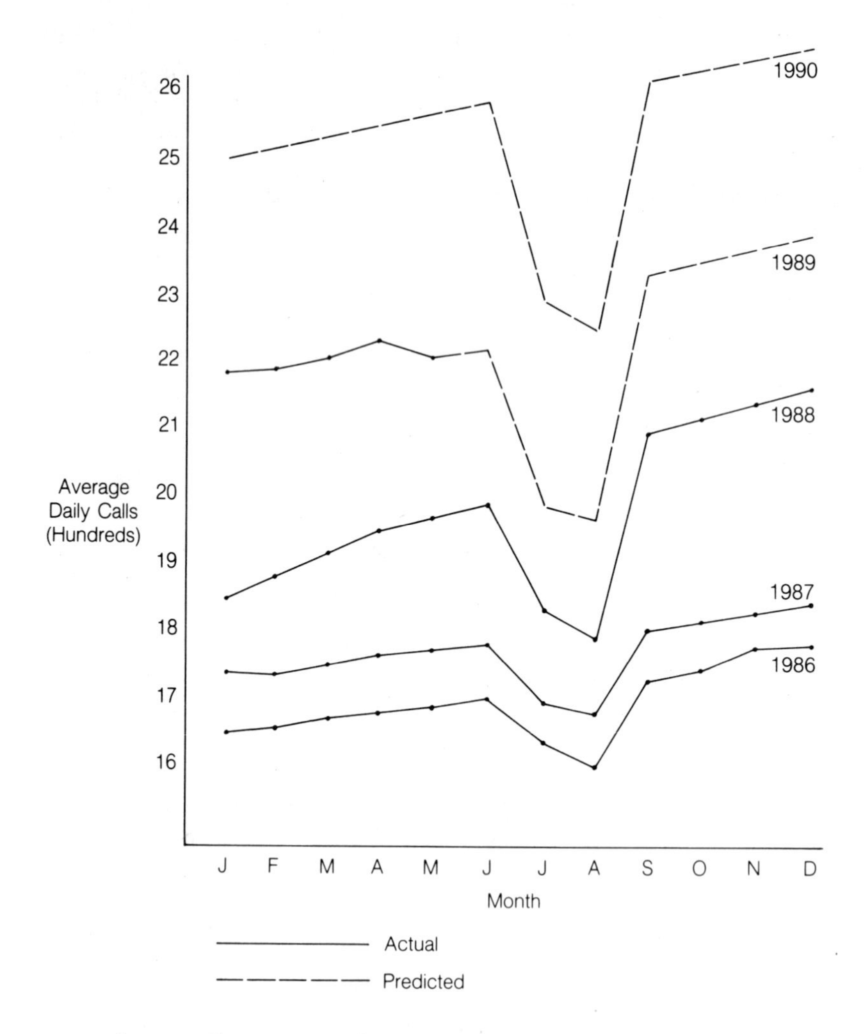

seconds per call, 300 CCS of usage would have been generated. (100 × 300 = 30,000; 30,000 ÷ 100 = 300.) Modern electronic switching systems provide direct readings in CCS, eliminating the need for this computation.

The CCS is one measure of usage. Another measure of usage is the Erlang. *One Erlang* equals *36 CCS*. Since there are 3,600 seconds in an hour (60 × 60), 36 CCS (3,600 ÷ 100) equals one hour of usage. Traffic engineers use both CCS and Erlangs in determining the quantities of equipment or trunks required. Figure 10.5 shows a sample conversion table between the two units of measurement.

The number of predicted busy-hour calls can be converted to CCS by multiplying the predicted number of busy-hour calls by the average hold-

Sample Conversion Table

Relationship between CCS and Erlangs

Call Seconds	CCS (Seconds/100)	Erlangs (CCS/36)
300	3	.083
1,200	12	.333
2,400	24	.666
3,600	36	1.000
10,000	100	2.777
18,000	180	5.000
36,000	360	10.000
72,000	720	20.000

Figure 10.5
Conversion from CCS and Erlangs

ing time. Thus, if in our example the average holding time (based on history) is 300 seconds per call, this figure is multiplied by 363 (the number of calls) and divided by 100, for a total of 1,089 CCS. To convert CCS to Erlangs, we divide by 36; i.e., 1,089 ÷ 36 = 30.25 Erlangs.

PREDICTING TELEPHONE GROWTH

The previous prediction of telephone usage is based on call volume data only. Another way to predict future usage is by using projections of anticipated telephone growth. The traffic engineer usually uses both methods, with one serving as a cross-check on the other.

In making projections, the traffic engineer needs to know the number of telephones to be served by the system, both at the time of system cutover and at the time the system reaches its maximum size. This information can be obtained from the telecommunications manager.

The following example uses the number of telephones to be served by the system in future years to calculate the usage that will be generated by these telephones when the system reaches its maximum capacity. In this example it is assumed that the busy-hour calling rate and the holding time per call will not change over the life of the system. In actual practice, estimates of these two items for future years would be required, although calling rates and holding times change very little, if any, from year to year.

Present number of telephones	200	
Growth in number of telephones per year	20	
Telephones in service at the end of one year	220	
Telephones in service at the end of five years	300	
Daily calls—present (from traffic study)	2,200	(Figure 10.1)
Percent calls in busy hour	15%	
Busy-hour calls	330	(2,200 × 15%)
Busy-hour calls per telephone	1.65	(330 ÷ 200)
Average length of calls	300 seconds	

Usage is equal to the number of telephones times the number of calls per busy hour per telephone times the average length of telephone calls.

Usage at Cutover

200 × 1.65 × 300 = 99,000 call seconds
99,000 ÷ 100 = 990 CCS
99,000 ÷ 3,600 = 27.5 Erlangs

Usage One Year from Now

220 × 1.65 × 300 = 108,900 call seconds
108,900 ÷ 100 = 1,089 CCS
108,900 ÷ 3,600 = 30.25 Erlangs

Usage Five Years from Now

300 × 1.65 × 300 = 148,500 call seconds
148,500 ÷ 100 = 1,485 CCS
148,500 ÷ 3,600 = 41.25 Erlangs

TELETRAFFIC THEORY

The mathematical description of message flow in a communication network is called *teletraffic theory,* a branch of applied probability. The traffic engineer uses teletraffic theory and engineering principles to design communication systems.

GRADE OF SERVICE

Every communication network is organized on the principle of sharing common equipment. When shared equipment is provided in insufficient quantity, there will be times when users will have to wait. Telephone systems could be designed so that users would never have to wait; however, such overprovision of equipment would be extremely costly and wasteful. Clearly, there must be a balance between the cost of providing service and the quality or grade of service provided.

A service delay occurs when a telephone number is dialed and a busy signal is received. The busy signal may occur for either of two reasons. First, the telephone line may be busy because a person is using it. Second, the dialed number is not busy but there are no available circuit facilities to reach the desired telephone. The second circumstance is referred to as a *network busy condition.* The term *blocking* describes a call that cannot be completed due to a network busy condition. When a call is blocked, the telephone user hears a faster busy signal than that received when the desired telephone line is busy. The standard signal to indicate a busy telephone line is a tone interrupted 60 times a minute (60 IPM). The standard signal to indicate a network busy condition is a tone interrupted 120 times a minute (120 IPM). The user is primarily concerned with getting the telephone call through and probably does not differentiate between these types of busy signals. Similarly, the term *blocked* is used, slightly incorrectly, by the public to denote a call that cannot be completed because of either a line busy or network busy condition.

Grade of service is the probability of a call being blocked expressed as a percentage, such as P.05. A P.05 grade of service means that there is a 5 percent chance of a call being blocked during the busy hour; in other words, 5 calls out of 100 will probably be unable to be completed because of a network busy condition.

TRAFFIC CAPACITY TABLES

The traffic engineer uses *traffic capacity tables* to determine the quantities of trunks or equipment needed for a telecommunications system. The two most frequently used traffic capacity tables are the *Poisson table* and the *Erlang B table*. Both tables are based on mathematical probability theory and are named after the mathematicians who developed them.

Simeon Poisson was a nineteenth-century French mathematician who developed a mathematical model to predict the outcome of events. His theory was adapted for use in telephony in the early 1920s by a Bell Laboratories scientist named E. C. Molina. The Molina formula is used to predict the number of trunks required to handle various volumes of usage. Poisson theory and the Molina formula are based on the assumption that the sources of telephone traffic are infinite and that all unsuccessful call attempts (blocked calls) are retried within a relatively short time interval.

A. K. Erlang, a Danish engineer and mathematician, is often referred to as the "father of teletraffic theory." The fundamental unit of traffic load, the Erlang, bears his name. In the second decade of this century, he developed a method of analysis used to predict the quantities of equipment required to handle a given volume of telephone traffic. The Erlang B traffic capacity tables are based on the assumption that the sources of telephone traffic are infinite but that all unsuccessful call attempts are abandoned.

Figure 10.6 shows a partial Poisson capacity table; Figure 10.7 illustrates a partial Erlang B capacity table. The tables have usage (CCS or Erlangs) as one dimension and grade of service as the other. Usage quantities are aligned under each of the desired grades of service. To use the tables, locate the predicted usage load in the appropriate "grade of service" column and read the number that appears on the same line in the "trunks" column at the extreme left. For example, to use the Poisson table (Figure 10.6) for 105 CCS of predicted usage to provide P.01 grade of service, find the quantity 105 in the P.01 column and read the corresponding number in the "trunks" column. In this example, the number of trunks required would be 8. The Erlang B table is used in the same way, except that usage must be expressed in Erlangs.

TRAFFIC QUEUING

Traffic engineering is based on assumptions and probabilities, not certainties. The Poisson theory is based on the assumption that when a call is blocked, the caller will try again within a relatively short time. Erlang B

Figure 10.6
Sample Poisson Capacity Table

Poisson Capacity Tables
Hundred Call Seconds at
Various Grade Levels

Trunks	Grade of Service At Indicated CCS Load			
	P.01	**P.02**	**P.05**	**P.10**
2	5.4	7.9	12.9	19.1
4	29.6	36.7	49.1	63.0
6	64.4	76.0	94.1	113.0
8	105.0	119.0	143.0	168.0
10	148.0	166.0	195.0	224.0
15	269.0	293.0	333.0	370.0
20	399.0	429.0	477.0	523.0
25	535.0	571.0	626.0	670.0
30	675.0	715.0	773.0	636.0
40	964.0	1012.0	1038.0	1157.0
50	1261.0	1317.0	1403.0	1482.0

To use this table find the usage in CCS in the appropriate grade of service column and read the number that appears on the same line in the trunks column on the extreme left.

Figure 10.7
Sample Erlang B Capacity Table

Erlang B Capacity Tables
Erlangs of Use at
Various Grade of Service Levels

Trunks	Grade of Service At Indicated Erlang Load			
	P.01	**P.02**	**P.05**	**P.10**
2	.153	.224	.382	.6
4	.870	1.093	1.525	2.0
6	1.909	2.276	2.961	3.8
8	3.128	3.627	4.543	5.6
10	4.462	5.084	6.216	7.5
15	8.108	9.010	10.63	12.5
20	12.03	13.18	15.25	17.60
25	16.13	17.51	19.99	22.80
30	20.34	21.93	24.80	28.10
40	29.01	31.00	34.60	38.80
50	37.90	40.25	44.53	49.60

To use this table find the usage in Erlangs in the appropriate grade of service column and read the number that appears on the same line in the trunks column on the extreme left.

theory assumes that the caller will not call again; that is, that the call is abandoned. There is no way to actually determine whether a caller will try again and, if so, when the retrial attempt will be made. There are so many variables that enter into the situation, such as the urgency of the call, time of day, and priority of other activities, that retrial attempts are impossible to predict accurately.

Telephone systems for some organizations are engineered so that many arriving calls experience some delay. These organizations include airlines, hotel reservation centers, service bureaus, and government agencies. Their systems are engineered on the assumption that the caller will be willing to wait, particularly if there is some assurance that calls are being processed in the order that they were received in the system. In these systems the cost of providing a better grade of service (less waiting time) is balanced against the likelihood of losing the calls by the inconvenience of waiting. Because of the nature of the calls to these organizations, it is assumed that callers will be almost certain to wait rather than abandon their calls.

In this system when calls are blocked, they are said to be *queued,* that is, waiting in the sequence in which they arrived in the telephone system. A *queue* is simply a waiting line. Customers are served when they arrive at the head of the line; newcomers enter the queue taking their place at the end of the line. Banks have refined the queuing procedure by combining many lines served by many tellers into a single line with access to the next available teller. This eliminates the possibility of the customer's selecting the wrong line and having to wait unduly long because of one customer's lengthy transaction. Supermarkets, on the other hand, generally require their customers to queue at each checkout station. Customers choose the queue in which they wish to wait, taking their chances on which line will provide the fastest service.

Most telephone systems do not have the capability to queue within the system. However, call distributing systems, such as those used by reservation centers and government agencies, and some least-cost routing systems do have queuing capabilities. In addition, distributed data processing systems are generally designed to provide queuing. Systems capable of queuing are generally engineered using a slightly different type of traffic capacity table, which is known as Erlang C or Crommelin tables. These tables add another dimension to the design process since they take into account the number of calls to be held in queue, as well as the grade of service and the predicted traffic load.

Telecommunication facilities are designed to transmit messages in either voice or data form. The principal differences between voice and data traffic are as follows:

1. Data traffic originates from a finite, or limited, source; voice traffic originates from an infinite source.
2. Data traffic has a longer holding time.
3. Data callers will accept a poorer grade of service. Computers do not mind waiting; humans do and become impatient.
4. Data traffic is capable of being stored for later transmission and is adaptable to queuing techniques. Voice traffic includes two-way interaction; thus it is not generally adaptable to queuing techniques.

Telecommunication facilities are engineered to handle a predicted amount of circuit usage, regardless of whether the traffic is in voice or data form. Thus, the engineering principles are the same for both types of transmission.

THE DECISION-MAKING PROCESS

Top management, the telecommunications manager, and the traffic engineer all participate in decisions concerning the telephone system. The telecommunications manager provides factual information to top management to assist them in making decisions regarding service and expenditure levels. The traffic engineer makes decisions regarding the quantities and types of equipment necessary to provide the desired grade of service.

COST VERSUS GRADE OF SERVICE

Telephone facilities can be engineered to provide any desired grade of service. The lower the probability figure, the better the service will be and the more the system will cost. Management determines the organization's service objective by balancing telephone system costs against the costs of the inconvenience to users when their calls are blocked. The traffic engineer translates management's service objective into a specific grade of service quantification.

In the United States, operating telephone companies (OTCs) provide public telephone service engineered to P.01 grade of service. Private telephone systems are often engineered to provide a slightly poorer grade of service, generally in the P.01 – P.05 range. Telephone users have become accustomed to the high-quality service provided by the United States OTCs and tend to use it as a frame of reference in judging all telephone service. Since users are particularly sensitive to changes in service rather than to absolute levels of service, any noticeable difference in service quality usually draws sharp criticism.

TRAFFIC ENGINEERING DECISIONS

The effective performance of a telecommunication system depends in large part upon the decisions made by the traffic engineer in the planning stages. If the system has been engineered to provide inadequate quantities of trunks and equipment, poor service is inevitable. If the system has been engineered for overprovision, the service will be good but excessively expensive.

■ ***CHOICE OF TRAFFIC CAPACITY TABLE*** After the traffic engineer determines the volumes of usage that the system will be required to handle, a decision can be made concerning the quantities of equipment and trunks to be provided. The traffic capacity tables (Poisson, Erlang B, and Erlang C) provide a guide to assist the engineer in making these decisions. The choice of table depends upon the characteristics of the traffic to be served

	Poisson	Erlang B	Erlang C
Usage Measurement	CCS	Erlangs	Erlangs
Traffic Source	Infinite	Infinite	Infinite/Finite
Blocked Calls	Prompt Retrial	No Retrial	Indefinite Wait
Type of Traffic	Random	Random	Random
Principal Users	OTCs	Private Systems and OTCs	Distributed Systems

Figure 10.8
Characteristics of the Leading Capacity Tables

by the system. Figure 10.8 summarizes the characteristics of the leading capacity tables.

Poisson, Erlang B, and Erlang C capacity tables were designed for public telephone systems with their corresponding large numbers of users. Therefore, none of the tables is perfectly suited for smaller private systems.

Communication theorists have developed other capacity tables designed specifically for use in private telephone systems. Dr. James Jewett of Telco Research Corporation, a consulting organization, developed tables called *Equivalent Queue Extended Erlang B (EQEEB)* to be used in the design of trunks that automatically route blocked calls to alternate routes.

The Center for Communications Management, Inc. (CCMI), a subsidiary of McGraw-Hill, has developed tables using a proprietary formula. These tables, known as *CCMI Pragmatist,* are based on the assumption that 70 percent of blocked calls will be retried during the busy-hour traffic period.

All traffic capacity tables are based upon statistical probability, and their use is based upon certain assumptions. There are two fundamental factors that must be kept in mind when an engineer uses any traffic capacity table. The first is the mathematical truth that the combination of a precise number and less precise number or approximate number can only be as accurate as the least precise number. The use of capacity tables involves at least two dimensions. One dimension is an estimate of usage, an imprecise number. Another number is grade of service, a number which contains an assumed number of retrials, thereby making it imprecise. Therefore, regardless of the degree of precision built into the traffic capacity tables, the figures resulting from their use can only be as accurate as the least precise quantities and the assumptions made.

The second factor is that the accuracy of a statistical sample varies directly with the size of the sample; the larger the sample, the more accurate the prediction will be. Since private telephone systems are designed for a relatively small number of users (compared with the public telephone system), estimates of usage tend to be less accurate.

Thus, traffic capacity tables are a guide in directing the judgmental process; they are not absolute. The quantities of equipment specified in the capacity tables should be tempered with judgment, taking into account such other considerations as the firmness of the grade-of-service decision and the extent of the busy-hour level of traffic. Before reaching a decision, it is wise to consider the quantities of equipment suggested by several appropriate traffic capacity tables. In spite of the theoretical shortcomings of traffic capacity tables, they provide a valuable guide for decision making. Their use contributes greatly to effective facilities engineering.

■ *DETERMINATION OF EQUIPMENT REQUIREMENTS* Returning to the previous example of usage for incoming and outgoing calls, consider how the traffic capacity tables are used as a guide in the decision-making process.

Since the traffic from the system under consideration is random, comes from an infinite source, and has no provision for queuing capabilities, the traffic engineer would use either Poisson or Erlang B traffic tables in determining the number of trunks required for incoming and outgoing calls. The following data summarizes the usage from the example and converts it into the number of trunks required at both P.01 and P.05 service levels using both Poisson and Erlang B capacity tables:

	Table	**Trunks Required**	
		P.01	P.05
Usage at Cutover			
990 CCS	Poisson	39	36
27.5 Erlangs	Erlang B	38	34
Usage One Year from Now			
1,089 CCS	Poisson	44	40
30.25 Erlangs	Erlang B	41	36
Usage Five Years from Now			
1,485 CCS	Poisson	57	53
41.25 Erlangs	Erlang B	54	47

In examining the differences between the numbers of trunks suggested by the Poisson and Erlang B tables, we see that although the two are very close, the Poisson estimate is slightly higher. The difference between the two tables is in handling blocked calls. The Poisson table assumes prompt retrial of blocked calls while the Erlang B table assumes no retrials. Since retrials are impossible to predict accurately, a possible decision might be to compromise halfway between the two figures.

In determining the number of trunks required, the traffic engineer would provide sufficient equipment to handle usage requirements for the life of the system (five years, in this example).

The trunks required to handle incoming and outgoing traffic are provided by the operating telephone company at a monthly rate. The engineer must provide enough equipment to terminate the ultimate number of trunks required, but the trunks do not need to be contracted for until they are needed. However, the ability to terminate the trunks must be foreseen and taken into account in the equipment provision.

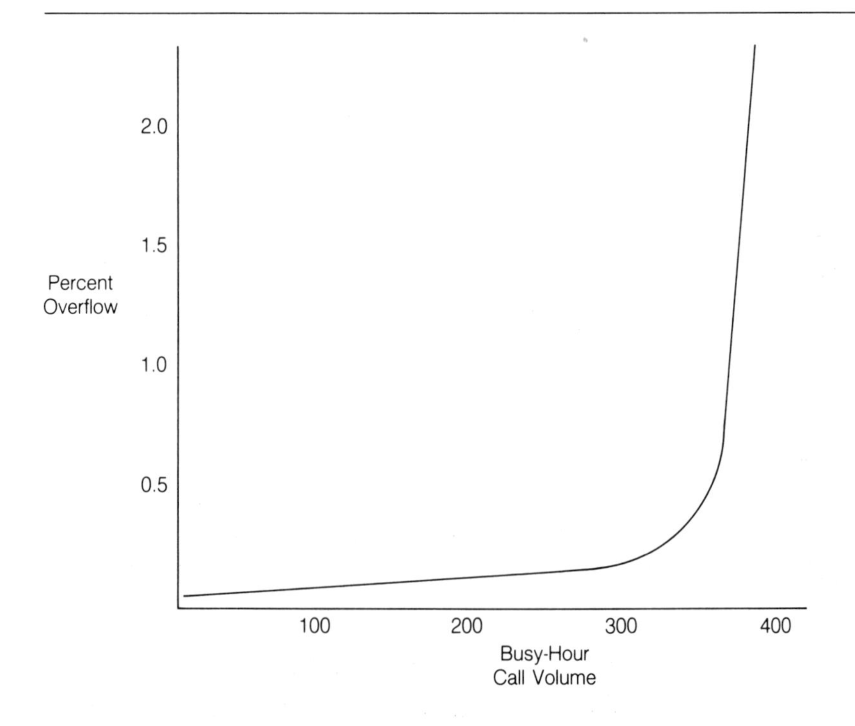

Figure 10.9
Load/Service Relationship of System Capacity

A common fault in the engineering of private systems is to underestimate future requirements, with the result that the system may not be able to meet user needs over its life expectancy. This results in poor service and unnecessarily high replacement costs. It is generally more advantageous to err on the high side of the prediction, since this merely postpones system replacement.

SYSTEM CAPACITY EVALUATION

After a new telephone system has been installed, the manager will want to know whether the system has been properly engineered to provide the desired grade of service. To do this, the manager conducts a *load/service relationship analysis,* designed to test the system.

To conduct such an analysis, data is collected on the number of calls handled by the system and the number of blockages encountered. This data is obtained by reading registers provided within the system. The data is graphed to show the load/service relationship by plotting the call volume on the X-axis and percentage of calls blocked on the Y-axis, as illustrated in Figure 10.9. In a fully loaded system, the graph shows a definite bend, indicating a critical area where the percent blockage increases exponentially. Thus, the graph depicts the number of calls that can be handled before excessive blockage occurs, indicating that system capacity has been reached. In Figure 10.9 the critical area occurs in the 350–360 call range.

If the system is new and thus underloaded, there is a probability that no blockages will be encountered. In this event, the manager can force a system overload to test the equipment capacity, causing some trunks to be inoperative (falsely busy) to obtain traffic load/service data. This technique provides a reliable check on the adequacy of the trunks and equipment.

SUMMARY

Telephone traffic is the flow of messages through a communication system. Traffic engineering is the science of designing facilities to meet user requirements.

User requirements are determined by conducting a traffic study, a count of telephone calls classified by types, such as incoming calls, internal calls, local calls, WATS-line calls, private-line calls, and any other types of calls handled by the system. In addition to the data on the number of calls, it is necessary to collect data on the average length of calls (holding time).

Present call volumes obtained in the traffic study are used to predict future call volumes. Estimates of telephone growth are also taken into account in predicting future call volumes. Present holding time data is used to predict future holding times. The product of these two quantities is described as usage (call volume $\times$ holding time $=$ usage). Usage may be expressed either in CCS or Erlangs.

A user receives a busy signal for one of two reasons: the line is busy because someone is using it, or there is no available circuit facility to reach the desired telephone. This second condition is referred to as a network busy condition. Grade of service refers to the probability of a call being blocked because of a network busy condition. It is expressed as a percentage. Thus, P.05 grade of service indicates that there is a 5 percent probability of receiving a busy signal because of a lack of circuit facilities during the busy hour.

Telephone facilities can be engineered to provide any desired grade of service. The lower the probability, the better the service will be and the more the system will cost. The traffic engineer balances the cost of providing service with the quality or grade of service to be provided.

The traffic engineer uses traffic capacity tables to determine the quantities of equipment required for the system. The two most frequently used traffic capacity tables are the Poisson table and the Erlang B table. Both tables are based on mathematical probability theory and contain certain assumptions. The Poisson theory assumes that when a call is blocked, the caller will try again within a relatively short time. Erlang B assumes that the caller will not try again; that is, the call will be abandoned. Traffic capacity tables have grade of service as one dimension and usage estimates as the other dimension. They are used as a guide in determining the number of trunks and/or units of equipment to be provided in the new system.

A queue is a waiting line in which customers are served in the order in which they entered the line. When telephone calls are blocked, they are said to be queued when they are waiting in the sequence in which they arrived. Most telephone systems do not have the capability to queue within the system. Call distribution systems, however, do have this capability.

A new telephone system may be evaluated to determine whether it has been properly engineered to provide the desired grade of service by conducting a load/service relationship analysis. This analysis shows the grade of service the system provides at various call volume levels.

REVIEW QUESTIONS

1. What is telecommunications traffic engineering, and what is its objective?
2. How is the average busy-hour traffic count computed? What is the purpose of obtaining this data?
3. Why is it important to identify busy-season traffic as well as busy-hour traffic? What are some types of businesses that experience seasonal fluctuations?
4. Why is it necessary to have holding-time data on calls?
5. If the average busy-hour call count is 312 calls and the average holding time 300 seconds, how many CCS does this represent? How many Erlangs?
6. If Jim calls Linda and receives a busy signal because Linda is talking to Dave, has Jim received a network busy condition? Explain your answer.
7. Which of the following figures represents the best grade of service: P.01, P.03, or P.05? Explain your answer.
8. Describe the condition of queuing. Which is more adaptable to queuing techniques — voice or data? Why?
9. What is the purpose of traffic capacity tables? Upon what factors does the choice of table depend?

VOCABULARY

telephone traffic
peak traffic
traffic study
busy hour
holding time
hundred call seconds (CCS)
Erlang
teletraffic theory
network busy condition
blocked call
grade of service
traffic capacity tables
Erlang B table
Poisson table
traffic queuing
load/service relationship analysis

CHAPTER

11

PRINCIPLES OF TELECOMMUNICATIONS MANAGEMENT

CHAPTER OBJECTIVES

After completing this chapter, the reader should be able:

- *To discuss the evolution of telecommunications management.*
- *To describe the basic functions of telecommunications management.*
- *To describe the skills required of a telecommunications manager.*
- *To discuss the need for a telephone equipment inventory.*
- *To discuss the need for corporate telecommunications policy.*
- *To describe the various ways to control telephone usage.*
- *To discuss the uses of the telephone cost history.*
- *To explain the meaning of system capacity and describe how it is evaluated.*

HISTORICAL PERSPECTIVE

Telecommunications in business and industry started out as a switchboard service managed by the telephone operator. The operator could receive calls, place outgoing calls, and handle internal calls using private branch exchange equipment. As telephone usage increased, organizations sought ways to improve the service and to control costs. The result was the development of the private automatic branch exchange, which allowed internal and outgoing calls to be placed directly by the user. The operator handled only incoming calls.

The next development in telephone technology was the Centrex system, which permitted callers to dial the desired telephone number directly. Since Centrex systems completed most calls automatically, the role of the operator was virtually eliminated.

The principal tasks to be performed in the traditional telephone system were checking the monthly bill and ordering repairs and equipment changes. In most offices, an office manager, secretary, or clerk performed these tasks.

THE TELEPHONE COMPANY AS MANAGER

Since the person responsible for the ongoing operation of an organization's telephone system rarely had specialized expertise in this area, the local telephone company served both as provider of services and consultant. In other words, the telephone company "managed" the telecommunications function of most organizations.

When an organization's telecommunications needs changed, it called upon the local telephone company for assistance. The telephone company made a study of the organization's call volume (time of day, destinations called, and length of calls), evaluated the data, and made recommendations. Because the telephone company furnished this service, many organizations found it unnecessary to establish internal management of their telephone system or to employ a consultant. There were few options in equipment or services and no options in price; therefore, few decisions had to be made.

TELECOMMUNICATIONS MANAGEMENT

The role of telecommunications management is to provide good telecommunications service to an organization and its employees at the lowest possible cost. Because data communications comprises a significant portion of the traffic on the telephone system, telecommunications management will include data communications requirements in the provision of telecommunications facilities.

NEED FOR TELECOMMUNICATIONS MANAGEMENT

Telecommunications costs represent a substantial and ever-increasing portion of a company's budget. In many organizations, they are exceeded only by costs of labor and office space. Before competition, telephone costs were considered an uncontrollable overhead expense. Now, however, there are many alternatives available for reducing telecommunications costs.

Rapid advances in technology, along with deregulation in the telephone industry, have resulted in many new types of equipment and services, a multitude of new suppliers, and aggressive price competition. A business has the following options in obtaining telecommunications equipment and services:

1. Renting or leasing equipment from AT&T or from any one of the equipment subsidiaries of the operating telephone companies.
2. Buying equipment from AT&T, from any one of the equipment subsidiaries of the operating telephone companies, or from an independent interconnect vendor.
3. Leasing equipment through a leasing company that, in turn, purchases the equipment from a vendor. (This is known as a third-party lease.)
4. Obtaining dial-up long distance service from AT&T or from any one of a number of long distance telephone companies.
5. Leasing private telecommunication circuits from the local telephone company for intra-LATA service and from AT&T or any one of the long distance telephone companies for inter-LATA service.

With the advent of competition in the telecommunications marketplace, it is imperative that customers be able to evaluate the various service and equipment offerings and to match them with their own business requirements on a cost-benefit basis. Choosing from the many services and vendors available is a difficult task. To make the right decisions, management must have comprehensive, up-to-date knowledge of telecommunications equipment, services, vendors, and prices. Effectively managed, the telecommunications system can make an important contribution to the profitability of any organization.

ORGANIZATION OF THE TELECOMMUNICATIONS FUNCTION

The relative position of telecommunications in the organization has changed considerably since the advent of competition. In days when the manager's responsibilities consisted primarily of ordering service and paying the bills, telecommunications was generally assigned as a minor responsibility to one of the main company departments. Now, however, management must be able to select the best equipment, services, and vendor(s) as well as exercise ongoing system controls, all in a highly competitive environment. As a result, the telecommunications function is becoming more important and consequently more visible. Many companies are addressing the increasing importance of telecommunications by establishing separate departments to administer this function. These departments are usually staffed with professionals trained in telecommunications.

The number of persons assigned to telecommunications is determined by the amount of work to be performed. This, in turn, is governed by the size of the organization. In a small organization, the telecommunications responsibilities may be performed by a manager or supervisor who is also responsible for many other tasks. If extensive changes in the telephone system are required, or if a new system is to be implemented, the small company would probably hire a consultant.

As a general rule, the telecommunications department budget should be 3 to 6 percent of the organization's telecommunications expenditures. An organization with an annual telecommunications expense in the $700,000 range needs a full-time communications manager, while an annual telecommunications expense of $2 – $3 million would generally warrant three specialists — a manager, an auditor, and a systems analyst.

THE MANAGEMENT PROCESS

Management has been defined in many ways; there is no universally accepted definition. For the purpose of our discussion, *management* is the process of directing the efforts of organization members toward stated organizational goals. The basic activities a manager performs are generally referred to as the "functions of management." These functions include:

1. planning — setting organizational goals and developing methods of achieving them
2. organizing — grouping activities, assigning activities, and providing the authority to carry out the activities
3. leading — directing and motivating people to perform tasks essential to goal achievement
4. controlling — setting standards, measuring performance against standards, and taking corrective action as required

FUNCTIONS OF TELECOMMUNICATIONS MANAGEMENT

The goal of telecommunications management is to provide good telecommunications services for an organization and its employees at the lowest possible cost.

Telecommunications management functions include:

1. administering the ongoing operation of the telecommunications system
2. preparing and administering the telecommunications budget
3. keeping abreast of changes in equipment, services, industry structure, and rates
4. implementing and administering strategies for usage control and instructing company employees in efficient usage procedures
5. assisting top management in developing corporate telecommunications policy
6. planning and implementing changes required in the telecommunication system, including an entirely new system if warranted

THE TELECOMMUNICATIONS MANAGER'S RESPONSIBILITIES

Within the four general functions of management, the telecommunications manager is specifically responsible for the following:

- ☐ Planning
  - Setting departmental goals
  - Preparing departmental budgets
  - Reviewing present telecommunications system
  - Validating telephone bill(s)
  - Keeping current on technical, regulatory, and corporate changes
- ☐ Organizing
  - Interpreting company policy
  - Identifying projects
  - Writing job descriptions
  - Selecting and training personnel
  - Developing specifications for services
  - Choosing a consultant
  - Writing requests for proposals
  - Defining service standards
  - Assigning projects and tasks
  - Conferring authority necessary for performance
  - Allocating costs to users
  - Maintaining the company telephone directory
- ☐ Leading
  - Motivating people and developing team spirit
  - Guiding and developing personnel
  - Communicating with management and staff
- ☐ Controlling
  - Establishing measurement standards
  - Measuring service performance
  - Measuring cost performance
  - Measuring target date performance
  - Identifying and correcting deviations

SKILLS REQUIRED OF A TELECOMMUNICATIONS MANAGER

To effectively perform the responsibilities of a telecommunications manager, a number of skills are required, including:

1. a knowledge of vendors and services
2. the ability to translate an organization's telecommunications requirements into the most economical configuration of equipment
3. a knowledge of traffic engineering principles
4. a knowledge of the legal and regulatory aspects of telecommunications

5. the ability to understand tariffs and rate changes
6. proficiency in applying human relations principles
7. the ability to apply sound management techniques that cut expenses without radically increasing equipment
8. the skill to plan moves and changes and predict future equipment needs
9. a knowledge of usage control techniques
10. the ability to communicate effectively
11. a knowledge of budgeting principles
12. a knowledge of cost accounting procedures to allocate telecommunications costs to appropriate departments

The remainder of this chapter discusses the administration of an established telephone system. The following chapter will examine the selection and implementation of a new telephone system.

MANAGING THE ESTABLISHED TELEPHONE SYSTEM

Telephone management is a service function. Effective management depends upon a thorough knowledge of all aspects of the system and a way to measure system performance.

REVIEWING THE PRESENT SYSTEM

The first responsibility of the manager is to become familiar with the existing telephone system. The manager personally visits each department of the company to verify the company's equipment records and discover how every telephone in the organization is used. During these visits the manager may locate equipment that is no longer necessary and can be removed from service. (Unfortunately, many departments request additional service as needed but neglect to request removal of unused equipment.)

These contacts establish communication between the manager and other departments and provide an opportunity to discuss their needs and solicit their suggestions. They also acquaint the manager with the company's business and the role that the telephone system plays in the successful operation of the business.

IDENTIFYING SYSTEMS COSTS

Any study of telephone costs begins with a review of two types of documents: the telephone bill and the telephone equipment record. Both documents are available from the telephone company on a monthly basis.

■ ***THE TELEPHONE BILLS*** One of the responsibilities of the manager is reconciling the telephone bills each month. To do this, the manager must understand how the bill is broken down. In the present post-divestiture era,

Figure 11.1
Sample Monthly Telephone Bill. The Business Service Monthly Charge does not include any charge for equipment because equipment is not provided by the telephone company. It must either be owned by the customer or provided by another vendor.

ANY TELEPHONE COMPANY, BOX 1234, ANYTOWN, MICHIGAN, 48207
SERVICE CHARGE FOR AUG 07–SEPT 06 * DUE AUG 27, 1989
CHARGES FOR (313) 358–9876

ABC CORPORATION
1234 NORTH 12TH STREET
ANYTOWN, MI 48705

BUSINESS SERVICE MONTHLY CHARGE (IN ADVANCE)	132.82
LOCAL AND ZONE CHARGES (STATEMENT ENCLOSED)	651.73
DIRECTORY ASSISTANCE USAGE (SEE DETAILS)	.22
ITEMIZED CALLS (SEE STATEMENT)	290.58
U.S. TAX 33.42 STATE TAX 36.01 TOTAL TAX	69.43
TOTAL	1,144.78
FEDERAL LINE CHARGE	3.60
TOTAL AMOUNT DUE	1,148.38

users receive at least two bills: one from their local telephone company and another from their long distance carrier. The typical telephone bill (Figure 11.1) contains nine basic items, at most. Each item pertains to a separate portion of telephone cost. These items include:

1. local service
2. local messages
3. directory assistance, toll calls, and telegrams
4. other charges and credits
5. total of current charges excluding taxes
6. taxes
7. total of current charges including taxes
8. balance from previous month
9. federal line charge

Local service is generally the first item on the telephone bill. It represents the total charge for all local telephone service. Local telephone service is service to telephones in the same zone and certain adjacent zones. Local service is billed one month in advance.

The *local messages* item is the *total* charges for calls made within the local calling area (Figure 11.2). These calls are generally bulk billed; no call details are available. Flat rate service, which is available in some areas for residential use, is only available for business in small communities. (With flat rate service, this line on the bill is blank.)

Directory assistance, toll calls, and telegrams shows the *total* charge for these items. A separate statement is also supplied that provides call details. The statement shows the following information on toll calls: the date on which the call was made, the type of call (such as customer dialed, calling card, person-to-person, collect, conference, mobile, ship to shore) the place called, the telephone number called, the time of day the call was made, the length of the call in minutes, and the total cost of each individual

LOCAL AND ZONE CALL DETAIL FOR (313) 358-9876

	CALLS	MINUTES	DISCOUNT	
LOCAL USAGE: 2,745 CALLS @ 8.2 CENTS				225.09
NEAR ZONE				
DAY RATE	1,656	5,946	00	386.27
EVENING RATE	149	499	9.80	22.87
NITE/WEEKEND	25	239	7.25	7.24
FAR ZONE				
DAY RATE	15	66	00	8.90
EVENING RATE	1	1	.05	.11
NITE/WEEKEND	2	19	1.25	1.25
TOTAL LOCAL AND ZONE CHARGES				651.73

Figure 11.2
Local and Zone Call Detail Billing

DETAILS OF ITEMIZED TOLL CALLS FOR (313) 358-9876

NO.	DATE	KEY	PLACE		AREA	NUMBER		TIME	MIN	
1	7/11	D4	PONTIAC	MI	313	338	XXXX	1245PM	1	.12
2	7/13	D5	UTICA	MI	313	732	XXXX	324PM	9	.68
3	7/13	D4	MONROE	MI	313	289	XXXX	237PM	3	.93
4	7/14	D4	PONTIAC	MI	313	981	XXXX	359PM	6	.47
5	7/15	D6	MONROE	MI	313	459	XXXX	910AM	4	1.21
6	7/15	D4	PLYMOUTH	MI	313	349	XXXX	1135AM	1	.06

KEY: B-BILL TO THIRD NUMBER, C-COLLECT, D-CUSTOMER DIALED STATION, E-ENTERPRISE, F-CONFERENCE, H-CALLING CARD, J-CIRCLE CALLING (30% DISCOUNT OFF RATE IN EFFECT IN STATE), K-PERSON CALL BACK, L-CUSTOMER DIALED PERSON, M-MOBILE, P-PERSON, T-TELEGRAM, X-BUDGET TOLL DIALING (30% DISCOUNT OFF RATE IN EFFECT IN STATE), 4-INSIDE STATE (FULL RATE), 5-INSIDE STATE (EVENING 30% DISCOUNT OFF FULL RATE), 6-INSIDE STATE (NIGHT AND WEEKEND 50% DISCOUNT OFF FULL RATE).

Figure 11.3
Details of Itemized Toll Calls (Toll billing provided by the local telephone company itemizes only calls that the local telephone company is authorized to handle. Other long distance telephone companies provide similar itemized statements for calls that are handled on their network.)

call (Figure 11.3). The statement also provides details of telegram calls that are handled by Western Union and billed through the telephone company. In addition, some long distance carriers charge for long distance directory assistance and include the total of these charges on the toll statement. The customer receives a separate toll statement from each long distance carrier that has been used.

The statement also provides details of directory assistance charges and telegrams (Figure 11.4). Directory assistance billing depends upon the charge plan of the serving telephone company.

Other charges and credits shows the *total* of all other charges and credits. A separate statement of these one-time charges and credits is supplied with call details. This section includes all charges that do not appear elsewhere, such as installations, moves, changes, and other adjustments.

Federal line charge shows the surcharge that is added to each bill as ordered by the FCC. The charge compensates the local telephone companies for providing long distance access for their customers. It replaces some of the revenues lost by the local telephone companies through the

Figure 11.4
Directory Assistance Billing

DIRECTORY ASSISTANCE USAGE DETAIL FOR (313) 358-9876	
211 CALLS TO 1-555-1212	
LESS 200 CALL ALLOWANCE	
11 BILLABLE CALLS @ 22 CENTS PER CALL	2.42
LESS MONTHLY CREDIT @ 22 CENTS PER LINE	2.20
NET DIRECTORY ASSISTANCE USAGE CHARGE	.22

discontinuance of former revenue-sharing practices. (This topic will be discussed further in Chapter 13.)

Most telephone companies charge for equipment items and flat rate service in advance. This means that the first telephone bill will include one entire month in advance plus the portion of the month in which the service was begun. Similarly, it will list a credit for any service discontinued during the month.

■ ***THE EQUIPMENT INVENTORY*** In order to maintain control of telecommunications costs, it is essential to have an accurate inventory of all telecommunications equipment. This inventory should include a detailed description of the equipment and circuits, including service features, and should classify each equipment item as to whether it is rented, leased, or owned by the organization. (See Figure 11.5.)

If equipment is provided by the telephone company, it is identified on the record in coded form. The record usually shows each item of equipment classified by telephone number and Universal Service Order Code (USOC) number. USOC codes were developed by the Bell System to provide uniformity in describing its services. Interpretation of the codes may be obtained from the telephone company representative.

ESTABLISHING CORPORATE TELECOMMUNICATIONS POLICY

A major responsibility of the telecommunications manager is controlling telephone expense. To do this, the manager establishes guidelines for the provision of telephones and controls on telephone usage. Without controls, telephone costs are almost certain to be unnecessarily high and continue to escalate.

The telecommunications manager is the key person in developing procedures and controls for telephone provision and usage. In developing these procedures and controls, the manager works closely with senior management. After approval by senior management, these procedures become the cornerstone of corporate communications policy.

Procedures are required for the following:

1. providing telephone equipment and services
2. controlling usage

Inventory of Telephone Equipment and Services
ABC Company, 9876 Main Street, Milltown, Michigan 48185

1. Main Telephone Number (313) 358–9876
2. 6 Business Lines provided by Milltown Telephone Company
3. All 6 lines served by AT&T for long distance service
4. 1 800 line (1–800–621–1ABC) provided by MCI
5. Two additional directory listings
6. Telephone equipment and telephones owned by ABC Company
   a. 3 six–line keysets — FS 256 Southwestern Bell Freedom Phones (each equipped with plug–in expansion units)
   b. 20 single line telephones — FS 800 Southwestern Bell Freedom Phones
   c. 1 speakerphone — Executive FS 900 Southwestern Bell Freedom Phone
   d. 1 single line telephone — FC 7 Southwestern Bell Freedom Phone (used to terminate 800 line)

Figure 11.5
Sample Inventory of Equipment and Service

3. allocating costs to system users
4. training system users

■ ***PROVIDING EQUIPMENT AND SERVICES*** A basic principle of the telecommunications policy should be that all orders for telephone equipment or service be placed through the telecommunications manager.

■ ***CONTROLLING TELEPHONE USAGE*** Telephone usage can be controlled either by manual or automatic (electronic) methods. One method is to require all calls for which a charge is made to be placed through an operator. The operator logs all such calls for later comparison with the telephone bill. Some companies modify this procedure by asking users to log their own calls; however, this provides a less reliable record.

As a general rule, placing calls through an operator acts as a deterrent to indiscriminate usage. However, if a more stringent procedure is desired, the operator can be instructed to ask whether the call is business or personal but to put it through irrespective of the answer. Just asking the question generally causes users to think twice before placing personal calls.

Most modern telephone systems can be equipped with automatic station identification capability. With this feature, call details are automatically recorded: identification of calling and called stations, the time the call was placed, the length of the call in minutes, and the cost of the call. Thus, the cost can be allocated to the extension making the call. This feature also acts as a deterrent to unauthorized usage. Two such features, Automatic Identification of Outward Dialing (AIOD) and Station Message Detail Recording (SMDR), were described in Chapter 9.

Some modern telephone systems are available with a service feature that requires user identification codes. This feature requires the caller to enter an identification code before any long distance call can be completed. Caller identification codes are used to restrict usage by unauthorized callers.

■ ***ALLOCATING COSTS*** There are several important principles of cost allocation:

1. The communications manager is responsible for providing the most economical service that will meet the needs of the various departments of the company.
2. Each department must bear the cost of the service that it uses.
3. The communications manager is responsible for identifying and calling attention to opportunities for improving telecommunication cost performance.

The telecommunications manager should receive and process all invoices from telephone companies and telephone vendors. Each bill must be compared with the corporate inventory record. Any discrepancies discovered must be reconciled with the carrier and refunds or credits negotiated. This function must be performed for each location for which the manager is responsible. After the accuracy of the bill has been determined or discrepancies reconciled, the manager has two important responsibilities. One is to make sure that the bill is paid, and the other is to allocate as many of the charges as possible back to the department responsible for the expense.

Charges for equipment and services can be prorated among departments on the basis of the equipment in each department. For example, if the company is composed of five departments, each of which has eight speakerphones, the expense for the speakerphones would be divided equally among the five departments. However, if one department has only four speakerphones while the others have eight, the portion of its expense for the speakerphones would be only half as much as that for the other departments.

Charges for local messages and zone call units cannot be identified and therefore must be prorated on an arbitrary basis. Prorates can be computed on the basis of the percentage of company employees in each department or percentage of the company's telephones in each department.

Toll calls can generally be identified from operator or user logs or automatic call detail recordings. Where this is the case, they can be billed directly to the department where the call originated. Other billed items that cannot be directly attributed to specific departments may be prorated on the basis of department size.

■ ***TRAINING SYSTEM USERS*** One of the manager's responsibilities is training system users. Training may be accomplished by any or all of the following methods:

1. intracompany training programs
2. instruction pages in the company telephone directory

3. in-house training by the vendor
4. intracompany letters or memos
5. bulletin board announcements

The telecommunications manager should develop an intracompany training program for new employees and assign responsibility for its execution. This program is often coordinated with the orientation program conducted by the personnel department.

The company telephone directory offers another means of instructing employees about the most efficient way of placing various types of calls.

The introduction of a new telephone system may require in-house training. Such training frequently consists of classroom sessions conducted by the telephone company or other vendor. Generally, written instructional material is available at these sessions.

Intracompany letters, handbooks, or memos are often used to define company policy for telephone usage. For example, company policy might require personnel away from the business location to direct-dial toll calls rather than use company credit cards. Reimbursement for such business calls might be handled through the employee's expense account.

Bulletin board announcements can be used to announce new services or changes in usage practices.

Training activities must be conducted on a continuing basis in order to make effective use of the telephone system.

CONTROLLING THE TELEPHONE SYSTEM

Control is the process that an organization employs to ensure that its activities are going according to plan. A basic premise of management is that effective management requires control.

The department's goals and objectives are established in the planning process. The control process measures progress toward these goals and alerts the manager to deviations from the plan that might require corrective action.

The control process consists of the following four basic steps:

1. setting standards
2. measuring performance
3. evaluating performance against standards
4. taking corrective action

Control addresses these basic questions: Where do we want to go? Where are we now? What actions are required to reach our goals?

■ ***THE ROLE OF STANDARDS*** A *standard* is a benchmark or point of reference against which performance can be compared. Without standards, there is no basis for measurement. Standards define desired performance and serve as the manager's operational goal. They are established in the planning process and reviewed periodically so that they can be modified as conditions change.

Standards should be as precise as possible. Ideally, they should be defined in quantitative terms. *Quantitative standards* have two principal advantages. First, they are specific, defining performance in clear-cut terms. For example, a quantitative standard for the length of a business long distance call is often established at 5.2 minutes. This tells the manager what the target length of business long distance calls should be.

A second advantage of quantitative standards is their reliability or consistency of measurement. Quantitative standards can be measured objectively. Thus, if two or more persons were comparing actual practice with standards, the decisions or ratings of each would be the same. Similarly, if the same person compared the same condition at different times, the results would always be the same.

All standards should be stated as precisely as possible, but some standards are difficult to describe in quantitative terms. Because setting quantitative standards for all activities is quite difficult, qualitative standards are also used. *Qualitative standards* are subjective. They are therefore of questionable reliability since reliability is directly affected by the degree to which judgments, biases, and emotions of the evaluator enter into the evaluative process.

Most telecommunication departments wish to provide "good service." However, "good service" is qualitative and subject to many interpretations. The current approach of managers is to try to translate qualitative standards into objective measures. For example, the goal specifying "good service" might by qualified to state, "good service as evidenced by no more than three blocked calls per hundred call attempts during the busy hour."

The manager must decide which areas of telephone service to control. The areas selected should be those most critical to the attainment of the department's overall goals and objectives. Since the overall goal of telecommunications is to provide good service at the most economical cost, the telecommunications manager would probably concentrate on controlling the following:

1. system capacity
2. costs
3. usage
4. special services

In setting standards in these four areas, the manager is guided by company policy. Most companies wish to provide sufficient service for efficient business operation. This policy is reflected in how they provide telephone service.

■ ***SYSTEM CAPACITY*** Telephone systems are available in many different sizes and configurations, ranging in size from only a few telephones to very large systems with several thousand telephones. A major concern of the telecommunications manager is how effectively the existing system serves the needs of the organization. This, in turn, is determined by the number of users, the number of calls to be handled, and the grade of service to be provided.

The utilization rate of the system can be computed by comparing the number of telephones installed to the maximum number of telephones that can be connected to the system. Thus, if a telephone system with a physical capacity of 400 telephones had 368 telephones installed, the utilization rate would be 92 percent. The manager uses utilization data to monitor the adequacy of the system. If the utilization rate is consistently low, it indicates probable waste of capacity. Conversely, 100 percent utilization indicates that there is no room for growth. In this case, the manager must decide whether to supplment the system (if this is possible), obtain a new system, or fail to provide new service to anyone.

Just as telephone systems differ in the number of telephones that they can physically serve, they also differ in how many calls they can handle simultaneously. If more calls are generated at the same time than the system can handle, some calls will be blocked and the service will probably be unsatisfactory.

In selecting a telephone system, the telecommunications manager must decide about the desired *grade of service*. The better the grade of service, the more the equipment will cost. Thus, top management should be involved in selecting the grade of service that is being purchased. Generally, telephone companies engineer the public telephone network to provide P.01 grade of service. Private telephone systems can be engineered to provide nearly any desired grade of service. However, most private systems do not meet the P.01 standard.

■ ***COSTS*** A *budget* is a plan describing how the financial resources of the organization will be used. It shows what portion of the total resources will be devoted to each expense item. The budget functions as a standard of financial control. Budget preparation and administration make up one of the manager's important responsibilities.

Preparation of the telecommunication budget starts with analyzing the telephone bill. The information contained in the telephone bill should be summarized in whatever way best demonstrates the history of the departmental expense. One way to analyze telephone expense is to maintain monthly records of the bills, summarizing them as follows:

1. local services
2. local message charges
3. long distance charges (including WATS)
4. other charges
5. other carrier charges
6. charges for directory advertising
7. charges for installations, moves, and changes
8. credits
9. taxes
10. federal line charges

Figure 11.6
Summary of Monthly Telephone Expense

Month	Service	Message Units	Long Distance (incl WATS)	Other Charges	Other Carriers	DIR Chrgs	Moves & Chgs	Credits	Taxes	Federal Line Charge	TOTAL

Figure 11.6 shows a form for summarizing monthly telephone expenses.

When these figures are posted monthly over a period of time, the resulting cost history serves as a basis for budget preparation. In addition to the costs for telephone services, the manager must add the expense of the telecommunications department, including the cost for any personnel needed to support the system.

The manager must be able to document past costs to convince corporate management that the expenditures specified in the budget are necessary and that the budgeted expense will produce the quality of service management wants.

In addition to serving as a basis for budget preparation, the telephone cost history may reveal trends that require corrective action. For example, if the cost history shows that long distance costs are escalating rapidly, the manager might want to consider implementing more stringent usage controls or using other common carriers, WATS lines, FX lines, or private lines.

The control process involves comparing actual telecommunications costs with projected costs set forth in the budget. Any discrepancies between actual and budgeted costs should be noted and carefully analyzed. An expense that exceeds the budgeted figure is an *overrun*. Any manager whose department overruns the budget is faced with a difficult situation. There are only two solutions — to cut expenses or to add money to the budget. Neither of these solutions is entirely satisfactory. Therefore, the importance of a realistic budget cannot be overemphasized.

An *underrun* of the budget may appear to be easier to handle, but underruns can present problems too. When the comptroller or top management sees an underrun, the usual reaction is to conclude that the department's budget figures were too high and to expect next year's figures to be reduced accordingly.

Costs are always of major concern to an organization. The high costs of communications as a percentage of the total business operating costs makes control of this area extremely important.

■ ***USAGE*** It is an accepted fact that telephone bills are escalating rapidly. While the manager cannot control telephone company rate increases, substantial savings can result simply from reducing system abuse. It is the manager's responsibility to establish rules and procedures for the use of business telephones. These rules should include:

1. a company policy concerning the use of business telephones for personal calls — both local and long distance
2. a company policy concerning special calls — person-to-person, calling card, charge to third party, and so forth
3. a company policy concerning the provision of special equipment and services, such as speakerphones, FAX machines, or data transmission equipment
4. a continuing program for training telephone system users
5. an effective procedure for allocating all telephone services and usage costs to the user
6. a follow-up procedure for dealing with system abuse

Monitoring both local and long distance usage is a continuing responsibility of the manager. Specific actions can be taken to reduce or eliminate unreasonable use. Probably the most effective procedure for controlling telephone usage is allocating costs to the users and making certain that the users are aware of this practice. Knowing that call details are being logged or recorded will do much to eliminate unnecessary usage and to shorten call duration.

The user training program can be a vehicle for educating system users about the various pricing mechanisms for telephone costs; for example, WATS line billing. When WATS service was initiated, the service was billed on a flat rate basis. This led many users to believe that if a WATS line was available, its usage would be "free." Although present WATS tariffs are based on usage, there are still people who regard WATS lines as a way to obtain "free" calls. Old ideas die hard (especially when people want to believe something), but education can do much to eliminate abuse based on misinformation regarding telephone costs.

■ ***SPECIAL EQUIPMENT AND SERVICES*** Technology has provided an abundance of new telecommunications equipment and services. While some of these features fill a real need, others may add prestige to an office but seldom be used. As the telecommunications marketplace continues to become increasingly competitive, we can expect to see more and more advertisements about these sophisticated new products and services. Once users are aware of their advantages, it is human nature to want these goodies. It is the responsibility of the telecommunications manager to evaluate service and equipment items realistically and to acquire them only when they will contribute to the total efficiency of the organization.

As we discussed earlier, the manager must verify the equipment inventory and evaluate the usefulness of special equipment. As changes are made in personnel, the manager should discuss special equipment with newcomers to be sure that the equipment will actually be used and will contribute to the effectiveness of the organization.

THE FULLY UTILIZED SYSTEM

If feedback obtained during the control process indicates that the system is operating at full capacity or is already overloaded, the manager should study the alternatives available to remedy the situation.

The first consideration should be whether to enlarge the present system by providing additional equipment and trunks. Where this is possible, it is often the most practical thing to do. However, it might not be possible because the system may already be configured at capacity. Or the system may be obsolete so it is impossible to obtain additional equipment or parts compatible with the system.

Still another reason why it might not be possible to enlarge the present system might be the lack of physical space. If the room where the equipment is housed is filled, rearrangement to include more equipment would be impossible.

When it is possible to enlarge the present system, the manager must investigate the feasibility of doing so. Even when it is possible to add to a given system, it may or may not be the best thing to do. The manager should evaluate conditions regarding the system before making a decision. One such condition is the commitment to the vendor. If equipment is leased, as is often the case, the length of time left to run on the existing contract is an important factor. When the contract with the vendor is soon to expire, any necessary changes should be considered before an organization signs a new contract. Once a contract has been signed, there is usually a substantial penalty for terminating it.

If the system is owned by the organization, the possibility of its reuse at another company location or its possible sale can be considered.

Future company plans must also be taken into account. For example, if the company plans to move the offices to another location, this certainly has a bearing on any decision regarding the enlargement of the telephone system.

Because telecommunications technology, regulation, and supplier offerings are changing at an incredibly fast pace, communications decisions are becoming increasingly complex. The effective manager must have regular briefings on these important aspects of the business in order to provide sound telecommunications management.

This discussion has described telecommunications management in the context of providing a service to an organization and its employees. Efficient management of the system entails providing the best possible service at the least possible cost.

When telecommunications service is being provided for external use—that is, for users who are not members of the organization—telecommunications can be managed not only to provide efficient service but also to generate a profit. Examples of this approach include systems in hotels, hospitals, and school/college dormitories.

SUMMARY

Prior to the advent of competition, most organizations relied upon the telephone company to manage their telephone services. After the competitive market developed, it became necessary for customers to evaluate the service and equipment offerings of various vendors and match them with their own requirements. Choosing among the proliferation of services and vendors available is a difficult task, requiring a comprehensive knowledge of telecommunications and the application of sound management principles. Effectively managed, the telecommunications system can make an important contribution to the profitability of any organization.

The goal of telecommunications management is to provide good telecommunications services for an organization and its employees at the lowest possible cost. The manager is responsible for the planning, organization, leadership, and control of the telecommunications system. Effective management requires a thorough knowledge of all aspects of the system and a way to measure system performance.

The first responsibility of the manager is to become familiar with the existing system, including system inventory and service and equipment costs. The manager should be a key person in the establishment of corporate telecommunications policy for the following:

1. providing telecommunications equipment and services
2. controlling usage
3. allocating costs to users
4. training system users

Another responsibility of the manager is to formulate standards for system capacity, costs, usage, and special telecommunications equipment. The manager monitors system performance and measures it against the established standards. When data obtained thorough the control process shows that the system is operating at or near capacity, the manager explores alternatives to remedy the situation.

REVIEW QUESTIONS

1. What is the objective of telecommunications management?
2. Historically, many organizations assigned the responsibility for telecommunications service to an office manager or clerical supervisor. Why was this possible?
3. What factors have been instrumental in bringing about the need for telecommunications management?
4. What are some of the options an organization has today in obtaining telecommunications equipment and services?
5. What are the functions of telecommunications management?
6. What is the first responsibility of the telecommunications manager?
7. What are the basic areas for which telecommunication policies are required?
8. What are the skills (knowledge, understanding, abilities) required of a telecommunications manager?
9. How does a telecommunications manager evaluate system adequacy?
10. What is telephone cost history used for?
11. What are some of the procedures a manager can implement to control telephone usage?
12. Why might it not be possible to enlarge an existing telephone system?

VOCABULARY

Federal line charge

telephone equipment inventory

telephone bill

Universal Service Order Code (USOC)

allocating costs

standard

system capacity

fully utilized system

CHAPTER

12

SELECTING AND IMPLEMENTING A NEW TELEPHONE SYSTEM

CHAPTER OBJECTIVES

After completing this chapter, the reader should be able:

- *To describe the role of vendors in telephone system selection.*
- *To describe the role of the telecommunications manager in telephone system selection.*
- *To describe the function of the consultant in telephone system selection.*
- *To explain how telephone system requirements are determined.*
- *To discuss the Request for Proposal (RFP).*
- *To discuss the evaluation of vendors' proposals.*
- *To compare the financial options available for obtaining a new telephone system.*
- *To describe the implementation of a new telephone system.*

A few years ago people would have laughed at the thought of "shopping" for a telephone system. A business needing a new system called the local telephone company representatives and left most of the decisions up to them. The manager's role was limited to casual concerns about details of the new system. Today, however, users can shop for telephone equipment from both the local telephone companies and interconnect companies. Businesses can select their own telephone system, "shopping around" in much the same way as they would for any other major purchase.

Many companies are convinced that they can save money by owning their own phone equipment rather than paying a monthly rental fee to the telephone company. Of course, organizations still pay the phone company every month for their phone calls, but they are relieved of monthly phone equipment bills. In any case, the management of an organization must now make informed decisions regarding telephone equipment, service, and financial arrangements. Thus, the job of the telecommunications manager is more complex than it was formerly was.

Telecommunications management is charged with providing telephone service that meets the needs of the organization. This requires continuing analysis and evaluation of the system's capability to meet service demands. Details of the analysis can identify possible shortcomings of the present system and help determine whether to improve, supplement or replace the system.

There are three situations in which a new telephone system may be required:

1. to provide service for a new location
2. to replace a system that lacks capacity to meet the needs of the organization
3. to replace a system that cannot provide desired service features, such as direct inward dialing or Touch-Tone capability for communication with computers

This chapter describes the options available to an organization selecting a new telephone system and guides the reader through the series of tasks necessary for successful implementation of the system.

SELECTION OF A TELEPHONE SYSTEM

The selection of a new telephone system is one of the most important, and probably one of the most difficult, responsibilities of the telecommunications manager. It is also one in which the manager has generally had little or no practice. The proliferation of vendors, each claiming to have the "best" system; the many attractive service features available; and the variety of financial arrangements available present the telecommunications manager with a diverse set of choices and make the selection more difficult.

There are three options available to organizations for acquiring a new telephone system: purchase, lease agreement, or rental agreement. Regardless of which option the organization chooses, it makes a long-term commitment of large sums of money, thereby precluding the opportunity to take advantage of further system advances for the duration of the commitment.

Rapid advances in telecommunications technology have resulted in a wide variety of improved service features, often at reduced costs. These advancements tend to build obsolescence into existing telecommunications equipment. Most organizations acquiring new telephone systems want the new system to have the latest capabilities. The implicit threat of system obsolescence poses a dilemma for the manager and acts as a deterrent in the decision-making process.

ROLE OF VENDORS IN SYSTEM SELECTION

The selection of a new telephone system must be based upon a careful determination of the organization's needs, extensive review of the various service offerings, and accurate projections of future needs. An examination of the system analysis procedures followed by the telephone companies prior to divestiture is helpful in understanding the process of selecting a new telephone system.

Telephone company marketing representatives served as the liaison between the telecommunications manager and the telephone company. They were trained in the communications needs of a particular industry. Very large nationwide organizations often required the services of many marketing representatives, headed by a national account manager. The Bell System's marketing department divided its customers into three broad categories: (1) government, education, and medical; (2) commercial; and (3) industrial.

These categories were further broken down into various industries within the categories. For example, the commercial category consisted of such firms as banks, law firms, automobile dealers, and department stores. Similarly, the industrial category included all types of manufacturing organizations.

The telephone marketing representatives were assisted by traffic and equipment engineers, who translated usage data into quantities of equipment and trunks and specific technical design details.

■ ***TRAFFIC STUDIES*** The telephone company conducted *traffic studies* (counts of calls classified according to types) to help an organization determine its telephone needs. Data obtained in these studies were then used to determine the adequacy of the existing system and provide a basis for the design of a new system.

Traffic studies include counts of calls that are blocked; that is, those which cannot be completed because of a network-busy condition. If the number of blocked calls is excessive, additional trunks will be required. If more trunks cannot be added to the existing system, a new system will have to be obtained.

Another dimension of traffic studies concerns long distance services. Traffic studies analyze long distance records to determine whether an organization does a substantial volume of calling to a locality that is not within the local calling area. If this is the case, there may be a need for special long distance facilities, such as WATS, foreign exchange, or tie lines, and the system should be designed to include this capability.

As a result of the traffic study, the telephone marketing representative was able to match system requirements with available types of systems and make an appropriate recommendation to the telecommunications manager. Each system was customized to meet the specific needs of the user organization.

In the post-divestiture marketplace, all vendors of telephone systems and services provide assistance in the form of traffic studies and specific recommendations similar to those formerly provided by the telephone company.

ROLE OF THE TELECOMMUNICATIONS MANAGER IN SYSTEM SELECTION

The major distributors of telephone systems have patterned their organization after the Bell System concept of industry specialization. As a result, the telecommunications manager deals with many vendors in addition to the telephone company, each competing for the organization's business. Their objective is to convince the telecommunications manager that their telephone system best meets the organization's needs.

In the new competitive environment, the manager receives just as much assistance as formerly. Now, however, much of the advice received is conflicting, and nearly all of it is prejudiced in favor of the providing vendor. The manager should learn as much as possible from the various vendors in order to evaluate the sometimes conflicting recommendations and make the best decision.

Today the evaluation of an existing telephone system begins with a traffic study made by the telephone company, by interconnect vendors, by consultants, or by the telecommunications manager (Figures 12.1 – 12.3). A good policy is to obtain traffic studies from several sources so that the various recommendations can be compared and evaluated. The telecommunications manager often begins the selection process by requesting a traffic study from the telephone company and/or other vendors.

■ ***SOURCES OF MANAGEMENT INFORMATION*** In addition to a basic knowledge of telecommunications concepts, the manager must be familiar with the latest technologies and vendor service and equipment offerings. Telecommunications technologies are changing at an incredibly fast rate. The manager's education must be continually updated to keep pace with new developments. Some of the sources of management information include periodicals, professional associations, seminars, and consultations with vendors.

Periodicals, seminars, and professional associations can provide the manager with information on state-of-the-art technology and industry and

(Courtesy of ROLM™ Corporation)

Figure 12.1
ROLM Analysis Center.™ The Analysis Center offers a wide variety of report formats for subscribers' call details.

(Courtesy of ROLM™ Corporation)

Figure 12.2
ROLM Telecommunications Analysis Center Library (TACL)™

Figure 12.3
ROLM CBX Analysis Center.™
CBX Analysis Center™ reports provide management with timely information to spotlight abuse, allow accurate cost allocation, and facilitate client billing.

(Courtesy of ROLM™ Corporation)

regulatory developments. Consultations with vendors can provide the manager with information on each vendor's service and equipment offerings. A list of professional associations, periodicals and newsletters, and organizations that sponsor seminars is found in Appendixes A through C.

ROLE OF THE CONSULTANT

Given the fact that communications decisions have become increasingly complex with advanced technology and myriad vendors, an outside consultant can often save an organization substantial time and money in system selection. The function of the consultant is to give unbiased advice. A good consultant should be well informed on state-of-the-art technology, knowledgeable about all types of available systems, and honest and impartial in system evaluation and recommendations.

The consultant's contract should detail the work to be done and the fee to be charged. Organizations should expect to pay a fair daily rate, which can be upwards of $500 a day. Managers should avoid retaining a consultant whose fee is contingent upon cost-saving; anyone can reduce costs by downgrading services. Organizations have a right to expect a consultant to improve service and/or save money.

The consultant's report should provide the following items:

1. an evaluation of the existing system
2. new system requirements and specifications
3. alternative solutions for a new system
4. conclusions and recommendations

As a general rule, the consultant should stay on the job until installation has been completed and the system checked to see that it has been installed according to specifications.

DETERMINING SYSTEM REQUIREMENTS

For the telecommunications manager, the next task after the traffic study is to determine the specifications for the new system. The areas to consider include:

1. Line capacity. How many telephone lines will be required over the lifetime of the equipment?
2. Call capacity. How many calls will the system be required to handle during its lifetime?
3. Service features. Which service features best serve the organization's needs and contribute to overall profitability?
4. Data communications requirements. What are the data communications requirements of the organization?
5. Costs. What are the cost considerations in selecting a new telephone system?
6. Financial arrangements. Should the system be purchased, leased, or rented? What are the terms of the various financial operations?
7. System maintenance. Are diagnostic procedures to be included in the system? What are the provisions for system repair?

There are two types of physical capacity for any telephone system: line capacity and call capacity.

■ ***LINE CAPACITY*** The physical *line capacity* is the maximum number of telephone lines that can be served by the system. For example, if the physical capacity of the system is 1,000 lines, it would not be possible to connect 1,001 lines to the system.

The telecommunications manager maintains records of the actual number of lines in service on the present system and combines this data with knowledge concerning the projected growth of the organization. These two sources of information are used to estimate the telephone lines required. When the existing line capacity has been exhausted, the organization requires a new system.

■ ***CALL CAPACITY*** *Call capacity* is the ability of the system to handle a specific number of telephone calls in the busiest hour of the day and still provide a specified grade of service. The manager must study call volume data on a continuing basis. If the busy-hour call volume exceeds the designed call capacity, calls will be blocked because of the equipment overload, and quality of service will deteriorate. The manager also uses call volume data to predict when call capacity will be reached and to estimate future requirements.

Sometimes call capacity can be increased by patching up the system — adding more trunks or switching equipment. Generally, however, a new system is required. Call counts are used to predict trends of future usage. These projections, in turn, are used to estimate the call volume that the system must be engineered to handle over its lifespan.

■ ***SERVICE FEATURES*** As has been noted, there are many service features available with modern telephone systems. A partial list of these features includes: least-cost routing, direct inward dialing, automatic caller identification, toll restriction capability, call transfer, call forwarding, speed dialing, camp-on, and group pickup. Some of these features produce cost savings, and others are just "nice to have." Research has shown that many of the "nice-to-have" features are, in fact, seldom used.

All digital telephone systems have certain switching and transmission advantages over older analog systems. Electronically controlled, programmable systems can often be upgraded to provide new service features. The need for some of these features could justify a new telephone system. All system features are not available from all vendors. Therefore, the choice of vendor may depend upon the availability of desired service features.

■ ***DATA COMMUNICATIONS REQUIREMENTS*** The data communications requirements are determined by the quantities of data to be transmitted, classified according to the following requirements: transmission speed, type of line (dedicated? conditioned?), type of transmission (realtime? delayed?), type of network (local area network? public network?), and other considerations specific to the organization.

■ ***COSTS*** Telephone costs represent a substantial and ever-increasing portion of an organization's budget. There are two underlying reasons for rising telephone costs. One reason is higher telephone rates. Telephone service and equipment, like most other elements in our economy, have been affected by inflation. Another reason for increased telephone costs is increased telephone usage. Today, many multinational corporations have established branch offices in widely dispersed locations, with the resultant increase in communications requirements. In addition, businesses keep finding more uses for the telephone. Telemarketing, to cite just one example, has greatly reduced the volume of personal sales visits but has caused a corresponding increase in telephone usage. While total telephone costs generally rise as an organizations's size and activities increase, the objective in obtaining a new telephone system is to reduce the costs per employee or per unit of production.

Opportunities for cost savings in a new telephone system exist primarily in the area of service features. There is no opportunity for cost savings on tariffed items, such as telephone trunks, provided by an operating telephone company. There is also very little, if any, opportunity for savings on telephone lines. Essentially, if an organization requires a specific number of telephone lines, that is the number of telephone lines that must be obtained.

The majority of new telephone systems are electronic rather than mechanical. Electronic systems combine computer capabilities and switching power to improve the performance of the telephone system. Electronic systems provide many service features unavailable in older telephone systems, including the following:

1. routing both incoming and outgoing calls over the least expensive route available
2. dialing numbers faster and more accurately than humans
3. recording details of telephone calls
4. restricting unauthorized toll call usage
5. identifying system malfunctions, thus facilitating system repair

These capabilities can provide substantial cost savings. They are combined in various ways in vendor system offerings.

■ ***FINANCIAL ARRANGEMENTS*** For many years, the only financial option available to a system user was rental from the telephone company. The telephone company tariff provided for a one-time installation charge and a monthly charge for equipment rental and service. Today, telephone systems may either be rented, leased, or purchased.

Under a rental agreement, the user receives telephone equipment for a monthly fee. The provider of the equipment is responsible for maintenance, taxes, and insurance. In addition to the monthly service fee, the user pays a one-time installation charge and also pays for any moves, changes, or rearrangements in the system. Rental agreements are made with any vendor of telephone equipment. The user signs a contract that specifies the terms and conditions, with a substantial penalty for cancellation of the service.

Many interconnect vendors sell their systems outright. When a system is sold, the vendor receives full payment. The purchase price includes the cost of installation. The vendor has no responsibility for maintenance beyond the system's warranty. Many vendors offer service contracts that provide system maintenance for a specified fee. These contracts generally include an escalation clause to protect the vendor against rising costs. The owner of a telephone system is, of course, responsible for insurance and taxes and is eligible for depreciation tax credit.

Another financial arrangement is leasing. To lease a telephone system, the organization enters into a contract with the vendor in which the organization agrees to pay a stipulated fee periodically in return for use of the system. The lease agreement specifies all the details of the arrangement, including the period of the lease, cancellation penalties, options to purchase, maintenance provisions, tax and insurance responsibilities, and depreciation credit.

Many vendors only sell telephone systems; they do not offer lease arrangements. When this is the case, organizations often obtain a lease through a third-party agreement. Financial institutions and leasing companies are in the business of providing financial services; they make a profit by purchasing equipment and leasing it to users.

The telecommunications manager should prepare an analysis of the financial impact of each of the three options on the organization's financial status. This financial analysis must be examined within the framework of corporate financial policy in order to select the best option for that particular organization.

■ ***SYSTEM MAINTENANCE*** An organization that rents its telephone system obtains system maintenance as part of the rental agreement. An organization that purchases or leases a telephone system is responsible for providing maintenance either through a service contract or by its own maintenance staff. When an organization elects to perform its own maintenance service, personnel must be trained in system maintenance. Such training is generally available from equipment vendors. The training prepares the trainees to accomplish system moves, changes, and rearrangements as well as maintenance.

An advantage of modern electronic systems is their capability to identify defective system components either from remote locations or on site. When potential system malfunctions are identified, the user can replace defective modules or circuit boards before system breakdown occurs.

COMPETITIVE BIDDING

After the requirements for the new telephone system have been determined, the telecommunications manager must find out how much the system will cost. The first step in this process is generally to issue a Request for Quotation (RFQ) inviting potential suppliers to submit an estimate of the cost of the system.

■ ***REQUEST FOR QUOTATION*** The RFQ is a document that describes the type of equipment desired, the approximate quantity and types of telephones required, the approximate number of incoming and outgoing trunks, the special features desired, maintenance arrangements, requested service date, and any other pertinent information.

The request asks vendors to provide an estimate of the cost of the system, alternative financial terms, and a description of the standard service features of the proposed system. The RFQ should also specify the deadline date for submitting quotations.

Although the ideal procedure is to issue a RFQ, occasionally the need for a new system is so urgent that it becomes necessary to bypass this step in the selection process and go directly to the next step, issuing the Request for Proposal (RFP).

■ ***REQUEST FOR PROPOSAL*** In obtaining a new system, it is generally desirable to obtain price quotations from several vendors as well as the telephone company. A convenient way to solicit such quotations is to issue a Request for Proposal, a formal document sent to potential vendors inviting them to submit bids on the new telephone system. The Request for Quotation differs from the Request for Proposal in that the former is merely asking for an estimate while the latter spells out precise system specifications and requests an exact price. Of course, the RFP must be very specific concerning the types and quantities of equipment required for the system so that vendors can accurately evaluate the cost. The following items are usually included in system specifications:

1. System capacity: number of incoming, outgoing, and internal calls the system must be capable of handling during the busy hour
2. Equipment: number of trunks (by types) required; number of telephones (by types) to be served by the system
3. Service features: types and quantities required
4. Changes: moves and rearrangement capabilities
5. Maintenance: diagnostic/repair capabilities of system and availability of maintenance service and parts
6. Training personnel: training for system users; training on repairs, moves, and changes
7. Proposal submission date: deadline for return of bids
8. Financial options: rental, purchase, direct lease, or third-party lease

Appendix D shows a sample RFP. This sample is general in nature and is for illustrative purposes only. The RFP should be customized to meet the organization's needs and approved by legal counsel before it is submitted to vendors. In contract law, the bid received from the vendor generally constitutes an offer; its acceptance constitutes a contract.

EVALUATING VENDORS' PROPOSALS The RFP responses should first be screened to be sure that they meet the system and service requirements of the organization in all respects. If they do not meet these requirements, they should be eliminated.

If all other things were equal, the proposals meeting system requirements could be ranked by price, with the one having the lowest price selected. However, no two organizations are identical in financial resources, credit ratings, cash flow, standards, and priorities. Thus, each proposal should be evaluated in terms of the organization's policies, preferences, and financial status. There are four basic factors to consider in evaluating vendors' proposals:

1. product
2. vendor
3. total system costs
4. financial options

PRODUCT When competition first entered the telephone marketplace, the service and equipment offerings of the interconnect vendors varied widely. Older telephone systems were chiefly mechanical. The development of solid-state electronics made a variety of new service features possible. Today all new telephone systems have essentially the same capabilities. As a result, nearly any service feature can be obtained from any vendor. The principal differences between telephone systems lie in the design arrangement of their components and in the physical characteristics of their telephone sets.

Different manufacturers arrange the functional components in different configurations, particularly in the number of components to serve the different types of trunks. The grouping arrangements must be matched with the requirements of the user. For instance, a new system requiring 12 trunks of a given type might have to be matched with manufacturers providing trunks in groups of 8, 10, or 12. Thus, the user requiring 12 trunks can select from one group of 12 trunks, two groups of 8 trunks, or two groups of 10 trunks. Since one group of 12 trunks would provide no expansion capability, this option can be ruled out. Therefore, two groups of either 8 or 10 trunks would be required. Since two groups of 8 trunks would provide sufficient expansion capability, this choice would be the most cost effective.

Telephone sets provided by different vendors also differ in physical characteristics, such as color, weight, and shape. The aesthetic qualities should be evaluated in terms of the organization's preferences and priorities. Another factor that should be considered in evaluating the product is reliability. Product reliability is measured by the number of times the system is out of service during a stated time interval and the average length of such occurences. This information is best obtained from other users of the system under consideration.

■ ***VENDOR*** Factors used to evaluate the vendor include reputation, support staff, experience, location, and repair service. These factors are best evaluated by questioning existing customers of the vendor's system. Some questions to consider include:

1. Does the vendor have a favorable reputation in the industry?
2. How long has the vendor been in business?
3. Is the vendor financially sound?
4. Does the vendor have a support staff of highly competent marketing representatives, customer service engineers, and training personnel?
5. Is the vendor's system marketed directly in the locality? In other words, can the organization deal directly with the vendor or must it deal with a third party to obtain the vendor's system in its locality?
6. How does the vendor provide maintenance service?
7. Are diagnostic and preventive maintenance procedures effective in avoiding serious system breakdowns?
8. What is the maintenance service response time?
9. What training does the vendor provide for system users?

■ ***TOTAL SYSTEM COSTS*** *Total system costs* are the amount an organization has to spend to get the system it needs. These costs include such items as purchase price, total rental costs, installation, maintenence, interest charges, depreciation, and taxes.

■ ***FINANCIAL OPTIONS*** Companies either purchase, lease, or rent telephone systems depending upon several factors, including the company's policies, preferences, and financial status. The telecommunications manager should make a detailed financial analysis comparing total system costs of the various financial options. Appendix E illustrates a telephone system financial analysis comparing rental, purchase, and lease options.

Under a rental agreement, the user pays a monthly equipment rental charge, which is subject to an excise tax. Additionally, the renter pays a one-time installation charge. The vendor pays taxes, insurance, maintenance, and other expenses; and is entitled to depreciation credit. The renter of a system can anticipate a yearly increase in rental costs because of inflation.

The *advantages* of *renting* include:

1. no large capital outlay
2. relative ease in breaking the rental agreement—an important advantage in view of rapid technological developments

The chief *disadvantages* of *renting* include:

1. the limited choice of equipment
2. the continued escalation of rates

In buying a telephone system, the purchaser pays a one-time purchase price, which is generally subject to a sales tax. Because the purchaser of a telephone system is responsible for system maintenance, maintenance charges (adjusted for inflation) must be included in the total costs. In addition, the purchaser is responsible for insurance and taxes. The purchaser, however, is entitled to an annual depreciation tax credit for the depreciable life of the system.

The principal *advantages* of *purchasing* a telephone system include:

1. the avoidance of continuing monthly payments that can amount to several times the purchase price over a period of years
2. the protection from continuing rate increases
3. the realization of substantial depreciation tax credits

The *disadvantages* of *purchasing* include:

1. the large capital outlay
2. the possibility of owning outdated equipment
3. the difficulty of moving the equipment if the organization relocates

Under a leasing contract, the leasing company provides a telephone system for the organization in return for a specified monthly lease factor.

This factor is determined by totaling all costs incurred by the leasing company — including insurance, taxes, and interest charges — and dividing this total by the number of months in the leasing contract. The system user is responsible for maintenance costs. In Appendix E, the user had the option to purchase the system after five years at 10 percent of the original purchase cost. After purchasing the system, the user is responsible for taxes and insurance.

Leasing a telephone system is similar in many ways to purchasing. The principal difference is that the large, immediate capital outlay is replaced by continuing monthly payments, with interest. An important *advantage* of *leasing* is that the lessee is protected from continuing rate increases. The principal *disadvantage* of *leasing* is that there is usually a substantial penalty for terminating the lease before its expiration. In most lease agreements, the lessee receives title to the system for a nominal sum of money after leasing it for a specified period of time, usually five years.

The financial study takes all expenditures in each alternative by periods (years), totals these amounts, and by using discount tables calculates the present value of the monies spent in each year. This information permits the manager to compare each of the options and to determine the most attractive one from an economic standpoint. The analysis also shows the effect of taxes on costs for each option. Microcomputers using electronic spreadsheet software make it possible for managers to compare the various financial options easily.

The financial analysis of total system costs should be reviewed with the organization's financial officers so that this expenditure can be evaluated within the framework of the organization's total financial picture.

IMPLEMENTING A NEW TELEPHONE SYSTEM

In the current deregulated environment of the telecommunications industry, the buyer and seller of a telephone system should enter into a written contract as protection against possible misunderstandings. The contract should specify all details of the agreement, including the rights and obligations of each party. Prior to industry deregulation, the public utility commissions protected both customers and providers of services by approving tariffs that defined terms, conditions, and rates for all telephone services. The commissions were empowered to enforce tariffs and to settle disagreements. Under deregulation, however, the public utility commissions have no power to settle disputes. Disagreements or performance defaults must be settled between the parties or in the courts.

CONTRACT PROVISIONS

The contract to acquire a new telephone system should contain the following provisions:

1. System capacity—the capability of the system to handle a specified number of calls in the busy hour
2. System configuration—the quantities of telephone sets by types
3. Registration requirements—conforming to FCC requirements regarding direct connection to telephone company lines
4. Building code conformance—all installation conforming to local building codes
5. Warranty—explicit details of the system's warranty
6. System failures—the seller's guarantee for restoration of service following system failures
7. System testing—the types of and times for system testing prior to system acceptance and authorization of payments
8. Operating manuals—the number of manuals to be supplied to the user organization
9. Installation schedule—for completion of each event comprising installation, including cutover date
10. User training—the vendor's responsibility to train system users in the system's service features
11. Payment—breaking the purchase price down into a series of payments that keep pace with system installation; customary to withhold a percentage of the total payment until system acceptance
12. Other charges—the vendor's responsibility for such charges as freight and insurance until the title passes
13. Title—when title passes to the buyer of the system

This list is not intended to be all-inclusive but to suggest the major items that should be included in the contract to protect the buyer's interests.

SYSTEM CUTOVER

The term *cutover* describes the process of putting a new system into operation. Before a new telephone system is ready to be implemented or placed in service, a number of procedures must take place:

1. Assigning telephone numbers. Each telephone number must be associated with a cable and wire pair assignment and the assignment details recorded. Wiring diagrams to show key telephone set assignments must be prepared.
2. Assigning user authorization codes. When the system incorporates user authorization codes, it is necessary to assign an individual identification code to each authorized user.
3. Selecting and implementing a computer program to allocate costs. The call detail data obtained from the telephone system must be interpreted so that each charge can be allocated to the person making the call. A separate computer program is required for this purpose.
4. Verifying the least-cost routing program. The system must be tested to ensure that the least-cost routing program is in operation.

5. Preparing traffic study procedures. The call count registers must be tested and appropriate data summarization forms prepared.
6. Preparing a system telephone directory. The names of all system users should be listed alphabetically, along with accompanying telephone numbers. Listings by departments should also be provided. The directory should also include detailed instructions for system usage.
7. Providing system training. System training should be provided for console attendants, system users, and personnel who may perform maintenance, moves, and changes.

The system cutover is scheduled for a specific time and date. The cutover should be a *turnkey* operation; that is, ready to go on an instantaneous basis at the turn of a key.

Experience has shown that call volumes on the first few days following a cutover are abnormally high because of curiosity calling. Therefore, judgments regarding system capacity should be withheld until a later data when the system has had a fair trial.

SUMMARY

There are three options available to organizations for acquiring a new telephone system: purchase, rent, or lease. Regardless of the option chosen, the organization makes a long-term commitment of large sums of money that precludes the opportunity to make changes without incurring a substantial penalty. The implicit threat of system obsolescence makes the choice difficult.

In addition to a basic knowledge of telecommunications concepts, the manager must be familiar with the latest technologies and vendor services and equipment offerings. A basic task in selecting a new telephone system is to determine the specifications for the system. Traffic studies are used to help identify system requirements. The areas to be considered include: line capacity, call capacity, service features, data communications requirements, costs, financial arrangements, and maintenance provisions.

The first step in obtaining a new system is to prepare a Request for Quotation describing the desired system in general terms. The RFQ requests vendors to provide an estimate of costs and financial terms of the system.

The next step is to request a proposal from vendors. The Request for Proposal document is very specific concerning the types and quantities of equipment required for the system. An important part of the selection process is the financial analysis comparing rental, purchase, and lease options. The evaluation of vendors' proposals is a critical step in system selection.

The contract to acquire a new system should contain provisions about the following: system capacity, system configuration, FCC registration requirements, building code conformance, warranty, system failures, system testing, operating manuals, installation schedule, user training, payment, and transfer of title arrangements.

The system cutover is scheduled for a specific time and date. It is necessary for the manager to plan and follow an orderly sequence of procedures in order to effect an instantaneous cutover of the new system.

REVIEW QUESTIONS

1. What are the factors that make the selection of a new telephone system difficult?
2. What was the role of the telephone company marketing representative in system selection?
3. What is the role of the consultant in selecting a new telephone system?
4. Why must the manager's telecommunications education be continually updated?
5. What are some sources of management information?
6. What are the considerations in determining specifications for a new system?
7. How does a request for quotation differ from a request for proposal?
8. Before evaluating a vendor's proposal, what preliminary screening should be conducted?
9. What are the details that should be specified in the contract to acquire a new telephone system?
10. What are the procedures that must be completed prior to the cutover of a new telephone system?

VOCABULARY

line capacity

call capacity

Request for Quotation (RFQ)

Request for Proposal (RFP)

total system costs

cutover

turnkey

CHAPTER

13

THEORY OF RATE MAKING

CHAPTER OBJECTIVES

After completing this chapter, the reader should be able:

- *To discuss the historical perspective of rate making.*
- *To describe the influence of regulation on rate making.*
- *To explain rate of return and distinguish between it and telephone rates.*
- *To explain the value-of-service concept as it relates to telephone rates.*
- *To describe vertical services and their relationship to rate-making theory.*
- *To describe the impact of industry competition on rate making.*
- *To define federal access charge and explain its relation to rate making.*
- *To describe usage-sensitive pricing and its relationship to rate-making theory.*

There are two times when people's attention focuses sharply on the telephone: when a telephone is inaccessible and when the cost of obtaining services changes dramatically.

Telephones have become so commonplace in our society that we take them for granted until something happens to remind us of their value. When our telephone is temporarily out of service, when we are away from our home or place of business and no telephone is available, or when service is denied temporarily for lack of telephone facilities, we are reminded of our dependence on the telephone—and impatient at not having instant service.

The other time we are reminded of the telephone is when we receive the monthly phone bill. As long as charges on the bill are reasonably in line with our expectations, we give the bill little thought. However, when a decided change in charges occurs, we immediately show interest in rates and pricing.

Fundamental changes are occurring within the telecommunications industry that recently have been reflected in our telephone bills and in our daily lives. Our telephone bills have become more complex, and we have to understand the new entries in order to interpret them. Many of us have begun to receive service from more than one telephone company and therefore receive two or more monthly telephone bills. Some of us have bought our telephones rather than continuing to rent them. Probably most of us have become more discriminating about when we call and over what telephone facilities. We all need education about obtaining telephone repairs. Finally, some of us have been overwhelmed by our options.

This chapter examines the two principal theories of rate making and describes the transformation of the telecommunications marketplace from one of regulation to one of competition.

HISTORICAL PERSPECTIVE OF RATE MAKING

When the Bell Telephone Company was organized in 1877, Alexander Graham Bell and his associate, Gardiner Hubbard, made a decision that proved to have far-reaching effects—the decision to rent telephone equipment while selling telephone service. This represented a departure from the accepted practice of public utilities, such as electric companies, which sold electricity but required their customers to own their own electrical equipment. As independent telephone companies entered the marketplace, they followed Bell's example and patterned their charging practices in the same way. This policy of requiring customers to rent their telephone equipment remained in effect for over a century.

THE COMMUNICATIONS ACT OF 1934

The Communications Act, passed by Congress in 1934, established the Federal Communications Commission and gave it the power to:

> regulate interstate and foreign commerce by wire and radio so as to make available, so far as possible, to all the people of the United States a rapid, efficient, nationwide and worldwide wire and radio communications service with adequate facilities at reasonable prices.[1]

This act resulted in the formulation of national policy regarding telephone rates, declaring that:

1. telephones be priced within the reach of everyone
2. telephone companies be entitled to a fair return on their investment
3. all telephone companies post tariffs in advance, and the state and federal commissions have authority to approve or suspend tariffs announced by the carriers
4. all telephone companies be permitted to interconnect to the long distance network and to share in long distance revenues in proportion to their investment and volume of business
5. state public utility commissions regulate intrastate telephone rates and the operation of intrastate activities of the telephone company

THE INFLUENCE OF REGULATION ON RATE MAKING

Three factors have helped bring the price of telephones within the reach of everyone:

- *cross-subsidization:* local services subsidized by long distance services, rural services subsidized by urban services, and basic services subsidized by deluxe products
- *low depreciation rates:* long life for telephone plant and equipment with accompanying long-term depreciation rates
- *low installation rates:* installation charges capitalized; that is, charged to the investor rather than to the subscriber

■ ***THEORY OF REGULATION*** The theory underlying regulation is that regulation serves the public interest by protecting customers from the arbitrary exercise of monopolistic power. This theory assumes regulated industries have special characteristics that make them a natural monopoly. Public utilities have been allowed to operate as regulated monopolies in order to prevent wasteful duplication of services. Regulation replaces market competition and is designed to protect the customer from excessively high service rates.

■ ***TARIFFS*** *Tariffs* are the published rates, regulations, and descriptions governing the provision of communications services offered by a regulated utility. They are prepared by the utility company and submitted to the regulating commission for approval. The commission studies the tariff and approves it when it is in accordance with the commission's policies relating to earning requirements and operating practices. Each common carrier has its own tariffs to describe its service offerings and rates. There are no tariffs for interconnect vendors because they are not common carriers.

Tariffs serve as a contract between the customer and the utility. They are written in language that is easy to understand. In the case of a disagreement, the commission is empowered to interpret the tariff and render a decision.

Both intrastate and interstate tariffs are required by law to be on file in the business office of the telephone company and available for public

Intrastate Tariffs

1. Local Telephone Exchange Service
2. Auxiliary Services and Equipment
3. Message Toll and Assisted Call Services
5. Wide Area Service
6. Directory-Assistance Service
7. General Regulations
8. Private-Line Services
9. DATAPHONE Digital Service
10. Facilities for Other Common Carriers
19. Mobile Telephone Service

The absence of some numbers indicates that tariffs have been withdrawn. To avoid confusion, the number corresponding to a withdrawn tariff is not reused.

Figure 13.1
Typical Intrastate Tariffs

Local Telephone Exchange Services

Exchange Service Offerings and Monthly Rates

EXCHANGE RATES					
Rate Groups:	A	B	C	D	E
Access Lines	1 to 20,000	20,000 to 50,000	50,000 to 200,000	200,000 to 500,000	Over 500,000
Business Services:					
PBX Trunks	$21.40	23.54	25.70	27.80	30.00
1 Party	10.10	10.10	10.10	10.10	10.10
Residence Services:					
1 Party Flat	$ 8.50	9.00	9.50	10.70	11.80
2 Party Flat	6.40	7.00	7.50	8.60	9.60
1 Party Measured	7.75	8.25	8.75	10.00	11.00
2 Party Measured	5.35	5.35	5.35	5.35	5.35

Figure 13.2
Typical Exchange Service Tariffs

inspection. To properly control communications expenses, the telecommunications manager should be familiar with the tariffs affecting the services obtained from the telephone company.

Figure 13.1 shows a list of typical intrastate tariffs for an operating telephone company. Figure 13.2 shows part of a typical exchange service tariff, illustrating service offerings and monthly rates for basic service, telephone access charges, and telephone sets. Figure 13.3 shows part of a typical message toll tariff.

■ ***RATE OF RETURN*** The Communications Act provides that regulated telephone companies are entitled to a "fair return on their investment." The term *rate of return* refers to the ratio of the funds available for distribution to investors to the total funds invested. It should not be confused with *telephone rates,* which are rates for telephone service.

Figure 13.3
Typical Message Toll Tariff

Message Toll Service

Basic Rate Schedule—applies to each Message Toll Service call with certain discounts as specified.

Rate Step	Rate Miles	First Minute or Fraction	Each Additional Minute or Fraction
1	1–10	$ 0.08	$ 0.04
2	11–15	0.10	0.06
3	16–20	0.12	0.08
4	21–25	0.17	0.12
5	26–30	0.23	0.17
6	31–50	0.30	0.23
7	51–100	0.37	0.29
8	101–200	0.44	0.35
9	Over-200	0.50	0.40

Discounts from the Basic Rate:

A 30% discount applies on each call placed Monday through Friday and Sunday during the period from 5:00 P.M. to, but not including, 11:00 P.M.

A 30% discount applies on each call placed during the period from 8:00 A.M. to, but not including 5:00 P.M. certain holidays (New Year's Day, Independence Day, Labor Day, Thanksgiving Day, and Christmas Day), or their resulting holiday when said holidays fall on a weekday (Monday through Friday).

A 50% discount applies on each call placed during the period from 11:00 P.M. to, but not including, 8:00 A.M. Sunday through Friday, all day Saturday, and Sunday to, but not including, 5:00 P.M.

Additional Charges

Collect Call	$0.90 per call
Calling Card	.35 per call
Bill to Third Number	.90 per call
Request for Time and Charges	.90 per call
Person to Person	1.50 per call
Calling Card Person to Person	1.50 per call

In the telecommunications industry, the FCC and the state public utility commissions determine the net revenue requirements necessary to compensate the investors for use of their money. This rate must be high enough to attract investment capital and to preserve the financial stability of the utility. At the same time, the revenue requirement must be as low as possible to protect the telephone customers.

Once the commission has established the net revenue requirements, the telephone company develops a *rate schedule* (in the form of tariffs) that will produce enough money to meet these requirements. The commission has the responsibility to approve—or disapprove—the proposed rate schedule.

The FCC has jurisdiction over interstate revenues, which are derived principally from interstate long distance services. The state public service commissions have jurisdiction over intrastate revenues, which are derived from local area services and intrastate long distance services.

THE VALUE-OF-SERVICE CONCEPT

Historically, telephone rates have been based on the *value-of-service* concept, which holds that rates for providing a service to a specific customer should be related to the value, or utility, of the service to the customer. Thus, the value of service for customers served by larger telephone exchanges should be greater than the value of service for customers served by smaller telephone exchanges because of the greater number of customers the former can call on a local basis.

The revenue requirements of the telephone company are converted into telephone rates. Rates that employ the value-of-service concept are designed to promote widespread telephone availability and usage.

■ ***SIZE OF EXCHANGES*** Telephone company exchanges are grouped according to the number of access lines they serve. For example, one group might include all exchanges serving up to 30,000 access lines. A second group might include from 30,001 to 40,000 access lines, while a third group might be comprised of all exchanges serving from 40,001 to 70,000 access lines. The local rates for customers in an exchange area serving a larger number of access lines are higher than those for an exchange area serving a smaller number of access lines. This pricing reflects the value-of-service concept.

■ ***BUSINESS TELEPHONE SERVICE*** Similarly, the rates for business telephone service are higher than residential service rates within the same exchange area. Since a business cannot conduct its affairs without telephone service, and since the service is capable of producing revenue for the business, the value of the service is deemed to be greater for the business customer.

■ ***INDIVIDUAL-LINE AND PARTY-LINE SERVICE*** Many local exchange areas offer individual- and party-line telephone service. *Individual-line service* provides a customer with exclusive access to the serving central office, while *party-line service* shares access to the serving central office among two or more customers. (The term *individual line* should not be confused with *private line,* which refers to private facilities connecting two points for the exclusive use of one customer, such as tie lines.) Telephone rates are highest for individual-line service, and they decrease for service with larger numbers of parties. This is another application of the value-of-service concept. Clearly, the exclusive access to the serving office provided by individual-line service is of greater value than the limited access provided by party-line service.

■ ***INSTALLATION SERVICE*** Historically, the cost of installing station equipment on customer premises has far exceeded the installation charge. This pricing has helped to promote the expansion of telephone service, in keeping with the national policy goal of universal service.

■ ***LONG DISTANCE SERVICES*** Traditionally, long distance charges have been based upon mileage, the distance from the point of origination to the point of termination. Both interstate and intrastate rates are designed to produce the amount of revenue that the respective commissions feel should be derived from long distance services. The mileage concept of rates is based on value of service, not costs of providing the service.

Since it would not be feasible to calculate mileage from one telephone to another, the country is divided into rate centers. A *rate center* is a geographically specified point used for determining mileage-dependent rates. The rate center of a telephone exchange is generally a point centrally located within the exchange area. All toll traffic originating within the exchange is assumed to originate from the rate center. The rate center is identified by locating its position in terms of vertical and horizontal (V – H) coordinates on a grid map. Similarly, terminating traffic is assumed to terminate at the distant rate center. The airline mileage between any two points is calculated by a mathematical formula using the V – H coordinates of both rate centers. The cost of each call can then be determined by use of a toll message rate table. These tables have mileage as one dimension and class of call (direct-dialed, person-to-person, collect, calling-card, etc.) per time interval as the other dimension.

The mileage method of computing toll charges averages the rates. It does not take into consideration the network facilities required to complete the calls. Thus, any two calls of the same type between any two equidistant points will cost the same.

■ ***VERTICAL SERVICES*** *Vertical services* are services over and above what is required for basic communications capability; e.g., speakerphones, deluxe telephone sets, deluxe bells, or custom calling services (such as call waiting, conferencing, and automatic dialing).

When a vendor decides to offer a new telephone service, a price for the service must be established. Under regulation, the approach to pricing vertical services differs from the approaches for local exchange service and toll service. The pricing of vertical services is based upon the principle of optimum pricing. The *optimum price* is one that will yield an optimum contribution to earnings and thus permit basic services to be offered at prices lower than would otherwise be necessary to meet overall earning requirements. The optimum contribution is the largest contribution that is prudent, taking into account market conditions.

The pricing of vertical services is another application of the value-of-service concept. By virtue of their deluxe aspects, vertical services have a special appeal, and therefore value, to many customers.

In summary, the value-of-service concept, which was promoted by national policy and which had the approval of the regulatory commissions, resulted in some services being provided below cost and others being priced substantially higher than cost. These practices were deemed to be "in the public interest" and to promote the affordability of the telephone for everyone. The result was that residential services were subsidized by business services, local services were subsidized by long distance services, and basic services were subsidized by vertical services.

THE IMPACT OF COMPETITION

The Carterfone decision permitted direct electrical connection of customer-provided equipment to telephone facilities, thereby opening the telecommunications marketplace to competition. Customers no longer had to rent telephones and associated equipment from the telephone company but could purchase this equipment from the interconnect vendor of their choice.

TARIFFS FOR INTERCONNECTION

As a result of the Carterfone decision, tariffs prohibiting interconnection of private systems and equipment with the public telephone network had to be rewritten. The FCC had ruled that interconnections would be permitted as long as there was no adverse effect on the telephone company's operations or on the utility of the telephone network to others. The telephone company argued that unlimited interconnection of terminal equipment directly to the public network could cause damage to the public network. This argument was considered by the FCC, with the result that a registration program for all types of terminal equipment was established. Registration of equipment certified that it complied with technical specifications issued by the FCC. Interstate tariffs were revised to reflect the registration program.

THE COMPETITIVE MARKET IN CUSTOMER-PREMISES EQUIPMENT

Competition was first introduced in customer-premises equipment because the regulated carriers were not sufficiently innovative. Competition was expected to promote creativity in the design and manufacture of interconnect equipment, and it did. It also acted as a stimulus for cost-cutting techniques and, in general, has worked well. Given the freedom of choice, many customers — both business and residential — purchased their own telephone equipment. The competitive marketplace also stimulated the development of communications devices with the capability to interact with other electronic equipment, such as computers and word processors.

THE COMPETITIVE LONG DISTANCE MARKET

Shortly after the Carterfone decision, the MCI case (1969) decided favorably on another aspect of competition — that of providing interstate long distance services. The decision permitted the MCI Corporation to provide long distance service over high-density telephone routes based on costs. The Bell System described the practice of competing in only the most profitable markets as "creamskimming" and contended that it was unfair. Bell argued that if the telephone company were to respond to creamskimming competition, it could result in the deaveraging of telephone rates and hinder the realization of universal telephone service. Uniform rates for both rural and urban subscribers would no longer be possible; rural subscribers would have to be charged more and urban subscribers less. The arguments of creamskimming were examined by the FCC and rejected.

As a result of these two decisions, regulated carriers with rates based on value of service had to compete with carriers whose rates were based on costs. Thus, regulated companies that subsidized local services by higher long distance rates and basic services by vertical services were at a disadvantage.

Today, the long distance segment of the telecommunications industry is highly competitive. Because it controls by far the largest part of the long distance market, AT&T is classified as a *dominant carrier*. As such, its rates are subject to regulation to ensure that they do not become excessive. The rates of other long distance carriers are not subject to this constraint.

RATES BASED ON COSTS

In a free enterprise economy, the price of a product or service is determined by adding a profit margin to the costs incurred in producing the product or service. Nonregulated entrants in the telecommunications marketplace base their rates and charges on the costs of providing the product or service. This is *cost-based pricing*.

Two major decisions caused a restructuring of telecommunications pricing policies: the Computer Inquiry II decision (1981) and the Modified Final Judgment (1982), which settled the United States Department of Justice antitrust suit of 1974 against AT&T.

■ ***DEREGULATION*** The Computer Inquiry II decision provided that customer-premises equipment supplied by the Bell System be deregulated. The Modified Final Judgment stipulated that the CPE segment of the business be transferred to an unregulated Bell subsidiary. Subsequently, AT&T organized the American Bell Company to market and service customer-premises equipment in competition with other vendors. These events necessitated Bells "unbundling" local service offerings by pricing their rates for switching services separately from prices for customer-premises equipment.

Since customer-premises equipment was no longer subsidized by other sources of revenue, it was inevitable that AT&T restructure its prices, basing them on costs.

■ ***NETWORK ACCESS*** The MFJ stipulated that the industry practice of AT&T sharing long distance profits with Bell and independent telephone companies must be discontinued. To compensate the operating telephone companies for their costs, the FCC directed that the unbundled rate for each business and residence telephone line be increased to cover the loss of the subsidy. To implement this policy, the FCC ordered that a *federal access charge* be imposed as a surcharge to each subscriber's line charge.

The FCC also directed that all long distance companies, including AT&T, be required to pay the operating telephone companies long distance access fees. In return for these fees, the operating telephone companies were to provide equal network access service to any long distance telephone company.

■ ***LOCAL ACCESS AND TRANSPORT AREAS*** The Modified Final Judgment ordered AT&T to divest itself of all its operating telephone companies. As a result of this divestiture, the spun-off BOCs were formed into seven regions. Their charter provided for monopoly carriage of local telephone traffic and its switching. Local calling areas were mapped into 165 Local Access and Transport Areas (LATAs) throughout the United States. The operating companies were empowered to handle intra-LATA calls and to charge all long distance carriers, including AT&T, for connecting calls to and from their LATAs.

Only long distance companies were empowered to provide telephone service between LATAs. No long distance company was required to serve any or all specific LATAs. Thus, if the costs of serving an area become so high that is unprofitable, long distance companies are not required to offer service.

These provisions resulted in two very important consequences:

1. Long distance services no longer subsidize local telephone services.
2. No long distance company—AT&T or other—is required to serve high-cost, low-density telephone routes.

AN INDUSTRY IN TRANSITION

Telecommunications is a dynamic, technology-driven industry. Prior to deregulation, operating telephone companies had moved in the direction of cost-based pricing by introducing *measured service,* a method of pricing which is *usage sensitive;* that is, based on the amount of usage. In many areas, measured service replaced *flat rate service,* in which a subscriber paid a fixed monthly charge for unlimited telephone service.

In an effort to reduce labor costs, telephone companies had promoted direct-dialing of long distance calls at reduced rates. To motivate more uniform traffic loading and hence use the telephone network more efficiently, the company had introduced discount evening, night, and weekend rates.

The year 1983 marked the beginning of deregulation in the telecommunications industry, the first step in what appears to be systematic, comprehensive regulatory reform. The divestiture of AT&T, with the division of responsibility among AT&T and the Bell operating companies, meant that users could no longer depend on "The Telephone Company" for everything. Users had to assume more responsibility for managing their own telephone systems.

Over the years, United States telephone users have experienced the best service in the world at comparatively low prices. The goal of universal telephone service became a reality. The rate policies of the regulatory commissions, along with technological advancements, protected telephone subscribers from the rampant inflation that pervaded most of the economy. In terms of constant dollars, telephone subscribes came out way ahead.

Recent changes in the FCC philosophy from one of protective regulation to one of market pricing have charted a different course for telecommunication rates. The termination of cross-subsidies as a result of the Modified Final Judgment has cause local telephone rates to rise. Compensating influences have been provided by advances in technology and competitive long distance pricing in a market-driven economy.

Another result of the FCC's change in philosophy was their decision to substitute *price-cap regulation* for *rate-of-return regulation* for AT&T. Under this decision, regulation that limited AT&T's earnings were traded for a ceiling on prices that AT&T charged for its long distance services. The new regulation procedure is expected to encourage innovation and motivate AT&T to reduce costs and improve efficiency.

It is anticipated that price-cap regulation will be extended by the FCC to apply to the intrastate services provided by local operating companies as well as AT&T.

SUMMARY

The Communications Act of 1934 established national policy regarding telephone rates. A basic principle of this policy was that telephones be priced within the reach of everyone. The following three factors helped to bring telephones within the reach of everyone:

1. subsidization of local telephone services by long distance services
2. long life for telephone plant and equipment, resulting in low depreciation rates
3. installation services subsidized by investors

The theory of regulation is that the public interest is served by regulating agencies that protect customers from the arbitrary exercise of monopolistic power. Public utilities have been allowed to operate as regulated monopolies to prevent wasteful duplication of services. The published rates, regulations, and descriptions governing the provision of communications services offered by a regulated utility are known as tariffs.

Historically, telephone rates have been based on the value-of-service concept, which holds that rates for providing a service be related to the value of the service to the customer. Thus, rates for business telephone service are higher than the rates for residential service within the same exchange area, because the value of the service is deemed to be greater for the business customer. Similarly, the value of urban services is deemed to be greater than the value of rural services because of the greater number of telephones subscribers can call on a local basis.

The telecommunications industry is presently in a state of transition. After operating for many years as a regulated monopoly, the industry now finds itself partially deregulated and in a highly competitive environment. Nonregulated entrants in the telecommunications marketplace base their rates on the cost of providing the products or service, with no subsidization of one service by another. Recent FCC rulings and court decisions have made it necessary for regulated telecommunications carriers to eliminate subsidies for equipment and local service. To compensate for loss of cross-subsidization, regulated carriers have found it necessary to increase local rates. Countervailing influences have been provided by advances in technology and competitive long distance pricing.

REVIEW QUESTIONS

1. What was the basic premise of the Communications Act of 1934?
2. What are the factors that have helped bring telephone service within the financial reach of nearly everyone?
3. What is the underlying theory of regulation? Why have public utilities traditionally been allowed to operate as regulated monopolies?
4. Explain the value-of-service concept and its role in telephone rate making.
5. What are vertical services, and how are they priced?
6. What two decisions opened the telecommunications marketplace to competition, and what two aspects of telecommunications were involved?
7. Briefly describe the changes in the telecommunications industry resulting from the Computer Inquiry II decision and the Modified Final Judgment.

VOCABULARY

cross-subsidization

rate of return

telephone rates

rate schedule

value-of-service pricing
individual-line service
party-line service
rate center
vertical services
dominant carrier
cost-based pricing
network access
federal access charge
usage sensitive
flat rate service
unbundling
rate-of-return regulation
price-cap regulation

ENDNOTE

1 *The Communications Act of 1934,* Title I, Section 1, p. 1, 47 USC 151.

CHAPTER

14

NEW DIRECTIONS IN TELECOMMUNICATIONS

CHAPTER OBJECTIVES

After completing this chapter, the reader should be able:

- *To discuss the factors that led to the creation of a highly competitive and uncertain marketplace for the providers and users of telecommunications services.*
- *To discuss research in progress and leading-edge developments that hold great promise for the advancement of the telecommunications industry.*
- *To define videotex, describe the technology, and discuss its role in providing information services.*
- *To describe voice processing.*
- *To describe telemetry, public communication centers, teleregistration, and telecommuting.*
- *To describe enhanced voice services.*
- *To discuss the major trends and predictions of the government study Telecom 2000.*

Rapid evolution of telecommunications technology and the break-up of the Bell System have significantly contributed to the creation of a highly competitive and uncertain marketplace for the providers and users of communication services. Capital investment decisions and strategic planning must be based upon a continually evolving network that considers the direction of state-of-the-art technology and the increasing level of demand for new products and services.

This chapter examines the technology underlying recent advances in telecommunications and the resultant marketplace for telecommunications services and applications.

THE TECHNOLOGY

The telecommunications industry has experienced many technological changes in the past, bringing innovative new services to telecommunications users. Research in progress and leading-edge developments suggest that the industry will see even more changes in the future.

COMPUTER TECHNOLOGY

Some of the areas in which ongoing research promises to contribute to the advancement of computer technology include photonic switching, microelectronics circuitry, optical interconnection, and microprocessor chip technology.

■ ***PHOTONIC SWITCHING*** The first photonic switch, a device which may become the primary building block of a totally new kind of computer or communication switching machine, has been developed by AT&T Bell Laboratories. The new switch, actuated by photons (packets of light), opens up possibilities for an optical computer that can carry on a large number of parallel operations at the same time. Consequently it will be much faster in operation than the current electronic computers, which work in serial fashion.

Photonic switching also improves transmission reliability and upgrades security, since photonic signals are far more difficult to intercept surreptitiously than are electronic signals.

■ ***GALLIUM ARSENIDE CIRCUITS*** AT&T Bell Laboratories, IBM, and others are working to develop high-performance gallium arsenide circuits for logic and memory functions in super high-speed computers and communications systems. Each chip will contain up to 5,000 logic gates and operate at a speed of more than 200 megahertz (200 million cycles per second). The circuits will use photons to transmit both voice and data faster and easier than currently used techniques.

■ ***GALLIUM ARSENIDE CHIPS*** Researchers also hope to use gallium arsenide to make chips for electronic computers. The chemical conducts electricity five times faster than silicon, the material traditionally used, but it is costly, brittle, and retains heat. Researchers at the University of Illinois have developed a hybrid chip made of alternating layers of gallium arsenide and silicon, thus combining the best qualities of each.

■ ***OPTICAL INTERCONNECTION*** With the shrinking dimensions of very large scale integration, a limit is being reached in the fabrication of metallic conductors to interlink chips and circuit boards. A technique of optical interconnection has been developed using a laser beam conducted along an optical waveguide and picked up by a photonic detector. Such a connection generates almost no heat and has high reliability, long life, and the same large bandwidth and freedom from electrical interference as lightwave technology.

COMMUNICATIONS TECHNOLOGY

Some of the areas in which ongoing research and development promise to contribute to the advancement of communications technology include optical communication, wideband packet transmission, and common channel signaling.

■ ***OPTICAL COMMUNICATIONS*** At laboratories in the United States, Japan, and Great Britain, research is being carried out on a new system of lightwave modulation and detection which may be the next great advance in optical communication. The new method, called "coherent detection," can provide many closely spaced multi-gigabit (billions of bits) channels over lightguide fibers by using multiplexing techniques. Experimental coherent lightwave systems can currently achieve a rate of 2 billion bits per second. In the coherent system, ten such channels can be derived and sent over a single optical fiber. With further development, such a system would be capable of sending 10 million telephone conversations or 10,000 television channels on one lightguide.

■ ***WIDEBAND PACKET TRANSMISSION*** Wideband packet transmission is a way of combining voice, data, and image transmission by converting the signals into digital form (bits) and assembling the pulses into small groups of "packets." The packets can then be sent individually, each to its own address over one or several high-speed digital lines. Packets are a more efficient and less costly way to move information. The technique, which is compatible with ISDN, will find early application in local exchange and interexchange networks with heavy traffic loads.

■ ***COMMON CHANNEL SIGNALING (CCS)*** A new interoffice signaling system allows local telephone companies to offer advanced telephone services. In *common channel signaling,* also known as *signaling system 7 (SS7),* all signaling functions are put on a separate packet-switched network and sent on a channel independent of the message channel. The system offers telephone users several new service features, such as tracing incoming calls by visually displaying the callers' phone numbers, forwarding calls, and blocking incoming calls from selected numbers. The CCS system will ultimately be linked to the nationwide long distance network.

BARRIERS TO PRACTICAL APPLICATIONS

The advances in telecommunications have outdistanced our ability to put them to practical uses. This phenomenon is not new; it has been true of nearly all discoveries throughout history. The internal combustion engine was available for over fifty years before it was put to practical use in the automobile. Gasoline was a waste product in the oil-refining process until it was used to fuel automobiles. Television technology was demonstrated some 25 years before it became available commercially.

There are several factors that tend to delay the implementation of new technology, namely:

- ☐ absence of creative ideas for practical use
- ☐ economic considerations
- ☐ legal and political deterrents
- ☐ attitudinal barriers

■ ***ABSENCE OF CREATIVE IDEAS FOR PRACTICAL USES*** Technological discoveries often occur as an offshoot of other research. Uses for these unexpected peripheral findings are not immediately apparent. Sometimes these types of discoveries are more important than the product of the basic research; however, the development of innovative ideas for their practical use depends upon the expensive trial-and-error process. Creative ideas cannot be developed upon demand; thus, there is usually a lag between any technological discovery and its practical application.

■ ***ECONOMIC CONSIDERATIONS*** Economic considerations play an important role in finding and developing practical applications for technological discoveries. Research activities are costly and time-consuming. The high cost of implementing new technologies must be weighed against the potential advantages they afford. Because the shape and dimension of the market are not clear, potential suppliers are reluctant to risk "technology-in-search-of-a-marketplace investments." Even when a cost-benefit analysis is favorable, the availability of investment capital can be a controlling factor.

In the case of chip technology, the initial development was very expensive. Yet once the circuitry had been designed, the chips could be mass-produced and sold in large quantities. The costs of research, development, and tooling had to be recovered before the chip could be replaced with newer technology. In other words, product obsolescence had to be balanced against economic considerations. At one time the chip industry had made twofold capacity increases standard, but the high cost of retooling to manufacture a new chip ruled out anything less than a fourfold jump. This criterion has been demonstrated by the recent increase in chip capacity from 256K to 1024K rather than to 512K.

In the telecommunications industry, many billions of dollars have been invested in analog transmission facilities. These facilities have formed the backbone of our communications system. They meet exacting maintenance standards and provide excellent communication service. However, digital transmission systems are definitely superior to analog. If the telecommunications industry were starting all over today, it would build all digital facilities. Unfortunately, the staggering amounts of work, money, and time required to convert the entire system mandate that the conversion be phased in over a period of time. Similarly, in spite of the superiority and widespread availability of electronic switching systems, many telephones will probably continue to be served by electromechanical systems for some time.

■ ***LEGAL AND POLITICAL DETERRENTS*** In a partially regulated industry such as telecommunications, legal and political forces are major determinants of industry policy. National policy was established by Congress in the Communications Act of 1934. Although the technological environment has undergone sweeping changes over the last fifty years, the legal environment has remained relatively constant. Regulatory agencies must work within a legal framework in interpreting the Communications Act. They prescribe the rates of return on invested capital, rates of depreciation, and methods of raising capital for telecommunications companies. This environment has generally slowed down the introduction of new products and services.

■ ***ATTITUDINAL BARRIERS*** People are creatures of habit. They generally resist change; it disrupts the comfortable status quo and evokes anxiety about their ability to master the new ways. The implementation of new technology involves change — in procedures, equipment, required skills and knowledge, organizational relationships, and/or work environments. Resistance to change is a natural phenomenon. Research studies show that attitudes about communications technologies are one of the most complex barriers to surmount in getting people to use new communications technologies.

THE MARKETPLACE: APPLICATIONS OF THE TECHNOLOGY

Today it is possible to accomplish more and more of our daily tasks electronically. There are electronic banking, electronic messaging, electronic shopping, and even electronic commuting, better known as telecommuting. These new telecommunications services are converting our homes and offices into an electronic environment and changing our work patterns, leisure activities, transportation, education, manufacturing, health care, news media, and government.

VIDEOTEX SERVICES

Videotex is a generic term used to describe a new group of consumer-oriented electronic information services that enable users to retrieve information stored in a computer data base on a television or microprocessor monitor screen. The term *videotex* includes both *teletext,* a one-way transmission system in which data signals are transmitted over the FM portion of a television signal, and *videotext,* a two-way interactive system that is transmitted over telephone lines. Videotex systems combine television, telephone, and computer technologies. They are sometimes regarded as an electronic version of the daily newspaper.

Videotex involves the transmission of text and high-resolution graphic information, which are prepared and delivered as single screens or "pages." The system is two-way; a user's request is sent upstream via coaxial cable or phone lines to a central computer, and the material requested is, in turn, sent downstream to the user. At the user's end, the digital data is either fed into a special terminal that translates the information into a television picture or processed via software decoders for display on a personal computer.

Videotex services can be classified in five general categories: retrieval (news, sports, weather, travel schedules, stock market reports), computing (computer-assisted instruction, management information services, income-tax aids), transactional (teleshopping, telebanking), messaging (sending and receiving), and downline loading (transmitting software programs to intelligent terminals).

■ ***VIDEOTEX SYSTEMS*** A *videotex system* is composed of three major segments: the service provider, the communications network, and the user. The service provider operates the computer system and maintains the data base and services that are accessed by the users. Service providers are often cable television operators, public data networks, and companies already in the business of providing information, such as newspapers.

The backbone of the videotex system is its communications network. This could be either a phone-based network or a cable TV network. Cable TV networks, however, have an advantage over phone-based networks because their wider bandwidth enables them to carry considerably more information than telephone lines.

To use videotex service, either a videotex terminal or a microprocessor is required. A *videotex terminal* consists of a control console connected to a television set and to a conventional telephone line. A microprocessor requires connection to a telephone line by means of a modem. The user accesses the service by dialing a local telephone number that connects the terminal to a computer system containing the information and programs. The user can select specific files, and the desired data is displayed on the screen of the television set or the microprocessor.

There are generally three charges for videotex service: a subscription fee, the telephone call (usually at local telephone rates), and a charge for the time the system is used.

■ ***HISTORICAL PERSPECTIVE*** The concept of videotex originated in the late 1970s. Early developments took place principally in Europe. The world's first public videotex service, Prestel, was formally launched by British Telecom in 1980, after a public trial lasting several years. Since it was the first public videotex service, Prestel has been a model for many other systems. However, Prestel has not lived up to the expectations of its founders, principally because there are so few users and growth has been slow.

By the end of 1985 there were more than two dozen public videotex systems operating around the world. Some of the systems operating in other countries include Blidschirmtext in West Germany, Viditel in Holland, Telset in Finland, Teletel in France, Prestel in Sweden, Telidon in Canada, and CAPTAINS in Japan.

In other countries the government exercises much greater control over the telecommunications system than in the United States. With the government paying the bills, the systems can be implemented with less concern about their ability to generate a profit. In this country, the government has not taken an active role in developing videotex, and the United States has therefore lagged behind other countries.

■ ***TELETEL: THE FRENCH VIDEOTEX SYSTEM*** The best known of the European videotex offerings is the French Teletel system, popularly known as Minitel. The service is operated by the Direction Generale des Telecommunications (DGT), France's government-operated telecommunications authority. Minitel is by far the largest videotex service in the world; however, it was achieved only through considerable state support.

Videotex service was introduced in France as part of the modernization and expansion of the French public network. The service was seen as the best way of modernizing directory assistance service for telephone subscribers. Existing paper directories were voluminous, making them difficult and costly to update. Additionally, they had to be supported by a directory inquiry service, a costly, labor-intensive procedure.

The Electronic Directory Service (EDS) was implemented to solve these problems. It is the only videotex service operated by the DGT. Inexpensive dumb terminals, known as "Minitels," were provided free of charge to telephone subscribers to replace the paper directory. By early 1987 over 2.5 million of these terminals had been placed in users' hands; by 1990 it is expected that every telephone subscriber will have one.

The electronic directory service has no equivalent anywhere in the world. The system stores directory details for 24 million telephone subscribers in France, and any system user can immediately obtain directory information on any one of these subscribers. Over 4,000 users can be connected to the system simultaneously, with a response time of less than 2 seconds. As an added benefit, the free distribution of Minitel terminals enables telephone subscribers to become familiar with the terminals. Once the terminals are in place, they can be used for other types of information retrieval, such as electronic banking, home shopping, and so forth, thus allowing users to explore the videotex system.

The data base for the electronic directory is set up and kept current by the DGT. Almost 4,000 other information providers offer people access to a wide range of services. A look at Teletel's directory gives an idea of how varied these services are. They include news reports, round-the-clock reports on sporting events, classified ads, banking information, airline and rail reservations, entertainment event and restaurant reservations, computer-aided teaching, armchair shopping, price comparisons between supermarkets, video games, and dating services. The information providers control and update their own data bases.

In addition to providing the videotex terminals, the French government has stimulated the use of videotex services by implementing an interesting marketing strategy — kiosk billing. This system allows end users to dial one number anywhere in the country to browse through and purchase electronic services on an impulse—without the need for prior subscription to that service. The bill for access is conveniently processed and collected by the telephone company, with about two-thirds of the amount forwarded to the appropriate information provider.

The electronic kiosk allows customers of information services the same ease of access that customers enjoy at a corner newsstand; that is, they can buy a publication without a subscription. They pay only for the specific item they choose.

When kiosk billing was first introduced as a network enhancement in 1982, it was an immediate success. Transpac, the French national data network, now carries more traffic than the rest of the world's public data networks combined.

■ ***VIDEOTEX IN THE UNITED STATES*** More than three dozen videotex experiments have been carried out in the United States. On the whole, they have been unsuccessful. One of the largest was Viewtron, which operated in the Miami area for nearly three years before closing for financial reasons.

The most successful videotex services operating in this country have been text-only services. The largest is CompuServe, which supports a full range of applications including news, weather, and sports information; shopping; messaging; financial services; travel information; and educational services.

Because of the large capital outlay required to implement a videotex system, private companies have found it advisable to form partnerships. Three of the larger partnerships are Prodigy (IBM, Sears), COVIDEA (AT&T, Chemical Bank, Bank of America), and CNR Partners (CITICORP, NYNEX Corporation, RCA).

■ ***THE TELEPHONE COMPANIES' ROLE IN VIDEOTEX*** Under the terms of the FCC's Computer Inquiry II, the Bell companies were allowed to provide enhanced services through separate subsidiaries. However, the MFJ agreement prohibited them from providing information services. The rules that came out of the FCC's Computer Inquiry III proceedings provided for replacing the separate subsidiary concept with *Open Network Architecture (ONA)*. ONA embodies a set of guidelines that permit the Bell companies to integrate their enhanced-services operations with their basic networks, in exchange for giving other enhanced-service providers access to the basic networks.

Subsequently, Judge Greene loosened the divestiture agreement's restrictions on information services by ruling that the Bells can provide information-transmission services—such as electronic mail, voice mail, and videotex—thus paving the way for the Bells to begin testing *information gateways* and other noncontent-based information services. In other words, the Bells can provide transmission service for information providers but they cannot provide the information itself.

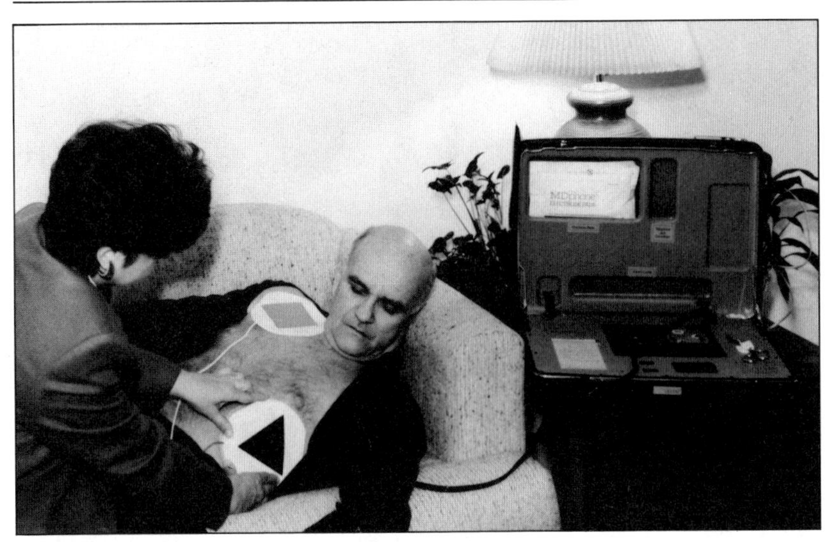

Figure 14.1
MDphone™ Patient Unit

(Reprinted with permission of MEDphone Corporation)

VIDEOTEX APPLICATIONS

Videotex has applications in many areas, including health care, banking, education, shopping, entertainment, and polling.

■ ***TELEMEDICINE*** Telecommunications technology makes it possible to provide health care services from a distance, a concept known as *telemedicine*. Under this concept a remote location is linked by two-way television and audio signals with a hospital or other health-care facility, allowing physician and patient to interact directly. The equipment enables doctors to conduct patient examinations, inspect X-rays and electrocardiograms, study microscopic specimens, diagnose medical conditions, and prescribe treatment. This technology also permits doctors to access medical reference sources and to consult with medical specialists.

Since there are many remote areas where doctors are in short supply and since medical equipment is extremely costly, telemedicine has economic potential for improving health care. It also makes more efficient use of medical personnel and equipment.

A recent development in telemedicine is the MDphone™, a transtelephonic defibrillator, which is said to be the first device of its kind that can remotely diagnose and treat patients exhibiting cardiac arrhythmia using standard telephone lines. The MDphone™ consists of a hospital-based display station and a portable unit contained in a briefcase. When the briefcase is opened at the scene of a cardiac emergency, it automatically dials the Base Station and establishes two-way communication. Two self-adhesive electrode pads, which are in the Patient Unit case, are placed on the patient's chest, which allows for the monitoring of the patient's electrocardiogram (Figure 14.1).

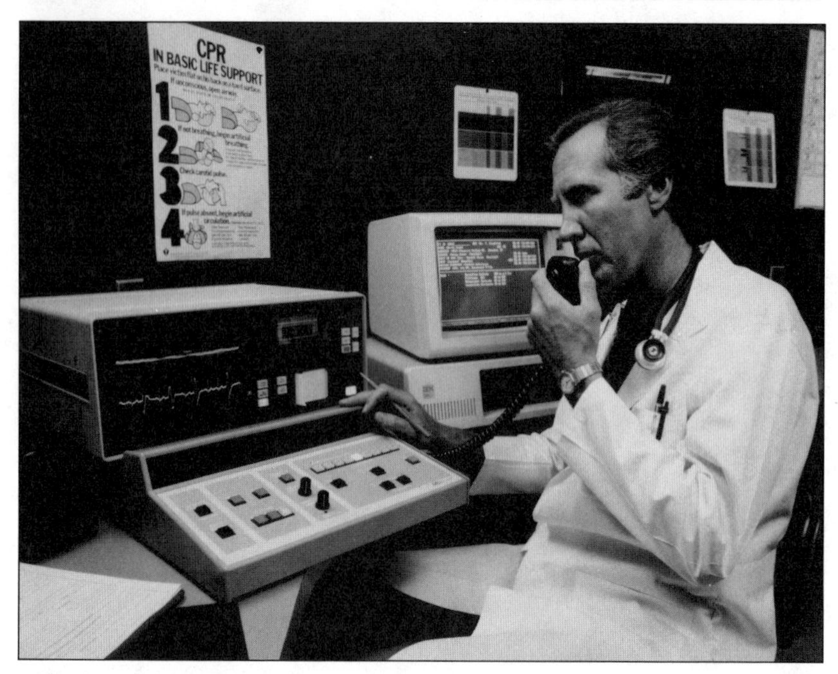

Figure 14.2
MDphone™ Base Station

(Reprinted with permission of MEDphone Corporation)

If the data transmitted to the hospital indicates cardiac failure, the clinician can remotely send a signal to charge and fire the patient defibrillator (Figure 14.2).

Other applications of the MDphone™ include fetal and geriatric monitoring. The device is the first in a family of interactive telephonic medical equipment that MEDphone Corporation is developing. The MDphone™ is currently being used by a number of hospitals in the United States.

■ ***HOME BANKING*** Home banking service allows customers with personal computers or videotex terminals to check the balances in their accounts, move funds from one account to another, pay bills, and do just about anything else other than make a deposit or a cash withdrawal.

The services reduce the amount of paperwork banks have to process. However, some retailers have refused to accept electronic transfers. When a home banking customer makes a payment to one of these retailers, the bank translates the customer's computerized order into a cashier's check and sends it along.

Banking institutions also use several other categories of electronic funds transfer that do not require videotex terminals or microcomputers:

1. Automated teller machines (ATMs), either locally or remotely attached to financial processing centers that connect to shared or switched ATM networks

2. Telephone bill paying service, which allows a customer to pay bills to participating merchants using a Touch-Tone telephone for data entry
3. Point-of-sale terminals and networks tied directly to a customer's bank account for purposes of check verification, credit authorization, and checkless (or debit card) purchases
4. Automated clearing houses, which facilitate the paperless dispensing and collecting of thousands of transactions automatically

Today, point-of-sale terminals represent the fastest growing segment of the electronic funds transfer category. In this application, customers use computer-coded bank cards to transfer funds from bank accounts to pay for merchandise. The transaction differs from credit-card sales in that the purchaser's bank account is debited when the sale is made, thus eliminating the need for sending out monthly bills and waiting for payment. Such systems are already online in a few states at supermarket and convenience store checkouts, gas stations, and department stores.

■ ***EDUCATIONAL SERVICES*** The potential of videotex in education stems from four important characteristics of this technology:

1. Two-way interactive capability supports individualization and leads to a better match between learner style and the instructional presentation.
2. Videotex instruction can be delivered to the home, where the learner can access and interact with a central computer any time on a 24-hour basis.
3. Videotex systems are easily used; knowledge of computer programming is not required.
4. Videotex systems can deliver high-resolution graphics.

The concept of *computer-assisted instruction (CAI)* has interested educators for some time, but until recently the high cost of computers delayed its implementation. With the advent of microprocessors, inexpensive microcomputers have made CAI feasible for use in the classroom and for delivering instruction to the home. CAI offers students the opportunity for high-quality, impartial, individualized instruction. Computers can identify learning problems and adapt instruction to the specific needs and learning pace of the students. They impose discipline upon learners, forcing them to attain mastery levels of competence before progressing to the next step.

Videotex systems can also be used to deliver instruction to remote locations, thereby permitting students to learn at home, a concept known as *distance learning.* This term refers to teaching via live audio and video transmission of lessons from a host classroom to distant receiving sites. The videotex aspect of distance learning stems from the fact that students interact with the source on a two-way basis.

Many urban and rural schools with low student enrollments have been able to continue providing education through distance learning. Students work with microcomputers that are connected to modems. They send their work over a telephone line to the teacher's site. After the teacher has corrected the work, it is sent back through the same channels.

Videotex systems are well suited for college independent study programs. Students enrolled in such programs have varying backgrounds and learner characteristics. He or she may be an on-campus student who is taking independent study because of schedule conflicts, or the profile may be that of an upwardly mobile working person taking courses for career advancement. Since independent-study students are removed from class structure and have no chance for interaction with other members of the class, they do not have the advantage of immediate feedback to questions. Thus, it is important that the instruction should be as self-explanatory as possible, should actively involve the learner, and should provide individualized feedback simulating a classroom environment. Videotex meets these requirements.

■ ***LIBRARIES*** Videotex service changes libraries into electronic information centers where users can access a central data base with a microcomputer. The data base may contain indexes, abstracts, citations, bibliographies, papers, or even full-text records. Computerized indexes permit users to search for items on a particular topic at a terminal screen; no card catalog is necessary. Computerized information systems can be accessed from anywhere in the world via telecommunication links, and many users can access the same data base at the same time. Homes equipped with videotex capabilities can also take advantage of the resources of electronic libraries.

■ ***TELESHOPPING*** Videotex home shopping service allows the user to see a product or service advertised on the television or microprocessor screen in his or her home, order it via terminal, and arrange for payment by credit card. The service enables shoppers to use comparative information on price and brand to hunt for bargains or find the best-quality products. Viewers select the merchandise they wish to purchase by pressing a key on the console to indicate their choice of the available options. The merchandise is then delivered to the purchaser's home.

Teleshopping is also available via cable TV; however, with cable TV the purchaser users the telephone to order the product rather than the videotex terminal or microprocessor.

■ ***ENTERTAINMENT INFORMATION SERVICES*** Entertainment information services provide information about restaurants, movies, plays, and tourism (places to go and transportation schedules). It is also possible to make reservations or purchase tickets for a specific activity through videotex terminals.

■ ***ELECTRONIC POLLING*** Videotex technology, with its two-way interactive characteristics, offers users the opportunity to voice their opinions about issues by participating in electronic polls. The subscriber can make and register choices by following instructions displayed on the terminal screen. This principle can be adapted for use in a variety of applications, such as conducting a survey or holding an election.

Electronic polling has the potential to stimulate interest in current affairs and make it easier for people to vote. However, its use is limited today because electronic polling requires the use of either a videotex terminal or a microprocessor and modem. Since the majority of voters do not have this equipment, participants in an electronic poll would not be a random sample of the general population. Therefore, the results could not be used for predictive purposes. Additionally, electronic polling would not be valid for general elections unless every qualified voter had access to an appropriate terminal.

VOICE PROCESSING

The technology of computers' speaking, storing human voices, and reacting to human speech is known as *voice processing*. The four types of voice-processing systems include:

1. voice response (synthesized speech)
2. voice recognition
3. voice store and forward
4. voice identification

Voice response is the conversion of computer output into spoken words and phrases that a human being can hear and understand. The computer combines various frequencies of electrical impulses stored in its memory to create vowels and syllables. Under the direction of a computer program, the vowels and syllables are synthesized into audible words, creating responses to program requests.

The systems present information via telephone in a natural voice form to callers. Messages are generated from a pre-recorded vocabulary of words, phrases, or sentences, and delivered in response to Touch-Tone dial input. This technology is currently in use by the financial industry (quoting account balances), by the telephone company (responding to directory assistance requests and reporting the status of telephone numbers not in service), and many other industries.

Voice recognition is the ability of a computer to understand and react to the human voice rather than being able to accept typed commands. Voice recognition is a new technology still undergoing laboratory testing; it is limited by the computer's ability to understand human speech.

One of the first consumer products to use voice-activation technology is the new FV 1000 Southwestern Bell Freedom Phone (Figure 14.3). The phone is unique in three ways: (1) It is totally voice activated—there are no dials or keypads; (2) it recognizes virtually any English-speaking voice (about 300 dialects); and (3) it does not need programming to recognize an individual's voice.

Figure 14.3
FV 1000 Voice-Activated Telephone

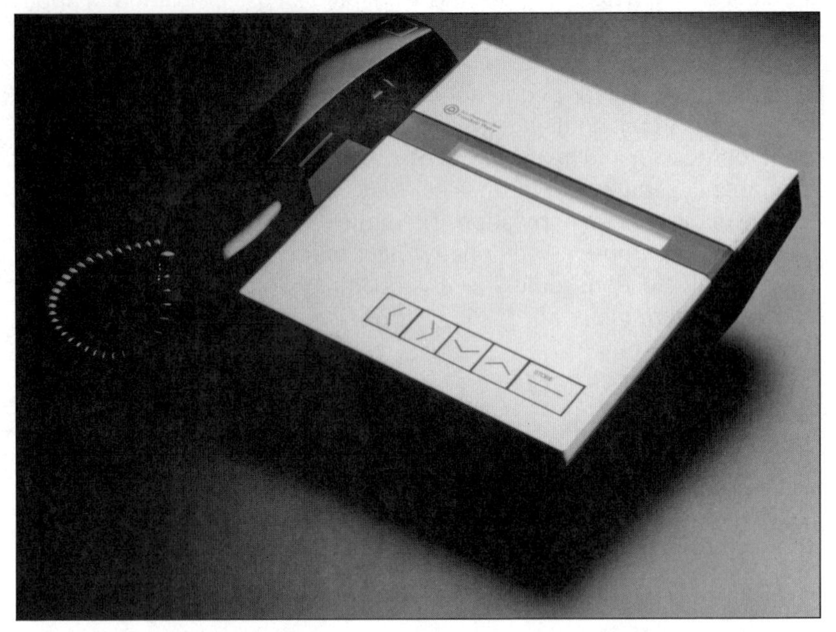

(Courtesy of Southwestern Bell Freedom Phone)

In all, this telephone recognizes about 40 words. These include digits, command words such as "call" and "dial," and function words like "flash" and "hold." There are basically three ways to dial:

1. Using key words — say "Call home" or "Call office."
2. Using generic memory — say "Memory: one, one."
3. Using digits that would be dialed on a conventional phone — say "Dial: five, five, five, one, two, one, two, one, two."

Today, computers with voice-recognition capabilities have limited vocabularies and are very expensive. As voice-recognition systems are further developed, their large-scale use is virtually assured because it is easier to speak than to keyboard. (Remember how the telephone has virtually replaced letter-writing as a means of keeping in touch with friends and relatives? It is easier to speak than to write!)

Voice store-and-forward systems enable a computer system to accept a message and store it until a transmission path is available or until the desired party calls to retrieve it. The latter concept is known as *voice mail.* An important advantage of voice mail systems is that messages are not missed because the recipient is unavailable at the time the message is delivered. Voice mail has become an integral part of office automation systems. Its cost has declined markedly during the last few years, resulting in a high rate of growth. (Voice mail is explained in more detail in Chapter 9.)

Voice identification (or speaker verification and/or authentication) is a technology used to verify the identity of a person accessing a system by comparing spoken passwords with computerized voice patterns. Voice identification offers a possible method for controlling and accessing secured locations or sensitive data. This technology is still under development.

TELEMETRY

Telemetry, the process of reading utility meters via telecommunications, uses ordinary phone lines to automatically transmit electric, gas, and water meter readings. The process uses encoders, supplied by the utilities, which are attached to meters to translate the readings on the meter dials into digital format. At the telephone switching center, a scanner and connecting equipment route all meter reading signals to and from a central host computer that directs the entire system.

The computer can read a meter in a fraction of a second, without ringing the phone. If someone is using the phone, the system waits until the line is free. The readings are accurate, timely, and virtually eliminate the need for visits by meter readers.

A trial program conducted by the Southern New England Telephone Company was extremely successful. A survey of the trial participants revealed an extremely positive response to the program. Benefits most often cited included the elimination of estimated bills and the virtual elimination of meter readers. Where the system is offered commercially, power and water utilities purchase and install the necessary equipment in their customers' homes. The utilities also pay the telephone company for each reading.

PUBLIC COMMUNICATIONS CENTERS

Public communications centers are the newest addition to airports, hotel lobbies, convention centers, truck stops, restaurants, courthouses, and airport executive lounges across the country. These public access terminals include a credit-card-enhanced telephone equipped with a computer keyboard, a video screen, and a fax machine for sending and receiving documents (Figure 14.4). Credit-card users can gain access to both public and private electronic-mail networks and can send messages by overnight letter couriers. These units can link into numerous data bases and networks, such as AT&T, CompuServe, Dow Jones Retrieval, Dialcom, GE Quick-Com, MCI Mail, Source Mail, and others.

In addition, users can connect to mainframe and personal computers. A secretarial service is also available, enabling users to dictate important correspondence and have it typed and faxed back to the user or anywhere else in minutes.

An important advantage of these centers is that they are available 24 hours a day, seven days a week. The service is helpful to the executive whose meetings often run late into the evening.

Figure 14.4
Public Communications Center

(Courtesy of Compugram™ Information Network)

A related service is *public facsimile-machine booths.* These booths are joining phone booths and vending machines in a variety of public-access locations across the country. For the traveling businessperson, a fax booth is the next best thing to a portable fax machine. The service is convenient for those who travel and, of course, for people who only occasionally need a fax machine.

To send a fax, the sender fills out a cover sheet with the recipient's fax number and the sender's name and credit card number, places the document on the feeder, and presses the "send" button. The fax is sent to the company's nearest computer center, where it is stored and forwarded to its destination.

TELE-REGISTRATION

College registration is probably one of the more unpleasant realities of student life. The long lines, visits to advisors, course cancellations, and course closings hardly add up to an enjoyable experience. But the use of new telecommunications technology, called *tele-registration,* can make the experience less burdensome.

A voice response system can automate the student registration process, letting college students register for classes from any Touch-Tone telephone. The availability and low cost of the necessary "terminals" (home and office telephones) make the registration process quick, simple, and convenient.

Voice response systems use digitized human voice—scripted into prompts, instructions, and information—to talk to incoming callers, who, in turn, "talk" back using the keypad of a Touch-Tone phone. When a student calls the registration number, he or she accesses one of a number of available ports on the computer. A human-sounding voice greets the caller and gives instructions on how to use the system.

The student then keys in an identification number, receives verbal confirmation, and proceeds to enter the desired course numbers printed in the catalog of classes. After each entry, the system checks the class's availability and notifies the computer accordingly. A verbal approval is given if the class is open, or the student is asked to make another choice if the class is full.

Each student's I.D. number has a personal curriculum code assigned to it that allows the system to check course prerequisites against the student's record. This feature prevents students from scheduling classes without having the proper academic background. Schedule conflicts are also eliminated by the system's checking abilities.

The entry-approval process continues until the student has selected all the desired courses. The system then gives verbal confirmation of the entire schedule as entered, affording the student the opportunity to make any necessary changes. When all is verified as correct, the student signs off, and the computer generates a printed confirmation of the schedule along with the associated billing, both of which are mailed to the student.

In addition to using the system for registration, faculty members and counselors can also access the computer to review a student's course load and make any appropriate changes. Administrative personnel can also use the system to broadcast special announcements, such as weather-related college closings. Such messages are particularly convenient to convey over the system because they can be entered remotely.

TELECOMMUTING

The sophisticated telecommunications systems of today have changed the way people do their work. They may also change where they do their work. Telecommunications hookups from the office to home allow people to work at home and still communicate effectively with colleagues. Since this system eliminates the need to travel to and from work, it is called *telecommuting*.

Telecommuting allows companies to tap into new labor pools, creating employment opportunities for groups often unable to hold office jobs, such as the handicapped and mothers with young children. Pilot studies have also discovered that working at home with fewer distractions tends to boost productivity, thereby providing significant financial benefits to companies employing the new working arrangements. However, before telecommuting becomes an accepted work practice, it must overcome resistance by managers, unions, and, in some cases, workers. Managers tend to shun the idea of trying to supervise personnel who work outside the office because they might lose control of them. Although unions do not think that the working arrangement will result in sweatship-type abuses, some suggest that it holds "a potential for exploitation." The chief drawback from the workers' standpoint is the lack of socialization. People generally want to be with people; they do not want to feel isolated.

About 450 companies have telecommuting programs involving about 100,000 telecommuters, who work as managers, programmers, marketers, researchers, claims processors, secretaries, data-entry operators, and typists. Among the companies participating in work-at-home projects are New York Telephone Company, AT&T Information Systems, Citibank, Digital Equipment Corporation, Equitable Life Assurance Corporation, Xerox Corporation, Control Data Corporation, McDonald's Corporation, Montgomery Ward & Company, Hartford Insurance Group, Marine Midland Bank, and J. C. Penney Company. The most successful telecommuting programs today are those that combine working at home with spending time in the office.

ENHANCED VOICE SERVICES

The continuing development of new technologies provides opportunities for the introduction of new telecommunications services. The competitive marketplace has been quick to capitalize on these opportunities, and the local exchange companies and nontelephone company providers are introducing services that heretofore were not possible.

■ ***AUTOMATIC NUMBER IDENTIFICATION (ANI)*** Automatic number identification, also known as *call identification (call ID)*, has been used by telephone companies for years as a means of identifying users of chargeable services so that bills could be issued. In this context ANI meant that when a customer dialed a long distance call, a record of the called telephone number, the calling telephone number, and the duration of the call was recorded. These call details were used to bill the customer for the call. This capability has resulted in the development of a wide range of new telephone services.

For example, enhanced 911 (described in Chapter 9) uses ANI capability to forward calling telephone numbers to an emergency reporting center. The calling number is displayed on a screen and is fed to a computer data base to locate the address of the calling number.

Similarly, a telephone equipped with a display screen can capture the telephone number of the caller if the telephone company provides this network capability. New network transmission techniques, such as ISDN and CCS, make it possible to provide ANI service across the telecommunications networks.

With ANI capability, a number of other new services can be offered because once the calling number is recorded in a properly equipped telephone, it is possible for calls to be selectively blocked, forwarded, returned, or traced.

■ ***DISTINCTIVE RINGING SERVICE*** Another service, which is marketed under various names such as "Ringmaster" or "Identa Call," provides distinctive ringing sequences on the called telephone. The ringing sequences can be identified by the person for whom the call is intended. This is accomplished by providing up to three different telephone numbers for the same telephone line.

■ ***RECORDER-ACTIVATED PAGING SERVICE*** Marketed under the name "Answer Anywhere," this service enables a person to receive incoming calls when away from his or her telephone. An unanswered call is transferred to an answering device, which, in turn, activates a pager. Upon receiving a paging signal, the called person goes to any telephone and calls his or her own number to be connected with the caller.

■ ***CUSTOM LOCAL AREA SIGNALING SERVICE (CLASS)*** Custom Local Area Signaling Service is a call management package comprising seven of the popular enhanced service features.

The package is built around call identification which, as previously described, lets users with special display attachments view the number of the person placing the incoming call. This feature enables the answering party to decide whether or not to take the call.

Another CLASS service, *call trace,* allows a customer to have the phone company trace a call for law enforcement purposes. The customer must hang up on the call and immediately dial prescribed digits.

Other services grouped under CLASS include the ability to let customers dial back the last incoming call whether or not the user answered it; redial the last outgoing call; key in up to six "priority" numbers that will give a distinctive ring to important calls; block unwanted calls; and forward calls.

CLASS services will become available gradually as telephone companies slowly deploy central office equipment that supports common channel signaling.

Since CLASS services are built largely upon the capacity to identify the number of the calling party, they open up seemingly endless communications possibilities. Certainly they could substantially reduce or even eliminate the telephone tag problem. However, the cornerstone of CLASS services, call identification, has come under fire from consumer advocate groups and the American Civil Liberties Union (ACLU). These groups claim that the feature constitutes an invasion of privacy of telephone customers who pay to have their telephone number unlisted. They also argue that the service would undermine the hot line and tip services that rely on caller anonymity. However, some law enforcement officials hail the plan, saying that it would let customers evade and even trace abusive and obscene phone calls. The telephone companies claim that call ID protects customers' privacy by allowing them to screen calls.

TELECOM 2000

Telecom 2000, a monumental study conducted by the National Telecommunications and Information Administration, takes a look at the next 20 plus years in telecommunications.

The study states that with the goals for basic telephone service accomplished and maintained, widespread access to information services must become the newly defined goal for the country. To accomplish this, the report calls for government regulatory and policy changes that will foster an environment to allow all interested companies, including the Bell regional holding companies, to contribute. They must move to modernize and to provide customers access to new services based on the customers' requirements.

The report goes on to find that there has been a positive trend in the telecommunications industry to deregulate at federal and state levels and that the less regulated a telecommunications sector is, the faster it grows.

As for the costs involved, the study states that the cost of providing new services should be borne primarily by the users.

MAJOR TRENDS AND CONCLUSIONS

The following major trends and conclusions of *Telecom 2000* relate to information services:

- *Competitive strength:* It is critical that this nation's telecommunications and information resources be made more fully available, especially to buttress the strength of the American economy.
- *Business information services:* Over the next three to five years, the business information services market will continue to flourish, growing at an annual rate of 20–30 percent. The business information market is currently at $5 billion annually, with projections of $15 billion annually in 1992.
- *Residential information services:* The residential market is about $115 million annually with slightly over one million subscribers, all of whom have personal computers, modems, and telecommunications software. Viewed in the context of the total telecommunications industry, the residential information services market reaches a rather small percentage of the U.S. population.

- *Future development:* The future development of the information services industry in this country is vital to the increased competitiveness of American companies both home and abroad and to the ability of all Americans to become part of the economic and social fabric of an electronic society, just as they are part of the voice telephone society.
- *The information services industry:* The U.S. online information services industry is strong and thriving for larger business users and those willing to purchase a personal computer, modem, and software, and to subscribe to particular information services.
- *The mass market:* A major component of the mass market is the household that is not likely to own a personal computer and that would benefit from the availability of a low-cost, user-friendly information services terminal.

Removal of regulatory impediments could accelerate growth of successful mass-market information service offerings in this country by allowing the local telephone companies to offer more information services and to produce suitable information services terminals.

The development and marketing of terminals, as well as of information services, should generally be left to the competitive marketplace, although appropriate government policies regarding the offering of information services by regulated common carriers can serve to further spur such growth.

SUMMARY

Rapid evolution of telecommunications technology and the break-up of the Bell System have significantly contributed to the creation of a highly competitive and uncertain marketplace for the providers and users of communications services.

Research in progress and leading-edge developments in photonic switching, microelectronics circuitry, optical interconnection, microprocessor chip technology, optical communications (coherent detection), wideband packet transmission, and common channel signaling hold great promise for important advances in the telecommunications industry.

Several factors have delayed the implementation of advances in telecommunications: lack of ideas for practical uses, economic considerations, legal and political deterrents, and attitude barriers.

Videotex is an interactive information service that allows users to converse with a remote data base, enter data for transactions, and retrieve textual and graphic information on a television set or microprocessor screen. Videotext services include telemedicine, home banking, educational services, libraries, teleshopping, entertainment services, and electronic polling.

The technology of computers' speaking, storing human voices, and reacting to human speech is known as voice processing. There are four types of voice processing: voice response (synthesized speech), voice recognition, voice store and forward, and voice identification.

Other innovative services made possible by telecommunications include telemetry, public communications centers, tele-registration, and telecommuting. Also a number of enhanced voice services are gradually becoming available. They include automatic number identification, distinctive ringing, recorder-activated paging service, call trace, automatic call back of last incoming call, last number redial (on outgoing calls), and call forwarding.

An extensive yearlong government study, Telecom 2000, takes a look at the next 20 plus years in telecommunications. A major conclusion of the study is that the nation's telecommunications and information resources must be made more fully available in order to buttress the strength of the American economy and that widespread access to information services must become the newly defined goal for the country.

REVIEW QUESTIONS

1. What two factors have contributed significantly to the creation of a highly competitive and uncertain marketplace for telecommunications vendors and users?
2. What are some of the areas in which research in progress promises to contribute to the advancement of the telecommunications industry?
3. What are some of the factors that tend to delay the implementation of new technology?
4. What is videotex service? What equipment is required to access this service?
5. What are some of the videotex services that can be used in the home?
6. What is voice processing? Name and briefly describe the four types of voice processing.
7. What are some of the other innovative services made possible by telecommunications?
8. What is automatic number identification? Why has it become a subject of considerable controversy?
9. What are some of the other enhanced voice services (other than ANI) that are gradually becoming available?
10. What are the major conclusions of the government study *Telecom 2000?*

VOCABULARY

videotex

telemedicine

computer-assisted instruction (CAI)

distance learning

teleshopping

electronic polling

voice processing

voice response (synthesized speech)

voice recognition

voice store and forward

common channel signaling (CCS)

signaling system 7 (SS7)

voice identification

telemetry

public communications center

tele-registration

telecommuting

automatic number identification (ANI)

distinctive ringing service

recorder-activated paging service

Custom Local Area Signaling Service (CLASS)

Telecom 2000

BIBLIOGRAPHY

GENERAL WORKS

Carne, E. Brian. *Modern Telecommunication.* New York: Plenum Press, 1984.

Dizard, Wilson P., Jr. *The Coming Information Age.* New York: Longman, Inc., 1982.

Doll, Dixon. *Data Communications Facilities, Networks and Systems Design.* New York: John Wiley and Sons, Inc., 1978.

FitzGerald, Jerry. *Business Data Communications.* New York: John Wiley and Sons, Inc., 1984.

Green, James Harry. *The Dow Jones-Irwin Handbook of Telecommunications.* Homewood, IL: Dow Jones-Irwin, 1986.

Gurrie, Michael L. and Patrick J. O'Connor. *Voice/Data Telecommunications Systems.* Englewood Cliffs, NJ: Prentice-Hall, Inc., 1986.

Martin, James. *Telecommunications and the Computer*, 3d ed. Englewood Cliffs, NJ: Prentice-Hall, Inc., 1989.

Meyers, Robert A., ed. *Encyclopedia of Telecommunications.* San Diego, CA: Academic Press, Inc., 1989.

Misra, Jay and Byron Belitsos. *Business Telecommunications.* Homewood, IL: Richard D. Irwin, Inc., 1987.

National Telecommunications and Information Administration. *NTIA Telecom 2000.* Washington, DC: U.S. Government Printing Office, NTIA Special Publication 88–21, October 1988.

Pool, Ithiel de Sola, ed. *The Social Impact of the Telephone.* Cambridge, MA: MIT Press, 1977.

Reynolds, George W. *Introduction to Business Telecommunications.* Columbus, OH: Charles E. Merrill Publishing Company, 1984.

Rowe, Stanford H., II. *Business Telecommunications.* Chicago: Science Research Associates, Inc., 1988.

CHAPTER 1

Bly, Robert W. and Gary Blake. *Dream Jobs: A Guide to Tomorrow's Top Careers.* New York: Wiley & Sons, Inc., 1983, 154–75.

Blyth, Mary. "Career Opportunities in Telecommunications," *Business Education Forum* (October 1985), 14–16.

Clemmensen, Jane M. "Telecommunications Curricula: The Status of Program Graduates in Industry," *Telecommunications* (September 1982), 99–101.

Fitzgibbons, Patrick. "Telecom Grads Welcomed By Industry." *Network World* (August 1, 1988), 22.

Harlow, Ronney. "Growing Choices Creating Urgent Need for More Telecom Professionals." *Communications News* (March 1983), 30–31.

Kander, Sharon L. "Women in Telecommunications," *Teleconnect* (May 1983), 30.

Kerr, Susan. "Telecom Pros Evolve," *Datamation* (January 16, 1989), 45–8.

Krepps, Karen A. "Automation and Its Effect on Telco Operations," *Telephony* (January 7, 1984), 30–32, 36, 40.

LaBlanc, Robert E. with Richard M. Wolf and Elizabeth A. LaBlanc. "Communications + Data = Compunications." *Telephone Engineer and Management* (May 1, 1983), 67–71.

Morley, Nik. "Advanced Degrees Can Speed Career Success." *Network World* (January 9, 1989), 1, 35, 37–9.

Thobe, Deborah J. "Telecom Compensation Issues for 1989." *Business Communications Review* (January 1989), 27–32.

Wiley, Don. "World Communications Year — 1983 — Will Be Landmark Year in USA." *Communications News* (January 1983), 26–41.

CHAPTER 2

Barret, R. T. *The Changing Years as Seen From the Switchboard.* New York: AT&T, 1936.

Bell Telephone System, *Alexander Graham Bell.* New York: Bell Telephone System, no date.

Boettinger, H. M. *The Telephone Book.* Croton-on-Hudson, NY: Riverwood Publishers, Ltd., 1977.

Brooks, John. *Telephone.* New York: Harper & Row, 1976.

Fagen, M. D. ed. *History of Engineering and Science in the Bell System.* Murray Hill, NJ: Bell Telephone Laboratories, Inc., 1975.

Harlow, Alvin B. *Brass-Pounders, Young Telegraphers of the Civil War.* Denver: Sage Books, 1962.

Jesperson, James, and Jane Fitz Randolph. *The Story of Telecommunications.* New York: Athenium Publishers, Inc., 1980.

Rhodes, Frederick Leland. *Beginnings of Telephony.* New York: Harper & Brothers Publishers, 1929.

Shippen, Katherine B. "Mr. Bell Invents the Telephone." New York: Bell Telephone System, 1955.

Thompson, Robert Luther. *Wiring a Continent, the History of the Telegraph Industry in the United States 1832–1866*. Princeton, NJ: Princeton University Press, 1947.

United States Independent Telephone Association. *The Ring of Success*. Washington, DC: U.S. Independent Telephone Association, 1979.

Watson, Thomas A. "The Birth and Babyhood of the Telephone." New York: AT&T, 1971.

CHAPTER 3

Brock, Gerald W. *The Telecommunications Industry*. Cambridge, MA: Harvard University Press, 1981.

Buckley, Linda M. "FCC Lays Down New Rules for Cost Allocation Game." *Telephony* (January 5, 1987), 13.

Brown, C. L. "AT&T and the Consent Decree." *Telecommunications Policy* (June 1983), 91–98.

Flax, Steven. "The Orphan Called Baby Bell." *Fortune* (June 27, 1983), 87–88.

Fowler, Mark. "We're Heading Ultimately Toward a Regulation-Free Telecom Market." *Communications News* (March 1983), 100.

Geller, Henry. "Regulatory Policies for Electronic Media." *Telecommunications* (May 1983), 128–33.

Iardella, Albert E., ed. *Western Electric and the Bell System*. New York: Western Electric Company, 1964.

Johnson, Ben and Sharon D'Amario Thomas. "Deregulation and Divestiture in a Changing Telecommunications Industry." *Public Utilities Fortnightly* (October 14, 1982), 17–22.

Kleinfield, Sonny. *The Biggest Company on Earth*. New York: Holt, Rinehart and Winston, 1981.

Lewin, Leonard, ed. *Telecommunications in the United States: Trends and Policies*. Dedham, MA: Artech House, Inc., 1981.

Rey, R. F., ed. *Engineering and Operations in the Bell System*, 2d ed. Murray Hill, NJ: AT&T Bell Laboratories, 1984.

Weber, Joseph H. "AT&T Restructure: 1982–1984, Its Causes and Effects." *Journal of Telecommunication Networks* (Spring 1983), 51–59.

CHAPTER 4

Annunziata, Robert. "What is a Teleport?" *Telecommunications Products + Technology* (November 1986), 24–28.

Badalato, Anthony. "Before Bypassing, Beware!" *Telephony* (December 3, 1984), 88, 92, 94, 95.

Booker, Ellis and Maribeth Harper. "MCI to Buy RCA Globcom," *Telephony* (September 7, 1987), 6.

Burstyn, H. Paris. "Teleports: at the Crossroads," *High Technology* (May 1986), 28–31.

Crockett, Barton. "Little-Known ITU Quietly Sculpts Industry Standards," *Network World* (August 8, 1988), 14.

GTE 1987 Annual Report. Stamford, CT: GTE, 1988.

Shooshan, Harry M. III, ed. *Disconnecting Bell: The Impact of the AT&T Divestiture*. New York, NY: Pergamon Press, 1984.

Simon, Samuel A. *After Divestiture: What the AT&T Settlement Means for Business and Residential Telephone Service*. White Plains, NY: Knowledge Industry Publications, Inc., 1985.

Smith, Tom. *Anatomy of Telecommunications*. Geneva, IL: abc Teletraining, Inc., 1986.

Telecommunications and You. Order #GE20–0790–01. White Plains, NY: IBM Corporation, 1987.

The Teleport Market in the U.S. Report #A1784. New York: Frost & Sullivan, Summer 1989.

United States Telephone Association. *Phone Facts 1988*. Washington, DC: U.S. Telephone Association, 1988.

Wallace, Bob. "AT&T to Cut Switched Service Rates," *Network World* (October 17, 1988), 1, 55.

CHAPTER 5

Castro, Janice. "Telephones Get Smart." *Time Magazine* (April 1987), 50–51.

Edwards, Morris. "Local Loop Bypass Paves Way for Wideband Services." *Communications News* (September 1983), 50–54.

Fike, John L. and George E. Friend. *Understanding Telephone Electronics*. Dallas: Texas Instruments, 1983.

LaCalzi, Pamela. "Voice-Activated Phone Gathers CES Attention." *Communications Week* (June 15, 1987), 26.

Lavoie, Francis J. "Today's Telephone System: Sophisticated and Affordable," *Modern Office Technology* (November 1988), 76–78.

Lohmann, Bill. "A Picture Phone That's More Than Just Talk," *The Detroit News* (June 10, 1986), D, 1.

Metzner, Kermit. "Telephone Instruments: Many Choices Await You." *Office Systems* (September 1988), 60–64.

Miller, Shelley. "Telephone Systems, Now and in the Future." *Office Systems* (March 1989), 72–6.

Noll, A. Michael. *Introduction to Telephones and Telephone Systems.* Norwood, MA: Artech House, Inc., 1986.

Rosen, David A. "Key Systems in the Large Organization," *TPT Networking Management* (November 1988), 46–48.

CHAPTER 6

Bates, Cyril P., Dennis H. Skillman, and Mitchell A. Skinner. "A New Microwave Radio System Expands Digital Transmission Capacity." *Bell Laboratories Record* (January 1986), 26–32.

Boettinger, Henry M. "A Global Network Through Cooperation." *Telephone Engineer & Management* (May 1, 1986), 14–17.

Bose, Keith. "What's Fiber?" *Teleconnect* (June 1987), 148–9.

Bowen, Terry. "Fiber Optics in the Age of Information." *The Office* (May 1987), 148–9.

Dzubeck, Francis X. "What is ISDN?" *Administrative Management* (April 1986), 55–56.

Gawdun, Michael. "Virtual Private Networks." *Telecommunications* (April 1986), 59–62.

Gershon, R. A. "Satellite and Fiber Optics: Redefining the Business of Long-Haul Transmission." *Business Communications Review* (March-April 1987), 22–25.

Heller, Michael. "Let's See What the Intelligent Network Can Do." *Telephone Engineer & Management* (June 15, 1988) 79–81.

Hinton, H. Scott. "Photonic Switching Connects to the Future." *Telecommunications* (May 1987), 79–83.

Landauer, Steve. "Data Over Voice Multiplexing," *Telecommunications* (April 1987), 82, 85.

Parsons, Arthur. "Why Light Pulses Are Replacing Electrical Pulses in Creating Higher-Speed Transmission Systems." *Communications News* (August 1987), 24–9.

Pyykkonen, Martin. "Local Area Network Industry Trends." *Telecommunications* (October 1988), 21–29.

Rosenwald, Jeffrey. "ISDN: Reinventing the Telephone," *Administrative Management* (October 1987), 41–42.

Scholes, William A. "Underseas Optics: Australia Readying Pacific Cable Link." *MIS Week* (January 30, 1989), 13.

Settanni, Joseph Andres. "Voice and Data Merging in Telecommunications." *Office Systems* (May 1987), 48–54.

Ulrich, Marc. "The Central Office: Cornerstone of the ISDN." *Telecommunications* (November 1985), 48k, 48l.

West, Fred. "Fiber Optics Technology Promises Dramatic Impact on Data Communications." *Data Management* (November 1986), 20–2.

Wienski, Robert M. and H. Charles Baker. "Getting Ready for ISDN." *Business Communications Review* (November-December 1986), 2–6.

Winfield, Marc. "FO for Maximum Security." *Telephony* (August 17, 1987), 65, 67, 70.

CHAPTER 7

Alexander, A. A., R. M. Gryb, and D. West. "Capabilities of the Telephone Network for Data Transmission." *Bell System Technical Journal* 39 (May 1960), 431–76.

Black, Uyless D. *Data Communications and Distributed Network,* 2d ed., Englewood Cliffs, NJ: Prentice-Hall, Inc., 1987.

Breslin, Judson, and C. Bradley Tashenberg. *Distributed Processing Systems.* New York: AMACOM, A Division of American Management Association, 1978.

Data Communications. *Executive Guide to Data Communications.* Vol. 5. New York: McGraw-Hill, no date.

Edwards, Morris. "Understanding Data Communication's Basics." *Infosystems* (July 1988), 52–54.

FitzGerald, Jerry. *Business Data Communications.* New York: John Wiley & Sons, 1984.

Forbes-Jamison, Doug. "Data Communication Basics." *Communication Age* (October 1984), 27–29.

Gentle, Edgar C., Jr. ed. *Data Communications in Business.* New York: American Telephone & Telegraph Company, 1965.

Greenwood, Frank, Mary M. Greenwood, and Robert E. Harding. *Business Telecommunications: Data Communications in the Information Age.* Dubuque, IA: Wm. C. Brown Publishers, 1988.

Housley, Trevor. *Data Communications and Teleprocessing Systems.* Englewood Cliffs, NJ: Prentice-Hall, Inc., 1987.

Kroenke, David M. *Business Computer Systems,* 2d ed. Santa Cruz, CA: Mitchell Publishing Inc., 1984.

Long, Larry. *Computers in Business.* Englewood Cliffs, NJ: Prentice-Hall, Inc., 1987.

Lusa, John M. "Fortune 1000 Telecom Spending to Get Increasing Share of Operating Budget in Next Five Years." *TPT Networking Management* (November 1988), 15.

Mullens, Richard W. "On-Line Transactions: Telcos in High Gear." *Telecommunications* (April 1987), 63–66.

Rhodes, Wayne L. and Perry S. True. "The 'Mythical' Datacomm Manager." *Infosystems* (March 1984), 28–30.

West, Fred. "Fiber Optics Technology Promises Dramatic Impact on Data Communications." *Data Management* (November 1986), 20–2.

CHAPTER 8

Asten, Kenneth J. *Data Communications for Business Information Systems.* New York: The Macmillan Company, 1973.

Data Communications. "Data Transmission Equipment." *Data Communications* (May 1985), 53–4.

DeNoia, Lynn A. *Data Communications: Fundamentals and Applications.* Columbus, OH: Merrill Publishing Company, 1987.

Friend, George E., John L. Fike, Charles H. Baker, and John C. Bellamy. *Understanding Data Communications.* Dallas: Texas Instruments, 1984.

Hanson, Kerry. "Datacom in the Integrated Office." *Telecommunications* (August 1984), 64, 68, 72, and 88.

NCC Publications. *Handbook of Data Communications.* Manchester, England: The UK Post Office, 1975.

Pyykkonen, Martin. "Computers in Communications." *Telecommunications* (January 1985), 67–70.

Sherman, Kenneth. *Data Communications, 2d ed.* Reston, VA: Reston Publishing Company, Inc., 1985.

Silver, Gerald A. and Myrna L. Silver. *Data Communications for Business.* Boston: Boyd & Fraser Publishing Co., 1987.

Techo, Robert. *Data Communications.* New York: Plenum Press, 1980.

CHAPTER 9

Caswell, Stephen A. "Electronic Mail — The State of the Art." *Telecommunications* (August 1988), 27–30.

Cline, Joe. "Voice Messaging Pioneer: SNET Gets Set for Residential Service." *Telephony* (November 14, 1988), 61–2.

Concannan, Larry. "How to Introduce Two-Way Videoconferencing to Your Organization." *Telecommunications* (September 1988), 57–60, 93.

Duke, David. "Teletex Tackles America." *Word Processing and Information Systems* (October 1982), 12–14, 38.

Evans, Sherli. "Fax in '88: More Models, More Features." *Modern Office Technology* (May 1988), BC3–6.

Green, James H. "Facsimile: On the Verge of Maturity." *The Office* (May 1987), 68–9.

Guyon, Janet. "Boosting Efficiency of Cellular Phones." *The Wall Street Journal* (November 3, 1988), 1.

Hanscom, Elizabeth A. "Electronic Mail: A Fivefold Message." *Office Systems* (March 1989), 27–30.

Hilton, Jack and Peter Jacobi. *Straight Talk About Videoconferencing.* New York: Prentice-Hall Press, 1986.

Hofferber, Michael. "Digital Advances Spur Sales of Fax Machines." *The Office* (November 1988), 68–70.

Kalow, Samuel Jay. "Introduction to Voice-Mail Systems Technology." *Journal of Data and Computer Communications* (Fall 1988), 28–33.

Kelleher, Kathleen, and Thomas B. Cross. *Teleconferencing.* Englewood Cliffs, NJ: Prentice-Hall, Inc., 1986.

Long, Gordon. "Fax Growth." *Office Systems* (September 1988), 43, 46, 48, 50.

Mallery, Dave. "The Northern Telecom Displayphone 220." *DEC Professional* (May 1988), 114–5.

Menkus, Belden. "Why Not Try Audio Teleconferencing?" *Modern Office Technology* (October 1987), 124–5.

Michigan Bell Telephone Company. "Phone Lines Become 'Post Office' of the Future." *Michigan Bell Tie Lines* (December 1988), 2.

Morris, John. "X.400 Breeds Third-Generation E-Mail Systems." *TPT/Networking Management* (March 1989), 34–7.

Miller, Shelley. "Telephone Systems Now and in the Future." *Office Systems* (March 1989), 72–6.

O'Bryan, Kathy. "Lifeline Works Toward Universal Service Goal." *Michigan Bell Tie Lines* (March 1989), 1–2.

Panko, Raymond R. "Electronic Mail: The Alternatives." *Office Administration and Automation* (June 1984), 37–43.

Pirani, Judy. "PC-Based Facsimile Transmission: A Guide for the Curious." *Telecommunications* (April 1989), 35, 36, 38.

Reiter, Alan A. "New Pagers Put a Mailbox in Your Pocket." *High Technology Business* (April 1988), 32–6.

Ryan, Donald J. "Making Sense of Today's Image Communications Alternatives." *Data Communications* (April 1987), 110–115.

Thomas, Jerry. "Get Used to It: Voice Mail is Here to Stay." *Telephony* (May 22, 1988), 62–3.

Voros, Gregory L. "Forecasting the Future for Facsimile." *Business Communications Review* (January 1989), 44–6.

Williamson, John. "Teletex Faces a Future Without Security." *Telephony* (February 4, 1985), 72–8.

Winkler, Gary. "Intelligent Facsimile: The Next Generation." *Office Systems* (May 1987), 36–8.

CHAPTER 10

Beckmann, Petr. *Elementary Queuing Theory and Telephone Traffic.* Geneva, IL: Lee's abc of the Telephone, 1977.

Booker, Ellis. "Taking Account of Traffic." *Telephony* (July 13, 1987), 42.

Datapro Research Corporation. "Call Distribution Systems." *Telemarketing* (February 1988), 40–3.

Ellis, Robert E. *Designing Data Networks.* Englewood Cliffs, NJ: Prentice-Hall, Inc., 1986.

Frankel, Theodor. *Tables for Traffic Management and Design.* Geneva, IL: Lee's abc of the Telephone, 1976.

Gunn, Howard J. *Principles of Traffic and Network Design.* Geneva, IL: Lee's abc of the Telephone, 1986.

Hoffman, Hugh. "Extended Erlang C: Traffic Engineering for Queuing with Overflow." *Business Communications Review* 113, No. 4 (July-August 1983), 28–33.

Jewett, James E., Jacqueline B. Shrago, and Bernard D. Yomotov. *Designing Optimal Voice Networks for Businesses, Government, and Telephone Companies.* Chicago: Telephony Publishing Corporation, 1980.

Lawson, Robert W. *Teletraffic Engineering and Administration.* Chicago: Telephony Publishing Corporation, 1983.

North American Telecommunications Association. *Industry Basics, 3d ed.* Washington, DC: North American Telephone Association, 1989.

Technical Staff, AT&T Bell Laboratories. *Engineering and Operations in the Bell System,* 2d ed. Murray Hill, NJ: AT&T Bell Laboratories, 1984.

Theis, Peter F. "Limits to Caller Tolerance When Inventorying Incoming Calls." *Business Communications Review* (March-April 1984), 30–4.

CHAPTER 11

Appleby, Jerry. "Managing Today's Technology." *Telecommunications* (September 1988), 51–2.

Avakian, Paul N. "Credit to Communications Managers." *Communications Week* (January 30, 1989), 13.

Bedrosian, Peter. "Managing Telecommunications Costs Is a Tricky Business." *The Office* (September 1987), 15–6.

Belitsos, Byron. "Profiles of Telecom Management Excellence." *Business Communications Review* (May-June 1987), 6–13.

Dickinson, Robert M. "Telecom Management: An Emerging Art." *Datamation* (March 1984), 120–30.

Frank, Howard. "Whose Problem Is Strategic Planning?" *TPT* (July 1988), 44–8.

Frasier, Rosalie Craven. "Selling Telecommunication to Management." *Telecommunications* (February 1986), 48s–v.

Fuhrman, John C. *Telemanagement.* Englewood Cliffs, NJ: Prentice-Hall, Inc., 1985.

Gantz, John. "How to Succeed: A White Paper to Management." *TPT* (October 1987), 35–54.

———."The Changing Face of Telecom Management." *TPT* (July 1987), 41–4.

———."The Growing Power of the Telecom Manager: A White Paper to Management." *TPT* (October 1986), 35–54.

James, Debbi. "Call Accounting Can Improve Your Bottom Line." *Telemarketing* (April 1989), 36–7.

McDonald, Harrison, "Check Phone Bill Services to Get What You Pay For." *Office Systems* (November 1986), 78–84.

Oldham, Paula. "The Changing Role of the Telecommunications Manager." *Telecommunications* (February 1986), 48q–r.

Thompson, Glenn R. "Telecommunications Competition: A New Challenge for Management." *Office Systems* (March 1986), 56, 58.

Trafton, Donald R. "Gain that Competitive Edge Strategically—Through Planning." *Communications Age* (January 1987), 38–41.

Zeibig, Robert A. "The Executive's Perspective on Telecommunications." *Business Communications Review* (March-April 1988), 20–1.

CHAPTER 12

Bennett, Duane. "You, Too, Can Avoid Those Frightening Pitfalls During Implementation of a New Phone System." *Communications News* (December 1986), 54–7.

Gordon, Jim, Maggie Klenke, and Karen Harrison. "Managing the Installation of a Telephone System." *Business Communications Review* (May-June 1985), 21–6.

Kuehn, Richard A. "When Growth and Technology Spur Phone-System Change." *Office Systems* (July 1986), 50, 52, 54.

McDonald, Harrison. "Telephone Selection: Everybody's Business." *Office Systems* (February 1987), 54–61.

Metzner, Kermit. "Telephone System Selection." *Office Systems* (August 1988), 38, 40, 42, 44.

Miller, Shelley. "How and When to Install a New Telephone System." *Office Systems* (May 1986), 46–51.

Morgan, James H. "The Telecom RFP; Three Pages or 50?" *TPT* (August 1987), 64–7.

Muller, Nathan J. "Telecom Comparison Shopping." *Infosystems* (August 1986), 46–51.

Petty, Harry. "How to Manage a Successful Cutover." *TPT/Networking Management* (November 1988), 42–6.

Waite, Andrew J. "Developing the RFP and Evaluating Responses." *Telemarketing* (March 1986), 8–11.

Whitney, Edward T. "Telemanagement Systems: Features and Procurement Issues." *Business Communications Review* (January-February 1987), 9–15.

CHAPTER 13

Brown, Bob. "FCC Details Price Cap Plan Promising Big Rate Cuts." *Network World* (May 16, 1988), 2, 4.

Buckley, Linda M. "Rate of Return Out, Price Cap in at FCC." *Telephony* (August 10, 1987), 10.

Fogarty, Joseph D. "Capital Recovery: A Crisis for Telephone Companies, a Dilemma for Regulators." *Public Utilities Fortnightly* (December 8, 1983), 13–18.

Fowler, Mark. "We're Heading Ultimately Toward A Regulation-Free Telecom Market." *Communications News* (March 1983), 100.

Jacobson, Andrew, Sam Rovit, Robert O'Brien, and Chris Vestal. "Report: USA Regulation and Policy — Access Charges." *Telecommunications* (February 1985), 26, 31.

Kahn, Alfred S. "Straight Talk About Local Rates: They Have Been and Are Today Much Too Low." *Telephony* (April 15, 1985), 68–70.

Killette, Kathleen. "FCC to Consider Price Caps." *Communications Week* (January 30, 1989), 4.

Lipman, Andrew D. "Price-Cap Proposal Stirs Lively Controversy." *Telephony* (November 2, 1987), 74–5.

Robinson, Teri. "FCC Price-Cap Plan: Regulatory Revamp." *MIS Week* (August 10, 1987), 1, 21.

Wallace, Bob. "Price Caps Take Center Stage at ICA Conference." *Network World* (May 23, 1988), 1, 54.

Wilson, Carol. "A Not-So-Fond Farewell to Rate of Return." *Telephony* (August 17, 1987), 39.

Wilson, John D. "Telephone Access Costs and Rates." *Public Utilities Fortnightly* (September 15, 1983), 18–25.

CHAPTER 14

Andrew, Tom. "Phones on Wheels." *Michigan Bell Tie Lines* (February 1989), 7.

Arden, Lynie. "Earning a Degree On-Line." *Home Office Computing* (November 1988), 102, 104–5.

Bennett, Margaret O., and Linda Noble. "The Revolution in Telecommunications." *Business Education Forum* (March 1989), 30–3.

Bushaus, Dawn. "Southwestern Bell Tests Meter Reading." *Telephony* (May 15, 1989), 16–7.

"Communications on the Fly." *Modern Office Technology* (September 1988), 40.

Desmond, Paul. "Bill Payment Network Benefits Utilities and Customers." *Network World* (January 9, 1989), 1, 9.

Diebold, John. "Videotex in the U.S.: An Assessment." *Telecommunications* (July 1988), 78–83.

Greenstein, Irwin. "Voice Processing Enters Public Net." *MIS Week* (March 13, 1989), 1, 54.

Harper, Maribeth. "New Directions in Telecommunications." *Nation's Business* (April 1987), 38–40.

Herman, Denise. "When Work Is on the Line—There's No Place Like Home." *Telemarketing* (April 1987), 35–9, 61.

Killette, Kathleen. "FCC Still Defining Enhanced-Services Market." *Communications Week* (January 30, 1989), 21.

"Management Misses the Bus on Telecommuting's Potential." *Network World* (January 23, 1989), 34.

McGowan, James. "Lessons Learned from the Minitel Phenomenon." *Network World* (December 5, 1988), 27.

Moore, Steve. "CLASS Controversies." *Network World* (October 5, 1987), 27.

National Telecommunications and Information Administration. *NTIA Telecom 2000*. Washington, DC: U.S. Department of Commerce, 434–5.

Nugent, Gwen C. "Videotex: A Delivery System for Educational Services." *T.H.E. Journal* (April 1987), 57–61.

Price, John. "Going to School via Fiber." *Telephony* (May 9, 1988), 28–9.

"Registration by Phone Benefits Students and Administrators Alike." *T.H.E. Journal* (November 1988), 56.

"Registration by Telephone Helps Make Community College More Accessible." *T.H.E. Journal* (February 1987), 46.

Robbins, Renee M. "French Success Can Guide Corporations." *Infosystems* (January 1988), 16.

Rockwell, Mark. "Nynex Plans to Offer Service with CLASS." *Communications Week* (March 6, 1989), 10.

Samuel, Jennifer. "Delaware Hospital Uses Phone Signals for Diagnosis." *Communications Week* (November 7, 1988), 32.

Shippen, Howard. "Phone Users' Rights Are Only Part of the Issue." *Network World* (April 10, 1989), 38.

Wallace, Bob. "Workers Punch Time Clock Over Phone Lines." *Network World* (January 30, 1989), 10.

Warr, Michael. "Talked to Any Nice Machines Lately?" *Telephony* (January 23, 1989), 38, 41.

Wilson, Carol. "New Jersey Bell Trial Moves to the Head of the CLASS." *Telephony* (September 28, 1987), 12.

Wu, Lisa. "A Public FAX Booth on Every Corner? *Home Office Computing* (November 1988), 10.

APPENDIX

PROFESSIONAL ASSOCIATIONS

The names of the current officers of most of the following organizations can be found in the *Encyclopedia of Associations,* published by Gale Research Company, Detroit, Michigan.

Associated Students for Career Orientation in Telecommunications (ASCOT)
Michigan State University
290 Communications Arts and Science Building
East Lansing, MI 48424
(317) 355-8312

Association of College and University Telecommunication Administrators
211 Nebraska Hall
Lincoln, NE 68588
(402) 472-2000

Association of Data Communications Users
P.O. Box 20163
Bloomington, MN 55420
(612) 881-6803

Competitive Telecommunications Association (COMPTEL)
120 Maryland Avenue, N.E.
Washington, DC 20002
(202) 546-9022

Computer and Communications Industry Association (CCIA)
666 11th Street, N.W.
Washington, DC 20035
(202) 783-0070

Data Processing Management Association (DPMA)
305 Busse Highway
Park Ridge, IL 60068
(312) 825-8124

Exchange Carriers Standards Association
Four Century Drive
Parsippany, NJ 07054
(201) 538-6111

Information Industry Association
555 New Jersey Avenue, Suite 800
Washington, DC 20001
(202) 639-8262

Institute of Electrical and Electronics Engineers (IEEE)
345 East 47th Street
New York, NY 10017
(212) 705-7866

International Communications Association (ICA)
12750 Merit Drive
Suite 710, LB-89
Dallas, TX 75251
(214) 233-3889

International Telecommunication Union (ITU)
Place des Nations
CH-1211 Geneva 20, Switzerland

International Tele/Conferencing Association
1299 Woodside Drive
McLean, VA 22102
(703) 556-6115

Joint Council on Educational Telecommunications (JCET)
1111 16th Street, N.W.
Washington, DC 20036
(202) 955-5278

Multi-Tenant Telecommunications Association
2000 L Street, N.W., Suite 200
Washington, DC 20036
(202) 822-9351

National Association of Radio and Telecommunication Engineers
P.O. Box 15029
Salem, OR 97309
(503) 581-3336

National Association of State Telecommunications Directors
P.O. Box 1190
Iron Works Pike
Lexington, KY
(606) 252-2291

National Association of Telecommunications Officers and Advisors
1301 Pennsylvania Avenue, N.W.
Washington, DC 20004
(202) 626-3250

National Communications Association (NCA)
404 Park Avenue
New York, NY 10016

National Payphone Association
1355 Beverly Road
McLean, VA 22101
(703) 556-3959

North American Telecommunications Association (NATA)
2000 M Street, N.W., Suite 550
Washington, DC 20036
(202) 296-9800

Organization for the Protection and Advancement of Small Telephone Companies
2301 M Street, N.W., Suite 350
Washington, DC 20037
(202) 659-5900

Satellite Broadcasting and Communications Association
300 N. Washington Street, Suite 310
Alexandria, VA 22314
(703) 549-6990

Society of Telecommunications Consultants (STC)
One Rockefeller Plaza
New York, NY 10020
(212) 582-3909

Tele-Communications Association (TCA)
1515 W. Cameron Avenue
West Covina, CA 91790
(818) 960-2849

Telecommunications International Union (TIU)
2341 Whitney Avenue
Hamden, CT 06518
(203) 288-2445

Telephone Pioneers of America
22 Cortlandt Street
New York, NY 10007
(212) 393-3252

United States Telephone Association (USTA)
900 19th Street, N.W.
Washington, DC 20006
(202) 835-3100

U.S. Telecommunications Suppliers Association (USTSA)
150 N. Michigan Avenue
Chicago, IL 60601
(312) 782-8597

Women in Telecommunications (WIT)
1827 Haight Street, #180
San Francisco, CA 94117
(415) 751-4746

APPENDIX

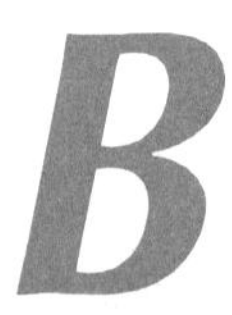

ORGANIZATIONS THAT SPONSOR TELECOMMUNICATIONS SEMINARS

abc Teletraining, Inc.
P.O. Box 537
Geneva, IL 60134
(312) 879-9000

Advanced Training Professionals, Ltd., Inc.
111 East Avenue
Norwalk, CT 06851
(203) 866-6060

American Institute
55 Main Street
Madison, NJ 07940
(201) 377-7400

Architecture Technology Corporation
P.O. Box 24344
Minneapolis, MN 55424
(612) 935-2035

AT&T
Knowledge Plus Center
55 Corporate Drive, Room 13K31
Bridgewater, NJ 08807
1 (800) 554-6400, Extension 1400

Bell Communications Research Technical Education Center
6200 Route 53
Lisle, IL 60532
(312) 960-6300

Business Communications Review
BCR Enterprises, Inc.
950 York Road
Hinsdale, IL 60521
(312) 986-1432

CAPE (Center for Advanced Professional Education)
11928 North Earlham
Orange, CA 92669
(714) 633-9280

Communications Solutions, Inc.
992 South Saratoga-Sunnyvale Road
San Jose, CA 95129
(408) 725-1568

Control Data Institute for Advanced Technology
1450 Energy Park Drive
St. Paul, MN 55108
(301) 467-8200

Data Communication Seminars
445 West Main Street
Wyckoff, NJ 07481
(201) 891-8405

Datamation Institute
Seminar Coordination Office, Suite 415
850 Boylston Street
Chestnut Hill, MA 02167
(617) 738-5020

Datapro Educational Services
McGraw-Hill Information Services Company
1805 Underwood Boulevard
Delran, NJ 08075
(609) 764-0100

Data-Tech Institute
Lakeview Plaza
P.O. Box 2429
Clifton, NJ 07015
(201) 478-5400

Friesen's School of Generic Telephony
Gerry Friesen
Building B, 1300 Chinquapin Road
Churchville, PA 18966
(215) 355-2886

George Washington University
Continuing Education Program
Washington, DC 20052
(202) 994-CEEP

GTE Telenet Communications
8229 Boone Boulevard
Vienna, VA 22180
1-(800)-835-3638

International Communications Association (ICA)
12750 Merit Drive
Suite 828, LB-89
Dallas, TX 75251
(214) 233-3889

Network Career Advancement Institute
202 Fashion Lane, Suite 113
Tustin, CA 92680
(714) 838-5458

Systems Technology Forum
10201 Lee Highway, Suite 150
Fairfax, VA 22003
(703) 591-3666

Telco Research Corporation
1207 17th Avenue, South
Nashville, TN 37212
(615) 329-0031

Telecom Library, Inc.
2205 West 19 Street
New York, NY 10011
(212) 691-8215

Telecommunications Research Associates
P.O. Box A
St. Marys, KS 66536
(913) 437-2000

TeleStrategies, Inc.
6842 Elm Street, Suite 102
McLean, VA 22101
(703) 734-7050

Tellabs, Inc.
4951 Indiana Avenue
Lisle, IL 60532
(312) 969-8800, Extension 241

United States Telephone Association
1801 K Street, N.W.
Washington, DC 20006
(202) 872-1200

APPENDIX

C

TELECOMMUNICATIONS PERIODICALS, REPORTS, AND NEWSLETTERS

This appendix lists the names and addresses of the most widely read telecommunications publications.

Auerbach Data Communications Management. Published bimonthly by Auerbach Publishers, Inc., 6560 N. Park Drive, Pennsauken, NJ 08109.

Business Communications Review. Published monthly by BCR Enterprises, Inc., Hinsdale, IL 00521.

Communications News. Published monthly by Harcourt Brace Jovanovich Publications, Inc., 124 S. First Street, Geneva, IL 60134.

Communications Week. Published weekly by CMP Publications, Inc., 600 Community Drive, Manhasset, NY 11030.

"Datacomm Advisor: IDC's Newsletter Covering Network Mangement—Products, Services, Applications." Published monthly by International Data Corporation, 214 Third Avenue, Waltham, MA 02254.

Data Communications. Published monthly by McGraw-Hill Publications Co., 1221 Avenue of the Americas, New York, NY 10020.

Datapro Reports on Data Communications. Published monthly by Datapro Research Corp., 1805 Underwood Blvd., Delran, NJ 08075.

Datapro Reports on International Telecommunications. Published monthly by Datapro Research Corp., 1805 Underwood Blvd., Delran, NJ 08075.

Datapro Reports on Management of Telecommunications. Published monthly by Datapro Research Corp., 1805 Underwood Blvd., Delran, NJ 08075.

Datapro Reports on Telecommunications. Published monthly by Datapro Research Corp., 1805 Underwood Blvd., Delran, NJ 08075.

"Information Age." Published monthly by TeleCom Management Associates, P.O. Box 2004, Columbus, IN 47202.

Information Week. Published weekly by CMP Publications, 111 East Shore Road, Manhasset, NY 11030.

LAN: The Local Area Network Magazine. Published monthly by Flat Iron Press, Inc., P.O. Box 40706, Nashville, TN 37204.

"The LOCALNetter Newsletter." Published monthly by Architecture Technology Corporation, P.O. Box 24344, Minneapolis, MN 55424.

"The MAPNetter Newsletter." Published monthly by Architecture Technology Corporation, P.O. Box 24344, Minneapolis, MN 55424.

MIS Week. Published weekly by Fairchild Publications, A Division of Capital Cities Media, Inc., 7 E. 12th Street, New York, NY 10003.

Modern Office Technology. Published monthly by Penton Publishing Inc., a subsidiary of Pittway Corporation, 1100 Superior Avenue, Cleveland, OH 44114.

Network Strategy Report. Published monthly by Forrester Research, Inc., P.O. Box 1091, Cambridge, MA 02230.

Network World. Published weekly by CW Communications, Inc., 375 Cochituate Road, Box 9171, Framingham, MA 01701.

"The PCNetter Newsletter." Published monthly by Architecture Technology Corporation, P.O. Box 24344, Minneapolis, MN 55424.

Satellite Communications. Published monthly by Cardiff Publishing Co., 3900 S. Wadsworth, Suite 560, Denver, CO 90235.

"Telecom Digest." Published twice a month by Telecom Publishing Group, P.O. Box 1455, Alexandria, VA 22313.

Telecommunication Journal. Published monthly by the International Telecommunication Union, Place des Nations, CH-1211 Geneva 20, Switzerland, in separate English, French, and Spanish editions.

Telecommunications. Published monthly by Horizon House, 610 Washington Street, Dedham, MA 02026.

"Telecommunications Alert." Published monthly by Telecommunications Alert, One Park Avenue, New York, NY 10016.

Teleconnect: A Monthly Telecommunications Magazine. Published monthly by Telecom Library, Inc. and G. A. Friesen, Inc., 12 West 21 Street, New York, NY 10010.

"Telephone Angles." Published semimonthly by United Communications Group, 4550 Montgomery Avenue, Suite 700N, Bethesda, MD 20814.

Telephone Engineer and Management: The Telephone Industry Magazine. Published semimonthly by Harcourt Brace Jovanovich, Inc., 402 W. Liberty Drive, Wheaton, IL 60187.

Telephone News. Published biweekly by Phillips Publishing, Inc., 7315 Wisconsin Avenue, Suite 1200N, Bethesda, MD 20014.

Telephony. Published weekly by Telephony Publishing Corp., 53 E. Jackson Blvd., Chicago, IL 60604.

TPT Networking Management. Published monthly by PennWell Publishing Company, 1421 South Sheridan, Tulsa, OK 74112.

APPENDIX

SAMPLE REQUEST FOR PROPOSAL

This specification describes a telephone system to be installed at the XYZ Company, 4321 Main Street, Milltown, Michigan 48185.

1.1 *General Information.* The following describes the specifications and features to be included in a telephone system to be installed for the XYZ Company in Milltown, Michigan. A seven-year life is anticipated for this installation. These specifications are intended to be functional. Any deviation from the stated specifications and features that the vendor proposes must be fully explained in the proposal.

1.2 *Proposal Submission Date.* Proposals will be received at the XYZ Company until 12:00 noon Eastern Standard Time on _____, 199_. The selected vendor will be notified of intent to negotiate within one hundred twenty (120) days from this date; therefore, all prices submitted with proposals will be considered to be firm for that period.

1.3 *General System Description.* Proposals should be accompanied with a General System Description, along with specifications for the system offered by the vendor. (Sales literature is not acceptable unless it is complete and accurate in all details.)

2.0 *Systems Specifications.* The system shall be a stored program control type system utilizing solid state electronic components. The system should be designed to provide P.01 grade of service for the specified number of lines and telephones.

2.1 *Attendant Features.* The attendant console(s) shall be capable of the following: camp on; automatic recall; conference set up for 2 trunks and 3 internal telephones, or 1 trunk and 4 internal telephones; transfer of incoming and outgoing calls; alphanumeric display of originating telephone number; call-waiting queue; trunk group busy indicators; line load control keys; and busy-station verification.

2.2 *Switching System Features.* The switching system shall be capable of the following: automatic dialing between internal telephones; dial "9" capability for automatic connection to outgoing trunks; dial "0" capability for connection to attendant console; rotary service groups (hunting) for selection of internal telephones in a group in a sequential basis; station transfer without operator assistance; add-on capability without operator assistance; pushbutton dialing; automatic route selection (least-cost routing); and call detail recording capability.

3.0 *System Administration.* Changes such as station class of service, hunting sequence, telephone number, and other features to be made from a terminal located on the customer's premises.

3.1 *Traffic Measurement.* The system should be equipped with the necessary devices so that traffic counts can be made. The following information, by trunk groups, will be required:

a. Number of calls
b. Number of blocked calls
c. Number of times individual features are used
d. Number of calls handled at operator consoles, and
 (1) average delay on answered calls
 (2) number of abandoned calls
 (3) average waiting time before abandonment

4.0 *Telephone Sets.* All telephone sets to be dual-tone multifrequency (pushbutton) dialing. All key telephone sets to be equipped with lamps to flash 60 interruptions per minute when ringing, 120 on hold, and remain lighted when in use.

5.0 *Installation.* All installation work shall be performed in an orderly business manner and in accordance with applicable building codes. Location of the switch on the premisses to be resolved with the selected vendor. All cabling to be concealed in conduit, building walls, ceilings, or similar areas. All feeder cable and distribution terminals to have a minimum of 30 percent spare capacity. Two (2) complete sets of "as installed" drawings, technical manuals, and spare part lists to be provided by vendor.

6.0 *Cutover Procedures.* Vendor will provide training materials for system users. Vendor will train XYZ key personnel, who will conduct training sessions for other XYZ employees. Testing of system to be performed by the vendor to the satisfaction of the XYZ Company.

7.0 *Maintenance.* Proposals to specify maintenance provisions and terms of proposed maintenance contract. If vendor does not provide maintenance service, recommend alternate contractor. Specify terms of warranty. Provide diagnostic capability at site.

8.0 *Financial Arrangement.* Specify financial terms under rental, sell outright, lease direct, or lease through third party.

XYZ Company
Milltown, Michigan 48155
System Call Data

	at Cutover	*Ultimate*
Internal Calls		
Total Day	3,400	4,600
Busy Hour	550	700
Incoming Calls		
Total Day	3,000	4,000
Busy Hour	425	600
Outgoing Calls		
Total Day	3,600	5,000
Busy Hour	575	750
Total Calls		
Total Day	10,000	13,600
Busy Hour	1,550	2,050

System Equipment Requirements

	at Cutover	*Ultimate*
Station Lines	825	1,120
Telephones		
Single Line	700	920
6 Line	100	150
10 Line	25	50
Speakerphones	10	20
Attendant Consoles	2	3
Trunks		
Incoming (DID)	55	80
Outgoing		
Local	60	80
WATS/OCC	20	30
Tie Lines	4	6

SUPPLIER'S QUESTIONNAIRE

This questionnaire is provided to assist in the evaluation of vendor responses to this Request for Proposal. Please list reference number in replying to each question. Use the following format in your answer:

1. Restate the question.

2. State answer, including cost sheets, if appropriate.

3. List reference.

1.1 Who is the manufacturer and what is the model number of the equipment to be supplied?

1.2 Does the system proposed comply with FCC registration requirements?

2.0 What features are available for the attendant consoles over those listed in the RFQ? List any features requested that cannot be provided.

3.0 What is the maximum traffic capacity per line at cutover, and what are the provisions to increase this capacity to meet growth to ultimate configuration? What is the maximum number for simultaneous conversations that can occur?

3.1 What types of calls can be recorded on the call detail recorder? What is the format and the capability of the system with respect to the preparation of the final cost allocation report? Please attach sample reports.

3.2 Can the least-cost routing feature handle calls through FX lines, OCC networks, WATS, and message toll? Does it provide time-of-day routing?

4.0 What will be the cable requirements for installation of the telephone sets? How much spare will be provided?

5.0 Please provide a diagram showing space requirements for the control unit. Specify power and environmental requirements.

6.0 What types of materials are available for training system users? What resources are available to train maintenance personnel?

7.0 What are the financial arrangements that you propose? In the event that a lease is requested, will you assign the lease to a third party? Please provide a proposed payment schedule.

APPENDIX

FINANCIAL ANALYSIS

COMPARISON OF MONTHLY RENTAL, PURCHASE, OR LEASE OF A TELEPHONE SYSTEM

RENTAL

Assumptions:

- A system that rents for $800 per month including taxes.
- Lines and trunks rent for $1,000 per year.
- One-time installation charge of $2,500.
- Maintenance, insurance and taxes included in rent.
- Income tax rate at 34%.
- Rental will escalate 6% per year due to inflation.

	1st yr.	2nd yr.	3rd yr.	4th yr.	5th yr.	6th yr.	7th yr.
Annual Rental	9,000	10,176	10,787	11,434	12,120	12,847	13,618
Rental Lines and Trunks	1,000	1,060	1,124	1,191	1,262	1,338	1,419
Installation	2,500						
Total Costs	13,100	11,236	11,911	12,625	13,382	14,185	15,037
Tax Credit (@34%)	4,454	3,820	4,050	4,293	4,550	4,902	5,113
Net Costs	8,746	7,416	7,861	8,332	8,832	9,283	9,924
Discount Factor (@15%)	1.000	.8691	.7561	.6575	.5718	.4972	.4323
Present Value	8,746	6,445	5,944	5,478	5,050	4,616	4,290

Total present value of all costs = $40,569

PURCHASE

Assumptions:

- A system that sells for $30,000 including tax.
- Lines and trunks to rent for $1,000 per year.
- Installation included in purchase price.
- First year under full warranty.
- Maintenance contract to cost $1,140 per year.
- Insurance to cost $100 per year.
- Property tax to be $1,200 per year.

- Depreciation to be over 5 years on a straight line basis.
- Income tax rate at 34%.
- Maintenance, insurance, and taxes will escalate 6% due to inflation.

	1st yr.	2nd yr.	3rd yr.	4th yr.	5th yr.	6th yr.	7th yr.
Purchase Price	30,000						
Rental Lines and Trunks	1,000	1,060	1,124	1,191	1,262	1,338	1,419
Maintenance		1,208	1,281	1,358	1,439	1,526	1,617
Insurance	100	106	112	119	126	134	142
Property Taxes	1,200	1,272	1,348	1,429	1,515	1,607	1,702
Total Expense	2,300	3,646	3,865	4,097	4,342	4,605	4,880
Total Costs	32,300	3,646	3,865	4,097	4,342	4,605	4,880
Depreciation	6,000	6,000	6,000	6,000	6,000		
Tax Credit (@34%)	2,822	3,280	3,354	3,433	3,516	1,566	1,659
Net Costs after tax	29,478	356	511	664	826	1,039	1,221
Discount Factor (@15%)	1.000	.8691	.7561	.6575	.5718	.4972	.4323
Present Value	29,478	309	386	437	472	517	528

Total present value of all costs = $32,127

LEASE

Assumptions:

- A system that sells for $30,000 including tax.
- Lines and trunks to rent for $1,000 per year.
- Installation included in the purchase price.
- Maintenance contract to cost $1,140 per year.
- Purchase options at end of five years at 10% of purchase price.
- Insurance and taxes to be assumed after title passes.
- Income tax rate at 34%.
- Maintenance, insurance, and taxes will escalate 6% due to inflation.

	1st yr.	2nd yr.	3rd yr.	4th yr.	5th yr.	6th yr.	7th yr.
Purchase Price						3,000	
Rental Lines and Trunks	1,000	1,060	1,124	1,191	1,262	1,338	1,419
Annual Lease	9,600	9,600	9,600	9,600	9,600		
Maintenance		1,208	1,281	1,358	1,439	1,526	1,617
Insurance						134	142
Property Taxes						1,607	1,702
Total Expense	10,600	11,868	12,005	12,149	12,301	4,605	4,880
Total Costs	10,600	11,868	12,005	12,149	12,301	7,605	4,880
Tax Credit (@34%)	3,604	4,035	4,082	4,131	4,182	2,586	1,659
Net Costs after Tax	6,996	7,833	7,923	8,018	8,119	5,019	3,221
Discount Factor	1.000	.8691	.7561	.6575	.5718	.4972	.4323
Present Value	6,996	6,808	5,991	5,272	4,642	2,495	1,392

Total present value of all costs = $33,596

GLOSSARY

Note: An asterisk after a definition indicates that it is an American National Standards Institute definition.

A

Access code The preliminary digits that a user must dial to be connected to a particular trunk or line.*

Access charges Fees paid for the use of a line provided by the local telephone company.

Acoustic coupler A portable device that performs the function of a modem, allowing a telephone handset to be used for access to the switched telephone network for data transmission.

Airfone A telephone in a commercial airliner used by passengers to place telephone calls.

Allocating costs Assigning responsibility for the charges for telecommunications equipment and services to various departments.

Alphanumeric pager A radio pager that can display numbers and letters.

Alternating current (AC) Electric current that travels first in one direction (+ to −) and then in the other direction (− to +).

American National Standards Institute (ANSI) An organization formed for the purpose of establishing voluntary industry standards.

American Standard Code for Information Interchange (ASCII) (Pronounced "ask-ee") An 8-bit code (1 bit is for parity check) developed by the American National Standards Association that has been adopted as the standard code for data transmission in the United States.

Ameritech One of the seven Regional Bell Operating Companies resulting from divestiture of AT&T.

Amplifier See *repeater.*

Amplitude The maximum variation from the zero position of any alternating current; the size or magnitude of an alternating wave form. It is sometimes described as volume, intensity, or loudness.

Amplitude modulation A form of modulation in which the amplitude of a carrier wave is varied in accordance with some characteristic of the modulating signal.*

Analog signal A continuous signal that varies in voltage to reflect variations in some quantity, such as loudness of the human voice.

Area code A 3-digit number identifying one of more than 150 geographic areas of the United States and Canada. It permits direct distance dialing on the telephone system.

ASCII See *American Standard Code for Information Interchange.*

Asynchronous transmission Transmission that is not related to a specific frequency, or to the timing, of the transmission facility; describing transmission characterized by individual characters, or bytes, encapsulated with start and stop bits, from which a receiver derives timing for sampling bits. It is known as start-stop transmission.

Attenuation The difference between the transmitted and received power due to transmission loss through communications equipment.

Audio frequencies Frequencies that can be heard by the human ear, about 20 to 20,000 Hz.

Audio teleconferencing Voice only teleconferencing; a telephone conference in which people at geographically distant sites carry on a discussion among themselves.

Automatic call back A feature of a telephone system that permits a caller to instruct the system to call back as soon as a busy station is free.

Automatic Identification of Outward Dialing (AIOD) A PBX Service feature that identifies the calling extension, thereby permitting the cost of the call to be allocated to the extension.

Automatic Number Identification (ANI) A system for tracking the details of a call, such as the calling telephone number; also called call identification or call ID.

Automatic route selection (ARS) A PBX service that permits automatic selection of the most efficient routing of a call in a corporate network. It is sometimes called least-cost routing.

B

Bandwidth The range between the lowest and the highest frequencies of a channel.

Baseline The starting point of a sine wave.

Basic services Telecommunication services that are limited to the transport of information; no processing of information takes place.

Batch processing A method of processing data in which the input records are collected in their original form over a period of time, transcribed onto an input medium that the computer can read, and then transported to the computer room in groups that are entered into the computer for processing.

Baud A unit of signaling speed derived from the reciprocal time of the shortest pulse width in the bit stream.

Baudot code A 5-bit, 32-character alphanumeric code used in asychronous teleprinter transmission.

Beeper A slang term used for a radio pager that signals the wearer that he or she has a message waiting.

Bell Associated Companies The 22 Bell Operating Companies that were owned by AT&T prior to divestiture.

Bell Atlantic One of the seven Regional Bell Operating Companies resulting from AT&T divestiture.

Bell Communications Research, Inc. (BELLCORE) An organization owned and funded by the seven Regional Bell Operating Companies. Its function is to evaluate new products, control the assignment of new area codes, and to serve as a standardization agency for the industry.

Bell Operating Companies (BOCs) The subsidiary telephone companies owned by the seven regional Bell operating companies.

BellSouth One of the seven Regional Bell Operating Companies resulting from AT&T divestiture.

Bell System The term used to collectively describe AT&T and all of its subsidiaries prior to divestiture. There is no Bell System since divestiture.

Binary A number system that uses only two characters ("0" and "1").

Bit A contraction of *BInary digiT*. The smallest unit of information in a code using the binary system. It represents one of two possible values, such as a *mark* or a *space*, a *1* or a *0*, or an *on* or an *off*.

Block A group of continuous characters transmitted as a unit.

Block character checking A method of error detection in data transmission based on the observance of preset rules for formation of blocks.*

Blocked call A call that cannot be completed because of a network busy condition.

Blocking The inability to complete a connection between two lines because of a network busy condition.

BOCs See *Bell Operating Companies*.

Box telephone Telephone instrument developed and patented by Alexander Graham Bell.

Broadband A synonym for *wideband*. A communications channel having a bandwidth broader than that of a voice-grade channel, thereby providing high-speed data transmission capability.

Buffer A temporary storage device used to compensate for a difference in rate of flow of data or time of occurence of events when transmitting data from one device to another.

Busy hour The two consecutive half-hour periods of a day in which the largest number of calls occur.

Bypass The use of private communications facilities or services to go around, avoid, or bypass the local telephone company exchanges of the public switched network.

Byte Generally an 8-bit quantity of information, used mainly in referring to parallel data transfer, semiconductor capacity, and data storage; also generally referred to in data communications as an octet or character.

C

Call capacity The ability of a telephone system to handle a specific number of calls to provide a specific grade of service.

Call-detail recording (CDR) A feature of private branch exchanges where each telephone call is logged, typically by time and charges, and retrievable by the network operator for cost charging by department; also called station message detail recording (SMDR).

Call distributing system A telephone system that has queuing capability, such as those used by reservation centers.

Call forwarding A telephone service feature that permits automatic forwarding of calls to another telephone number.

Call pick-up A telephone service feature that permits a person receiving a telephone call to have access to the incoming call on any telephone in the system by entering a code. This feature enables a telephone system to operate effectively without having all lines appear on each telephone station.

Calling Rate The number of calls per telephone; it is determined by dividing the count of busy-hour calls by the number of telephones in the system.

Call waiting (camp on) A telephone service that permits a call to a busy telephone to be held while an audible tone notifies the busy telephone that a call is waiting.

Carrier-based message systems Systems that offer electronic mail service to the general public.

Carterfone decision An FCC decision striking down tariff restrictions that had prohibited attachment or connection to the public telephone system of any equipment or device not supplied by the telephone company. The decision ended an antitrust suit against AT&T and GTE which had been filed by Carter Electronics, a Texas company that wanted to interconnect private two-way radio by means of an acoustic device.

CCITT See *Consultative Committee on International Telegraphy and Telephone.*

CCS Abbreviation for hundred call seconds.

Cell A subdivision of a mobile telephone service area; it contains a low-powered radio communicating system connected to the local telephone network.

Cellular radio Technology employing low-power transmission as an alternative to local loops for accessing the switched telephone network; users may be stationary or mobile — in the latter case, they are passed, under the control of a central site, from one cell's transmitter to an adjoining one with minimum switchover delay.

Central office A synonym for *switching center,* also referred to as a telephone exchange.

Central office switching equipment The mechanical, electromechanical, or electronic equipment that routes a call to its ultimate destination.

Central processing unit (CPU) The component of a computer that does the actual processing of data.

Centralized processing A data processing configuration wherein the processing for several divisions, functional units or departments is centralized on a single computer, with input/output devices located in the same area as the computer.

Centrex A type of PBX service. The PBX facilities for Centrex are not located on the customer's premise, but are an integral part of a telephone company central office.

Channel A communications path for electrical transmission.

Character Standard bit representation of a symbol, number, or punctuation mark; generally means the same as byte.

Character checking A method of error detection in data transmission using preset rules for checking of characters.

Check bit One noninformation-carrying bit added to characters being transmitted that enables the computer to run its own check on every character that it processes. Also called a parity bit.

Circuit The complete path between two end-terminals over which one-way or two-way communication can be provided.

Circuit-switching The process of establishing and maintaining a circuit between two or more users on demand and giving them exclusive use of the circuit until the connection is released.

Class-of-service restriction A feature that limits the use of a telephone station to certain types of calls.

Coaxial cable A cable consisting of one or more hollow cylinders with a single wire running down the center of each cylinder. It can carry a much higher bandwidth than a wire pair.

Code Any system of communication in which arbitrary groups of symbols represent units of plain text of various lengths.*

Code set The complete set of representations defined by a code.*

Coding The process of converting information into a form suitable for communications.*

Coin telephones Telephones that provide service on the public network requiring the use of coins deposited in coin slots on the telephone to pay for each call.

Common-battery A DC power source in the central office that supplies power to the central office switching equipment and to all subscribers connected thereto.

Common carrier An organization in the business of providing communications services to the public, and which is subject to regulation by the appropriate state or federal agency.

Common Control Switching Arrangement (CCSA) A private network provided by the telephone company that shares switching facilities with the public network.

Communicating word processors Word processors equipped with electronic components that permit the sending of documents from one location to another over telecommunication channels.

Communications control unit (CCU) See *front end processor.*

Communications satellite An orbiting vehicle that relays signals between communications stations.

Communications Workers of America (CWA) A labor union that represents members who are employed by communications companies in the United States and Canada.

Computer-Assisted Instruction (CAI) Using a computer for individualized instruction.

Computer-based message system General purpose computer that has the capability to store messages and forward them to terminals having access to the CPU of the computer.

Computer conferencing A visual form of conference telephone call, in which conferees talk to each other by keyboarding messages and transmitting them over telecommunications facilities to other computers connected to the system.

Computer Inquiry I (CI-I) A study conducted by the FCC, concluded in 1971, that examined the relationship between the telecommunications and data processing industries to determine which aspects of both industries should be regulated for the long term.

Computer Inquiry II (CI-II) A study conducted by the FCC, concluded in 1981, that accelerated the deregulation of the telecommunications industry.

Computer Inquiry III (CI-III) A study conducted by the FCC to determine to what extent the AT&T and the BOCs are allowed to provide enhanced (data processing) services in the network.

Conditioned line A private line that has been specially treated to reduce distortion and improve transmission quality.

Consent decree of 1956 An agreement entered into between AT&T and the United States Department of Justice, and approved by the Court. The agreement permitted AT&T to retain Western Electric but limited AT&T to activities related to the telephone business and government projects, thereby excluding them from the data processing business.

Consultative Committee on International Telephony and Telegraphy (CCITT) A committee of the International Telecommunication Union, which serves as the medium for recommendations for international communications systems.

Control character A character used to define a subsequent series of characters until the next control character appears.

Control unit A component of modern telephone instruments that allows callers to place calls directly, without the assistance of an operator. The control unit can be either a rotary dial or pushbutton keys.

Country code The second set of digits a customer dials to place an international call (following the international access code).

Cost-based pricing Rates for telecommunication services that are based upon the cost of providing the service.

Creamskimming The practice of competing in only the most profitable markets.

Crossbar system A type of common-control switching system using switches that have vertical paths and horizontal paths interconnected to form a communications link.

Cross-subsidization A pricing method wherein some rates are much higher than cost while others are below cost. The excess revenues from the overpriced rates are used to compensate for the revenue shortage resulting from the below cost rates.

Crosstalk The phenomenon in which a signal transmitted on one circuit or channel of a transmission system creates an undesired effect in another circuit or channel.*

Current The amount of electrical charge flowing past a specified circuit point per unit of time, measured in amperes.

Cursor A blinking symbol that indicates the current location on the CRT screen.

Custom Local Area Signaling Service (CLASS) A call management package comprised of seven popular service features.

Custom-premises equipment (CPE) Terminal equipment installed on the customer's premises that is connected to the telephone network. It may be obtained from any supplier.

Cutover The activation of a newly installed telephone system that either replaces an older system or is a new installation.

D

Data Digitally represented information, which includes voice, text, facsimile, and video.

Data base A comprehensive collection of data that relates to a function.

Data communications The movement of coded information by means of electrical transmissions systems.

Data set A Bell System synonym for a *modem*.

Dedicated circuit or line A point-to-point telecommunications channel used exclusively by a single subscriber. Also called private line or leased line.

Delivery time The time from the start of transmission at the transmitting terminal to the completion of the reception at the receiving terminal, when data is flowing in only one direction.

Demodulation A function of changing the band pattern of a message on a carrier wave back into the form of the original message signal after transmission.

Dial pulse The signal that is transmitted by a rotary dial telephone by momentarily opening a direct circuit a number of times corresponding to the decimal digit which is being dialed.

Dial-tone first service A coin telephone that permits customers to reach the operator and to dial certain calls, such as directory assistance or 911, without depositing a coin.

Dibits A method used by some transmission systems to combine groups of bits and transmit two bits at a time, thereby doubling the transmission speed.

Digital signal A nominally discontinuous electrical signal that changes from one state to another in discrete steps.*

Direct distance dialing (DDD) A long distance service that permits customers to dial their own long distance calls without assistance from an operator.

Direct current (DC) Electrical current that travels in only one direction in a circuit (+ to −).

Direct Inward Dialing (DID) Feature of some telephone switches and PBXs that allows an external caller to call an extension without going through an operator.

Directory assistance A bureau operated by the telephone company that provides telephone numbers upon request.

Distance learning Learning at home through the use of videotex systems.

Distributed processing The processing of data at remotely located sites using communication lines to interconnect microcomputers or intelligent terminals with the central computer. See also *centralized processing.*

Divestiture of AT&T The breakup of AT&T mandated by the federal courts, based upon an antitrust accord reached between AT&T and the U.S. Department of Justice, effective January 1, 1984; most notable effects include the separation of 22 AT&T-owned local Bell operating companies (BOCs) into seven independent regional Bell holding companies, the requirement that AT&T manufacture and market customer premises equipment through a separate subsidiary, and use of the Bell name and logo only by the divested BOCs (RBOCs).

Dominant carrier A carrier that has such a large share of the market that it virtually controls the market.

Downlink The rebroadcast of a microwave radio signal from a satellite back to earth.

Drop wire The wire running from a residence or a business to a telephone pole or its underground equivalent.

Dual-tone multifrequency (DTMF) A method of signaling a desired telephone number by sending tones on a telephone line.

E

Earth station In satellite communications, a terrestrial communications center that maintains direct links with a satellite.

EBCDIC An acronym for *Extended Binary Coded Decimal Interchange Code.* An 8-bit data transmission code used in IBM systems.

Echo A type of distortion; an electric wave that has been reflected back to the transmitter with sufficient magnitude and delay to be perceived.

Echo canceler Device that performs the same function as an echo suppressor, but unlike that device does not clip speech of the speaker and can work during two-way transmissions.

Echo suppressor A device installed by telephone companies to reduce echo to a negligible level.

Echo suppressor disabler A device that transmits a tone that can be heard on the telephone as a high-pitched whistle. The tone disables the echo suppressor until there has been no signal on the line for about 50 milliseconds.

800 service A form of long distance telephone service wherein the caller places the call using "800" as the area code and the caller does not pay for the call.

Electronic mail The delivery of mail, at least in part, by electronic means.

Electronic polling Voting or voicing opinions at home through the use of videotex systems.

Electronic switching system (ESS) Any switching system whose major components utilize semiconductor devices.*

Enhanced services Telecommunication services that involve computer processing of the transmitted information and which are provided on a competitive basis by a number of vendors.

Equal access A provision of the MFJ which mandated that local telephone companies provide access from the local exchanges to any long distance company, and that the quality of the access must be equal for all long distance carriers. Thus, customers could dial carriers of their choice without having to dial extra digits.

Equivalent Queue Extended Erlang B Tables developed by Dr. James Jewett to be used in the design of trunks that automatically route blocked calls to alternate routes.

Erlang A measurement of telecommunication traffic usage. One Erlang equals 3600 seconds of usage.

Erlang, Erlang B, and Erlang C tables Tables used to predict quantities of equipment required to produce a desired grade of service at a given level of traffic.

Error detection A systematic method to detect whether a data message transmitted over telecommunication channels is valid or if something has gone wrong that might have caused an error in the transmission.

ESS See *electronic switching system*.

Exchange A specific geographical area served by one or more telephone offices, including the physical plant and equipment necessary to provide communications service in the area.

Extended Area Service (EAS) A service that permits a subscriber to make calls to a designated area beyond the local exchange area and be charged local rates instead of toll rates.

Extended binary coded decimal interexchange code (EBCDIC) An 8-bit code with 256 characters, generally transmitted in synchronous systems.

F

Facility A transmission path between two or more locations.

Facsimile (FAX) The process of transmitting text, pictures, diagrams, etc. via a telecommunication system to a remote location where hard copy of the transmitted material is reproduced.

Federal access charge A surcharge ordered by the FCC directing that an amount be added to every local line charge to compensate for the loss in revenue caused by the discontinuance of the subsidy of local service by long distance service.

Federal Communications Commission (FCC) A board of five commissioners appointed by the president of the United States under the Communication Act of 1934, having the power to regulate interstate and foreign electrical communication systems originating in the United States.

Federal line charge Another name for federal access charge.

Fiber optics Hair-thin filaments of transparent glass or plastic that use light instead of electricity to transmit voice, video, or data signals.

Flat rate service Service wherein the user is entitled to an unlimited number of telephone calls within a specified local service area for a fixed monthly rate.

Foreign Attachment A historical term used to describe equipment not provided by the telephone company that was attached to telephone lines.

Foreign equipment See *foreign attachment.*

Foreign Exchange Service (FX) A service providing a circuit connecting a subscriber's main station or private branch exchange with a central office of an exchange other than that which normally serves the exchange area in which the subscriber is located.

Franchise The right or license granted to an indivudual or group to market a company's goods or services in a particular territory.

Freeze-frame television A telecommunication service that transmits "snapshot" images over standard unconditioned telephone lines.

Frequency The number of cycles or events per unit of time. When the unit of time is one second, the measurement unit is the hertz (Hz).*

Frequency-division multiplexing One of the two basic multiplexing techniques, in which the channel frequency range is divided into narrower frequency bands. See also *time-division multiplexing.*

Frequency modulation (FM) A process in which the intelligence of a signal is represented by variations in the frequency of the oscillation of the signal.

Front-end processor A programmed-logic or stored-program device that interfaces data communication equipment with an input/output device or memory of a data processing computer;* also called a communications control unit (CCU).

Full-duplex (FDX) A type of operation in which simultaneous two-way conversations, messages, or information may be passed between any two given points.*

Fully utilized system A system that is operating at full capacity.

G

Gateway A network station that serves to interconnect two otherwise incompatible networks or devices; performs a protocol conversion operation across numerous communications layers.

Geosynchronous orbit The orbit where communications satellites will remain stationary over the same earth location; about 23,300 miles over the earth's equator.

Grade of service The probability that a call will be unable to be completed due to a network busy condition, expressed as a percentage.

H

Half-duplex (HDX) A circuit that affords communication in either direction but in only one direction at a time.*

Handshaking An exchange of predetermined characters or signals between two stations to provide control or synchronism after a connection is established.*

Hard copy Printed copy of the output of a computer.

Hard-wired A link (remote telephone line or local cable) that permanently connects two nodes, stations, or devices; describes electronic circuitry that performs fixed logical operations by virtue of fixed circuit layout, not under computer or stored program control.

Harmonic telegraphy The transmission of a number of telegraph messages over a single wire simultaneously, using interrupted tones of different frequencies.

Hertz (Hz) A unit of measurement formerly called cycles-per-second.

Holding time Telephone conversation time plus the time that equipment is used to establish connection.

Home banking Banking services performed at home through the use of Touch-Tone telephones, personal computers, or videotex terminals.

Hookswitch See *switchook*.

Hush-A-Phone A device attached to the handset of a telephone designed to reduce interference from room noise.

Hundred call seconds (CCS) A measurement of telecommunication usage. One hundred seconds of usage equals one CCS.

I

Identified ringing A telephone service feature that provides distinctive ringing tones for different categories of calls.

Image communication The transmission of an exact image of a document or an event by means of telecommunication equipment and facilities.

Independent telephone company Prior to AT&T's divestiture, a company providing common carrier telephone service independent of any Bell affiliation.

Individual-line service One telephone line to serve one subscriber.

Information Processed data (as opposed to raw data).

Information utility A commercial firm that rents computer time.

In-house system A computer within an organization that is usually timeshared among several departments.

Intercept service A telephone service provided by telephone companies to inform customers of the status of telephone numbers that are not in service. Typical information relates to disconnected telephone, number change, and no such number.

Integrated Services Digital Network (ISDN) An international concept whose objective is a digital public network. When in place, a totally digital network will extend from the user's terminal to the user's destination for the transmission of voice, data, and video information.

Intelligent terminal A terminal that contains a processing unit and can perform data processing and storage functions.

Interactive system A realtime communication system that provides immediate, two-way communication between terminals and a computer, processing transactions as they occur.

Interconnect equipment The equipment at each end of a communication channel; also called terminal equipment and customer-premises equipment.

Interconnect company A company that provides telecommunications terminal equipment for connection to telephone company lines.

Interexchange carrier (IXC or IC) A carrier engaged in the provision of interexchange (long distance) telecommunications services.

International access code The first set of digits a customer dials to place an international telephone call.

International Communications Association (ICA) An association of large users of telecommunications.

International Record Carrier (IRC) One of a group of common carriers that until a few years ago exclusively carried data and text (record) traffic from gateway cities in the United States to locations abroad and overseas; with recent FCC rulings, there is no longer rigid IRC monopoly, and several new carriers have been allowed to service points domestically.

International Standards Organization (ISO) An organization established to promote the development of standards to facilitate the international exchange of goods and services, and to develop mutual cooperation in areas of intellectual, scientific, technological, and economic activity.

International Telecommunication Union (ITU) The telecommunications agency of the United Nations, established to promote standardized telecommunications on a worldwide basis.

International Telegraph Alphabet (ITA) A code with the same characteristics as the Baudot code that is used in international telex transmission.

ISDN See *Integrated Services Digital Network.*

ISO See *International Standards Organization.*

K

Key telephone set A telephone with buttons or keys located on or near the telephone. It is used with associated equipment to provide features such as call holding, multiline pickup, signaling, intercommunication (intercom), and conferencing.

Key telephone system An arrangement of key telephone sets and associated circuitry located on a customer's premise that permits more than one telephone line to be terminated on one telephone instrument.

Kingsbury Commitment A 1913 letter from AT&T vice-president Nathan C. Kingsbury to the U.S. attorney general in response to accusations that AT&T was a monopoly. Kingsbury committed AT&T to interconnecting its facilities with those of the independents, disposing of its stock in Western Union, and refraining from acquiring independent telephone companies without approval of the ICC.

L

LAN See *local area network.*

Large scale integration (LSI) The integration of thousands of circuits onto a single chip.

Laser An acronym for *Light Amplification by Stimulated Emission of Radiation.* Used to generate very high frequency beams of light with tremendous information capacity.

LATA An acronym for *Local Access and Transport Area,* one of the 161 local telephone serving areas in the United States, generally encompassing the largest standard statistical metropolitan areas; subdivisions established as a result of the Bell divestiture that now distinguish local from long distance service; circuits with both ends within a LATA (*intra-LATA*) are generally the sole responsibility of the local telephone company, while circuits that cross the LATA boundaries (*inter-LATA*) are the responsibility of an interexchange carrier.

Leased line See *private line.*

Least-cost routing See *automatic route selection.*

LEC See *local exchange carrier.*

Light-emitting diode (LED) A device that serves as a source of light for use in fiber optic systems.

Lightwave Referring to electromagnetic wavelengths in the region of visible light; wavelengths of approximately 0.8 to 1.6 microns; referring to the technology of fiber optic transmission.

Line capacity The capacity of a telephone system expressed in terms of the maximum number of lines that can be physically served.

Line conditioning Communication channels that are specially treated to reduce distortion and improve transmission.

Line discipline The sequence of operations involving the actual transmitting and receiving of data; sometimes synonymous with protocol.

Line of sight Characteristic of some open air transmission technologies where the area between a transmitter and receiver must be unobstructed.

Line privacy A telephone service feature that prevents a person at an extension telephone from listening in on a conversation.

Link The communication facilities existing between adjacent nodes.*

Load/service relationship analysis A test to determine whether a telephone system is actually providing the grade of service it was engineered to provide.

Local Access and Transport Area See *LATA.*

Local area network (LAN) A configuration of telecommunications facilities designed to provide internal communications within a limited geographical area.

Local exchange carrier (LEC) A carrier that provides telecommunications services within a local exchange.

Local exchange service Public telephone service to points within the designated local service area (exchange area) for a telephone.

Local loop The transmission channel between a customer's premise and the telephone company central office.

Local service area A geographical area that has a single, uniform set of charges for telephone service.

Lockout A telephone service feature that prevents any interference with a call that is in progress.

Long distance access code A code used to gain access to the long distance network of a specific long distance carrier.

Longitudinal redundancy checking (LRC) See *block character checking.*

Loop See *local loop.*

M

Magneto A component of early telephones that was a hand-operated electrical generator. Callers cranked the handle on the generator to activate a bell or light that signaled an operator.

Mainframe computer A large computer capable of processing large amounts of data with very fast processing speeds, but requires a special environment and staff with data processing skills. See also *microcomputer and minicomputer.*

Management information system (MIS) The formal management of the flow of information throughout an organization, usually coordinated by a management information system department.

MCI decision An FCC decision that permitted Microwave Communications, Inc. (MCI) to provide private, leased line communication service between Chicago and St. Louis via microwave facilities.

Measured local service Telephone service for which a charge is made in accordance with a measured amount of usage, billed as message units.

Medium The path through which information flows.

Message switching A method of handling message traffic through a switching center, either from local users or from other switching centers, whereby a connection is established between the calling and the called stations or the message traffic is stored and forwarded through the system.*

Message systems Electrical transmission systems, such as TWX and Telex, that send messages (as opposed to conversation) in data form.

Message toll service (MTS) Synonymous with *long distance.*

Message unit A unit of measurement used in charging for local telephone calls. Criteria used are the length of the call and the distance involved.

Microcomputer The smallest general-purpose computer. Often it serves as a special-purpose computer or single-function computer on a single chip. See also *minicomputer and mainframe computer.*

Microelectronics The branch of electronics that deals with the miniaturization of electronic circuits and components.

Microprocessor An electronic device consisting of a central processing unit, memory circuits, and input-output devices.

Microwave radio Line-of-sight radio transmission using very short wavelengths, corresponding to a frequency of 1,000 megahertz or greater.

Minicomputer A medium-sized class of computers that are larger and more expensive than microcomputers but smaller and less expensive than mainframes. See also *mainframe computer and microcomputer.*

Mobile telephone service Telephone service between stationary telephones and moving vehicles that uses both the telephone network and a radio circuit to establish communication.

Modem Acronym for *MOdulator/DEModulator.* An electronic device used for converting digital signals into analog signals for transmission and reconverting the analog signal into digital signals.

Modular jack An interface device that permits easy interconnection of various telecommunications equipment and circuits.

Modulation The process, or the result of the process, of varying certain characteristics of a signal in accordance with a message signal (e.g., amplitude, frequency, and phase).

Modified Final Judgment (MFJ) The document approved by the court that ordered the divestiture of the Bell Operating Companies from AT&T, and that set up the conditions under which AT&T and the Regional Holding Companies could operate.

Multidrop circuit or line See *multipoint circuit or line.*

Multiplexer A device that combines a number of low-speed channels into one higher speed channel at one end of a transmission system and divides it back into low-speed channels at the other.

Multiplexing Use of a common channel to make two or more channels, either by splitting the frequency band transmitted by the common channel into narrower bands, each of which is used to constitute a distinct channel, or by allotting this common channel to multiple users in turn, to constitute different intermittent channels.*

Multipoint circuit or line A circuit providing simultaneous transmission among three or more separate points.*

Multiprocessing A process wherein two or more CPUs are interconnected into a single system and one control program operates both processors.

Multiprogramming The execution of two or more programs on the same computer simultaneously by interweaving their operations.

N

Narrowband Describing sub voice-grade channels whose bandwidth is between 0 to 300 Hz.

National Association of Regulatory Utility Commissioners (NARUC) An organization composed of Public Utility Commissioners whose purpose is to develop uniform regulatory policies and techniques.

National Exchange Carriers Association (NECA) An association of local exchange carriers, mandated by the FCC upon the divestiture of AT&T.

National Telecommunications and Information Administration (NTIA) An organization formed in March 1978, combining the Office of Telecommunications Policy and the Office of Telecommunications of the Commerce Department to provide advisory assistance in telecommunications and information issues for the Department of Commerce.

Natural monopoly An economic concept which holds that certain services can be provided more efficiently by one vendor in a given market.

911 service A service provided by local telephone companies to municipalities for the purpose of receiving emergency reports for police, fire, or ambulance service.

Network A series of points, nodes, or stations connected by communication channels.

Network access The capability of interconnection with a network.

Network busy condition A condition that is encountered when the network has received a greater volume of traffic than it can process. A caller is notified of this condition by a fast-busy signal (120 IPMs).

Node In network topology, a terminal of any branch of a network or a terminal common to two or more branches of a network.*

Noninteractive system A system (such as an offline system) where no interaction takes place between the user at a terminal and the computer during the execution of a program.

North American Telecommunications Association (NATA) An organization comprised of manufacturers, suppliers, and service companies engaged in interconnect activities, and/or furnishing equipment for interconnect facilities, and private users.

NTIA See *National Telecommunications and Information Administration.*

Numbering plan area (NPA) A geographic division within which telephone directory numbers are subgrouped. A three-digit code is assigned to each numbering plan area.

NYNEX One of the seven Regional Bell Operating Companies resulting from divestiture of AT&T.

O

Offline That condition wherein devices or subsystems are not connected into, do not form a part of, and are not subject to the same controls as an operational system. These devices may, however, be operated independently.*

Off-hook In telephony, a condition indicating the active state of a subscriber's telephone circuit; a line state that signals a central office that a user requires service; opposite of on-hook.

ONA See *Open Network Architecture.*

On-hook Deactivated condition of a subscriber's telephone circuit, in which the telephone or circuit is not in use; opposite of off-hook.

Online That condition wherein devices or subsystems are connected into, form a part of, and are subject to the same controls as an operational system.*

Open Network Architecture (ONA) A set of provisions imposed by the FCC on the BOCs and AT&T to ensure competitive availability of and access to unregulated enhanced network services.

Open-wire pairs A transmission facility comprised of pairs of bare (uninsulated) conductors supported on insulators, which are mounted on poles to form an aerial pole line.

Optical fiber See *fiber optics.*

Originating restriction A telephone service feature that restricts the telephone station from being used to place outgoing telephone calls.

Other Common Carriers (OCCs) Common carriers other than Bell System carriers.

Outgoing restriction A telephone service feature that provides the capability to restrict certain telephones from making specified types of outgoing calls, such as long distance.

Overflow A condition that prevents a call from being completed because the system is unable to complete the calls at the rate they are being offered.

Overnight telegram A telegraph service with following morning delivery.

P

Pacific Telesis Group One of the seven Regional Bell Operating Companies resulting from AT&T divestiture.

Packet A sequence of data, with associated control information that is switched and transmitted as a whole.

Packet Switching A data transmission technique whereby user information is segmented and routed in discrete data envelopes called *packets,* each with its own appended control information for routing, sequencing, and error checking; allows a communication channel to be shared by many users, each using the circuit only for the time required to transmit a single packet.

Pager A small radio receiving device carried or worn on the person used to receive a signal that a message is waiting.

Panel switching equipment An early type of electromechanical switching equipment in which groups of numbers are arranged on frames resembling panels.

Parity In binary-coded systems, a condition obtained with a self-checking code such that in any permissable code expression the total number of 1's or 0's is always even or always odd.*

Parity bit A single bit used to detect errors in data transmissions.

Parity checking A classical method of error detection during data transmission.

Party-line service Local telephone service wherein two or more customers share the same line to the telephone company central office.

PBX Private branch exchange. A private telephone exchange located on the user's premises and connected to the public network.

Peak traffic The highest volume load of traffic offered to a telecommunications system.

Phase The relative timing of an alternating signal.

Phase modulation (PM) A form of modulation in which the phase or timing of a signal is shifted to respond to the pattern of the intelligence being transmitted.

Photons Packets of light; used in lightwave communications.

Point of presence (POP) The geographic location where an interexchange carrier's facilities interconnect those of the local exchange carrier.

Point to point Describing the simplest type of network, in which lines directly connect two points in a communication network.

Poisson table Traffic capacity table developed by Simeon Poisson; based on the assumption that the sources of telephone traffic are infinite and that all unsuccessful call attempts are retried relatively soon.

Polling The process of calling up terminals in sequence to request the terminal to transmit a message. It is usually performed automatically by a central control unit.

Post, Telephone, and Telegraph (PTT) The department responsible for operating the telecommunications system in many countries.

POTS Plain Old Telephone Service. A term used to describe the basic service of supplying a single telephone set and access to the public-switched network.

Price-cap regulation A form of regulation that sets a maximum rate a company can charge for given services rather than regulating the rate of return of the company.

Primary carrier The long distance carrier chosen by a customer to carry long distance traffic when the customer dials "1" plus the area code and called telephone number.

Private branch exchange See *PBX.*

Private line See *dedicated circuit.* (Not to be confused with individual or one-party line.)

Private network A configuration of private lines and related switching facilities that are provided for the exclusive use of one customer.

Protocol The rules for communication system operation that must be followed if communication is to be effective.*

Public communications center Public access terminals equipped with credit-card-enhanced telephones, computer keyboards and monitor screens, and faxsimile machines, provided to meet the needs of travellers.

Public Service Commission (PSC) An agency charged with regulating communication services, as well as other public utility services, within a state. Called Public Utility Commission in some states.

Public switched network See *public network.*

Public Utility Commission (PUC) Same as Public Service Commission.

Pulse code modulation That form of modulation in which the modulating signal is sampled, the sample quantified and coded, so that each element of information consists of different kinds of numbers of pulses and spaces.*

Q

Queue Any group of items, such as computer jobs or messages, waiting for service.

R

Radio paging The broadcast of a special radio signal that activates a small portable receiver carried by the person being paged.

Rate case A formal procedure conducted by a regulatory agency to determine the level of rates that a utility will be permitted to charge its customers.

Rate center A geographically specified point used for determining mileage-dependent telephone rates.

Rate of return The ratio of net profit to the total invested capital.

Rate-of-return regulation A method of regulating public utilities that specifies the maximum rate of return a utility is permitted to earn.

Rate schedule A list of rates to be charged for specific services.

Raw data Unprocessed data (as opposed to *information,* which is processed data).

RBOC See *Regional Bell Operating Company.*

Realtime processing Processing that occurs at the same time a transaction is taking place.

Regional Bell Operating Company (RBOC) One of the seven regional companies (Ameritech, Bell Atlantic, BellSouth, NYNEX, Pacific Telesis, Southwestern Bell, and US West) created by the breakup of the Bell System.

Regulation A rule or order having the force of law, issued by an executive authority of a government.

Regulatory agency An agency with the legal power to control.

Remote access Pertaining to communication with a data processing facility through a data link.*

Remote job entry (RJE) The submission of data processing jobs via a data link.

Remote terminal The source in a data communications system; a computer device with keyboard, used for entering data.

Repeater A device that amplifies an input signal or, in the case of pulses, amplifies, reshapes, retimes, or performs a combination of any of these functions on an input signal for retransmission.*

Request for Proposal (RFQ) A formal document sent to interested vendors, inviting them to submit bids on new telecommunication systems or new services that meet user's requirements. The document is very specific concerning the requirements of the new systems or services.

Request for Quotation (RFO) A formal document sent to potential vendors, asking them to provide an estimate of the costs of equipment and services described in the document.

Resale carriers (resellers) A communications carrier engaged in selling long distance services of other carriers.

Response time In a data system, the elapsed time between the end of transmission of an inquiry message and the beginning of the receipt of the response message, measured at the inquiry originating station.*

RFP See *Request for Proposal.*

RFQ See *Request for Quotation.*

Rotary Dial A rotary mechanism on a telephone that, when wound up and released, creates an interruption of the line current that causes the central office equipment to operate in accordance with the digit dialed.

Routing The process of selecting the circuit path for a message.

Routing code The area code that comprises the third group of digits a customer dials to place an international call, or the first set of digits a customer dials for a long distance call within the same nation.

S

Satellite An object or vehicle orbiting, or intending to orbit, the earth, moon, or other celestial body.*

Satellite communications The use of geostationary orbiting satellites to relay transmissions from one earth station to one or more other earth stations.

Satellite earth terminal The portion of a satellilte link that receives, processes, and transmits communications between the earth and a satellite.

Satellite relay An active or passive satellite repeater that relays signals between two earth terminals.*

Simplex A circuit using a group return and affording communications in either direction, but in only one direction at a time.*

Sine wave An undulating wave used to represent the frequency of oscillation of an alternating current.

Sink A component of a data communication system that functions as the receiver of the information.

Soft copy A visual display on a CRT screen that provides no permanent record of the information displayed.

Source A component of a data communication system that functions as the originator of the information.

Southwestern Bell One of the seven Regional Bell Operating Companies resulting from AT&T divestiture.

Specialized Common Carrier (SCC) A common carrier that specializes in a specific type of telecommunications service such as long distance. Also called Other Common Carrier (OCC).

Specialized Common Carrier decision A 1971 decision of the FCC that expanded the MCI decision to permit specialized carriers to offer new, innovative long distance services.

Speed calling A telephone service feature that permits a caller to reach certain frequently called numbers by using abbreviated telephone codes in place of the conventional telephone number. Also called automatic dialing.

Standard A benchmark or point of reference against which performance can be compared; also an agreed-to specification for equipment and circuit design.

Station One of the input or output points on a communications system.

Station Message Detail Recording (SMDR) See *Automatic Identification of Outward Dialing (AIOD).*

Step-by-step switching equipment An automatic switching system in which a call is advanced progressively step-by-step to the desired terminal under the direct control of pulses from a customer's dial.

Store-and-forward A communication service in which messages are received at intermediate points and stored for later retransmission to a further point or to their ultimate destination.

Stored-program control Electronic switching equipment that can be programmed to perform a variety of functions in addition to conventional call completion.

Strowger switch The first automatic switch; a step-by-step switch named after its inventor, Almon B. Strowger.

Subscriber loop See *local loop.*

Switchhook A switch on a telephone set that signals the central office that the telephone is either idle or in use. It is operated by the removal or replacement of the receiver or handset on the support mechanism. (Sometimes referred to as hookswitch.)

Switching The process of transferring a connection from one telephone circuit to another by interconnecting the two circuits.

Switching center An installation in which switching equipment is used to interconnect communication circuits on a message or circuit switching basis.*

Synchronization The process of determining and maintaining the correct timing for transmitting and receiving information.

Synchronous transmission Data communications in which characters or bits are sent at a fixed rate; the rate is maintained by electronic clocking devices at both transmitting and receiving ends of the circuit.

System capacity The fullest extent to which a telecommunications system can be used.

T

T-1 carrier A time-division multiplexed digital transmission system that provides 24 voice-grade digital channels on one pair of copper wires.

Talking paths A network of interconnected paths forming a communication link in a switching system.

Tandem office A high-level switching center in the local exchange or serving area.

Tariffed item A service item that is described in an approved tariff, along with the approved rate for the provision of the service.

Tariffs The published rates, regulations, and descriptions governing the provision of communications services.

Telco A generic abbreviation for telephone company.

Telecommunication Any transmission, emission, or reception of signs, signals, writing, images, and sounds or information of any nature by wire, radio, visual, or the electromagnetic system.*

Telecommuting Use of a computer system in the home that allows an employee to communicate with the office without actually traveling to and from work.

Teleconferencing A conference between persons remote from one another but linked by a telecommunications system.*

Telecopier Facsimile machine.

Telecourse Instruction delivered to the home via telecommunications.

Telegraph A system of communication using coded signals.*

Telematics The marriage of telecommunications and computer technology. The word is derived from *telematique,* a French term that describes the merging of telecommunications with computers and television.

Telemedicine Provision of health care from a distance, linking a remote location by two-way television and audio signals with a hospital or health-care facility.

Telemetry The process of reading utility meters via telecommunications.

Telephone equipment inventory A detailed description of all the telecommunications equipment and circuits, including service features.

Telephone exchange A room or building equipped so that telephone lines terminating there may be interconnected as required.*

Telephone traffic The flow of messages through a communications system.

Telephony The science and practice of transmitting speech or other sounds over relatively large distances, i.e., distances normally greater than earshot range, and rendering the sound audible upon receipt.*

Teleport A collection of earth stations oriented to carry telecommunications to and from communication satellites. Usually it has transmission facilities connecting directly to the end-user, thereby bypassing the local exchange carrier.

Teleprinter See *teletypewriter.*

Teleprocessing The overall function of an information transmission system that combines telecommunications, automatic data processing, and human-machine interface equipment and their interaction as an integrated whole.*

Tele-registration Registering for college classes through the use of Touch-Tone telephones or other telecommunications equipment.

Teleshopping Shopping at home through the use of personal computers or videotext terminals.

Teletex A high-speed version of ASCII Telex, intended eventually to replace Telex.

Teletext A one-way transmission system in which data signals are transmitted over the FM portion of a television signal.

Teletraffic theory The mathematical description of message flow in a communications network; a branch of applied probability.

Teletypewriter A printing telegraph instrument having a signal-actuated mechanism for automatically printing received messages. It may have a keyboard similar to that of a typewriter for sending messages. (The term *teleprinter* may be applied to a receive-only unit having no keyboard.)*

Teletypewriter Exchange Service (TWX) The earliest public teletypewriter network, now owned by Western Union.

Telex A worldwide switched message-exchange service.

Terminal A point in the network at which data can either enter or leave; a device, usually equipped with a keyboard, often with a display, capable of sending and receiving data over a communications link.

Terminal equipment Communications equipment at each end of a circuit to permit the stations involved to accomplish the mission for which the circuit was established.* (Sometimes called *terminal.*)

Tie line Same as dedicated or private line.

Time division mutliplexing One of the two basic multiplexing techniques, in which the channel frequency is assigned successively to several users at different times. See also *frequency-division multiplexing.*

Time sharing An online realtime computer system that allows several users to share the facility for different purposes on what appears to be a simultaneous basis.

Tone dialing See *dual-tone multifrequency (DTMF).*

Total system costs The amount an organization has to spend to obtain and maintain the system it needs (including purchase price, rental costs, installation, maintenance, interest charges, depreciation, and taxes).

Traffic The flow of messages through a communications system.

Traffic capacity tables Tables used to determine the quantities of trunks or equipment needed for a telecommunications system.

Traffic engineering The science of designing facilities to meet user requirements.

Traffic queuing See *queue.*

Traffic study A study to determine the levels of traffic that a system is presently handling. It consists of a count of calls classified by types (incoming, outgoing, local, long distance, WATS, etc.). The data obtained is used to forecast future traffic, which, in turn, is used in determining new system requirements.

Transmission The dispatching of a signal, message, or other form of intelligence by wire, radio, telegraphy, telephony, facsimile, or other means.

Transmission link See *channel.*

Transaction terminal A terminal designed for use in a particular industry to communicate transaction details directly to the computer as the transaction is completed.

Transponder A combination receiver-transmitter that receives a signal, amplifies it, and retransmits it at different frequency. Communications satellites serve as transponders.

Trunk A single transmission channel between two points, both of which are either switching centers or nodes, or both.*

Trusteeship Supervisory control of property by legal person(s).

Turnkey The activation of a system by a single act, such as the turning of a key.

U

Unbundling Separation of charges by pricing charges for switching services separately from prices for customer-premises equipment.

Universal Service Order Code (USOC) A code system developed by the former Bell System used for items to be billed by the local telephone company. Also used on service orders and equipment records.

Uplink Describing the earth-station transmission and the carrier signal used to transmit information to a geosynchronous satellite; complement of downlink.

Usage sensitive Describing a communications charge that is based on usage (number of calls).

US West One of the seven Regional Bell Operating Companies resulting from AT&T divestiture.

V

Value-added carrier (VAC) A specialized common carrier that provides a service over and above the transmission of voice or data. The added value is usually computer-oriented.

Value-of-service pricing A method of pricing telecommunications services that is based upon the anticipated value of the service to the users rather than on the costs of providing the service.

Vertical services Services over and above what is required for basic communications capability, such as deluxe telephone sets or custom calling services.

Video teleconferencing A form of teleconferencing in which conferees can both hear and see each other.

Videotelephone A telephone set that is capable of receiving and transmitting images.

Videotex An interactive communications application that allows users to converse with a remote data base, enter data for transactions, and retrieve textual and graphics information for display on subscriber's television set or microprocessor screen.

Viewdata A form of videotex service offered by the British Post Office using telephone lines for transmission.

Virtual network A carrier-provided service in which the public switched network provides capabilities similar to those of private lines, such as conditioning, error testing, and higher-speed, full duplex, four-wire transmission with a line quality adequate for data.

Voiceband A communications channel with a bandwidth appropriate for audio transmission, generally with a frequency range of about 300 to 3,000 Hz.

Voice identification A technology used to verify the identity of a person accessing a system by comparing spoken passwords with computerized voice patterns.

Voice mail service An advanced form of telephone answering service that permits a caller to send a one-way spoken message to a service user; the message is stored in the voice message equipment for retrieval by the intended recipient.

Voice processing The technology that allows computers to speak, store human voices, and react to human speech.

Voice recognition system A telephone service using speech recognition to activate equipment that dials telephone numbers automatically.

Voice response The conversion of computer output into spoken words and phrases that a human being can hear and understand; it is a combination of various frequencies of electrical impulses.

Voice store-and-forward system A system that enables a computer to accept a message and store it until a transmission path is available or the desired party calls to retrieve it.

W

WATS See *Wide Area Telecommunications Service.*

Waveguide Specially constructed metal pipe for containing, directing, and focusing microwave electromagnetic radiation transmission.

Wide Area Telecommunications Service (WATS) A service that permits customers to make (OUTWATS) or receive (800) long distance voice or data calls at reduced rates.

Wideband A communications channel offering a transmission bandwidth greater than a voice-grade channel; data transmission speeds on wideband facilities are typically in excess of 9.6 kbit/s and often at rates such as 56 kbit/s and 1.544 Mbit/s.

Wide-pair cables Insulated copper wires, twisted and packed into a covering of lead or plastic to form a cable.

Z

Zone call A call to an exchange that is contiguous to a defined local exchange area.

FREQUENTLY USED TELECOMMUNICATIONS ACRONYMS

AIOD	Automatic Identification of Outward Dialing
ANI	Automatic Number Identification
ANSI	American National Standards Institute
ARS	Automatic Route Selection
ASCII	American Standard Code for Information Interchange
AT&T	American Telephone & Telegraph Company
BASIC	Beginners All-purpose Instruction Code
BELLCORE	Bell Communications Research, Inc.
BOC	Bell Operating Company
CAI	Computer-Assisted Instruction
CCITT	Consultative Committee on Telegraphy and Telephony
CCS	Hundred Call Seconds
CCU	Communications Control Unit
CLASS	Custom Local Area Signaling Service
CPE	Customer-Premises Equipment
CRT	Cathode Ray Tube
DDD	Direct Distance Dialing
DDS	Digital Data Systems
DTMF	Dual-Tone Multifrequency
EAS	Extended Area Service
E-Mail	Electronic Mail
ESS	Electronic Switching System
FAX	Facsimile
FCC	Federal Communications Commission
FDX	Full Duplex
FX	Foreign Exchange Service

HDX	Half Duplex
IC	Interexchange Carrier
ISDN	Integrated Services Digital Network
ISO	International Standards Organization
ITU	International Telecommunication Union
LAN	Local Area Network
LATA	Local Access and Transport Area
LEC	Local Exchange Carrier
LED	Light Emitting Diode
MCI	Microwave Communications, Inc.
MFJ	Modified Final Judgment
MIS	Management Information Systems
NPA	Numbering Plan Area
OCC	Other Common Carrier
OLRT	Online Realtime
ONA	Open Network Architecture
PABX	Private Automatic Branch Exchange
PBX	Private Branch Exchange
POP	Point of Presence
PSC	Public Service Commission
POTS	Plain Old Telephone Service
PUC	Public Utility Commission
RBOC	Regional Bell Operating Company
RFP	Request for Proposal
RFQ	Request for Quotation
SCC	Specialized Common Carrier
SMDR	Station Message Detail Recording
TWX	Teletypewriter Exchange Service
VAC	Value Added Carrier
VDT	Video Display Terminal
WATS	Wide Area Telecommunications Service

INDEX

D

E

F

G

H

I

J

K

L

M

N

O

P

R

S

T

U

V

W